T0226113

Materials Science of High Temperature
Polymers for Microelectronics

MATERIALS RESEARCH SOCIETY SYMPOSIUM PROCEEDINGS   VOLUME 227

Materials Science of High Temperature Polymers for Microelectronics

Symposium held April 29-May 2, 1991, Anaheim, California, U.S.A.

EDITORS:

D.T. Grubb
Cornell University, Ithaca, New York, U.S.A.

Itaru Mita
Dow Corning Japan Ltd., Kanagawa, Japan

D.Y. Yoon
IBM Almaden Research Center, San Jose, California, U.S.A.

MⓇⓈ MATERIALS RESEARCH SOCIETY
Pittsburgh, Pennsylvania

CAMBRIDGE UNIVERSITY PRESS
Cambridge, New York, Melbourne, Madrid, Cape Town,
Singapore, São Paulo, Delhi, Mexico City

Cambridge University Press
32 Avenue of the Americas, New York NY 10013-2473, USA

Published in the United States of America by Cambridge University Press, New York

www.cambridge.org
Information on this title: www.cambridge.org/9781107409859

Materials Research Society
506 Keystone Drive, Warrendale, PA 15086
http://www.mrs.org

© Materials Research Society 1991

First published 1991
First paperback edition 2012

Single article reprints from this publication are available through
University Microfilms Inc., 300 North Zeeb Road, Ann Arbor, MI 48106

CODEN: MRSPDH

ISBN 978-1-107-40985-9 Paperback

Contents

Preface

A Symposium on Materials Science of High Temperature Polymers for Microelectronics was held from April 29th to May 2nd 1991 in Anaheim, California, as Symposium J of the 1991 MRS Spring Meeting. Previous meetings concentrating on high temperature polymers have primarily considered their use as structural materials. High temperature polymers are finding use in aerospace as matrix and fibers for composites. They are also finding use in microelectronics, as thin films, where the problems are somewhat different. There has also been a tendency to focus more on the engineering or end-use application of these novel materials.

In this symposium, the intention was to have a balance between the end-use materials engineering and a more scientific and general approach. This shows itself in the balance between industrial and academic authors of the papers given. Although the topic of the symposium is very closely focused, it is one of great current interest. Sixty-four papers were presented at the meeting, to an audience sufficient to make the organizers consider moving to a larger room. Of these papers, 50 are included in these proceedings.

The organization of this volume follows that of the meeting. The first two sections are on synthesis and new materials and characterization of structure and properties. They are the largest sections, and there is considerable overlap; most synthetic papers measure some properties; many characterization papers are looking at very new materials. Sections on specific applications and on thermotropic liquid crystalline polymers follow, and finally a more distinct section on surface properties, surface modification and adhesion.

The corporate support of this symposium by IBM Corporation, Hoescht Celanese Corporation, and E.I. dupont de Nemours & Co. enabled us to invite a number of speakers from a wide range of locations. This made the symposium the success that it was, and we are grateful for their support. Many of these invited papers are included in this volume.

D.T. Grubb
I. Mita
D.Y. Yoon

August 1991

Volume 201—Surface Chemistry and Beam-Solid Interactions, H. Atwater, F.A. Houle, D. Lowndes, 1991, ISBN: 1-55899-093-3

Volume 202—Evolution of Thin Film and Surface Microstructure, C.V. Thompson, J.Y. Tsao, D.J. Srolovitz, 1991, ISBN: 1-55899-094-1

Volume 203—Electronic Packaging Materials Science V, E.D. Lillie, P. Ho, R.J. Jaccodine, K. Jackson, 1991, ISBN: 1-55899-095-X

Volume 204—Chemical Perspectives of Microelectronic Materials II, L.V. Interrante, K.F. Jensen, L.H. Dubois, M.E. Gross, 1991 ISBN: 1-55899-096-8

Volume 205—Kinetics of Phase Transformations, M.O. Thompson, M. Aziz, G.B. Stephenson, D. Cherns, 1991, ISBN: 1-55899-097-6

Volume 206—Clusters amd Cluster-Assembled Materials, R.S. Averback, J. Bernholc, D.L. Nelson, 1991, ISBN: 1-55899-098-4

Volume 207—Mechanical Properties of Porous and Cellular Materials, K. Sieradzki, D. Green, L.J. Gibson, 1991, ISBN-1-55899-099-2

Volume 208—Advances in Surface and Thin Film Diffraction, T.C. Huang, P.I. Cohen, D.J. Eaglesham, 1991, ISBN: 1-55899-100-X

Volume 209—Defects in Materials, P.D. Bristowe, J.E. Epperson, J.E. Griffith, Z. Liliental-Weber, 1991, ISBN: 1-55899-101-8

Volume 210—Solid State Ionics II, G.-A. Nazri, D.F. Shriver, R.A. Huggins, M. Balkanski, 1991, ISBN: 1-55899-102-6

Volume 211—Fiber-Reinforced Cementitious Materials, S. Mindess, J.P. Skalny, 1991, ISBN: 1-55899-103-4

Volume 212—Scientific Basis for Nuclear Waste Management XIV, T. Abrajano, Jr., L.H. Johnson, 1991, ISBN: 1-55899-104-2

Volume 213—High-Temperature Ordered Intermetallic Alloys IV, L.A. Johnson, D.P. Pope, J.O. Stiegler, 1991, ISBN: 1-55899-105-0

Volume 214—Optical and Electrical Properties of Polymers, J.A. Emerson, J.M. Torkelson, 1991, ISBN: 1-55899-106-9

Volume 215—Structure, Relaxation and Physical Aging of Glassy Polymers, R.J. Roe, J.M. O'Reilly, J. Torkelson, 1991, ISBN: 1-55899-107-7

Volume 216—Long-Wavelength Semiconductor Devices, Materials and Processes, A. Katz, R.M. Biefeld, R.L. Gunshor, R.J. Malik, 1991, ISBN 1-55899-108-5

Volume 217—Advanced Tomographic Imaging Methods for the Analysis of Materials, J.L. Ackerman, W.A. Ellingson, 1991, ISBN: 1-55899-109-3

Volume 218—Materials Synthesis Based on Biological Processes, M. Alper, P.D. Calvert, R. Frankel, P.C. Rieke, D.A. Tirrell, 1991, ISBN: 1-55899-110-7

Volume 219—Amorphous Silicon Technology—1991, A. Madan,
Y. Hamakawa, M. Thompson, P.C. Taylor, P.G. LeComber,
1991, ISBN: 1-55899-113-1

Volume 220—Silicon Molecular Beam Epitaxy, 1991, J.C. Bean, E.H.C. Parker,
S. Iyer, Y. Shiraki, E. Kasper, K. Wang, 1991, ISBN: 1-55899-114-X

Volume 221—Heteroepitaxy of Dissimilar Materials, R.F.C. Farrow,
J.P. Harbison, P.S. Peercy, A. Zangwill, 1991, ISBN: 1-55899-115-8

Volume 222—Atomic Layer Growth and Processing, Y. Aoyagi, P.D. Dapkus,
T.F. Kuech, 1991, ISBN: 1-55899-116-6

Volume 223—Low Energy Ion Beam and Plasma Modification of Materials,
J.M.E. Harper, K. Miyake, J.R. McNeil, S.M. Gorbatkin, 1991,
ISBN: 1-55899-117-4

Volume 224—Rapid Thermal and Integrated Processing, M.L. Green,
J.C. Gelpey, J. Wortman, R. Singh, 1991, ISBN: 1-55899-118-2

Volume 225—Materials Reliability Issues in Microelectronics, J.R. Lloyd,
P.S Ho, C.T. Sah, F. Yost, 1991, ISBN: 1-55899-119-0

Volume 226—Mechanical Behavior of Materials and Structures in
Microelectronics, E. Suhir, R.C. Cammarata, D.D.L. Chung,
1991, ISBN: 1-55899-120-4

Volume 227—High Temperature Polymers for Microelectronics, D.Y. Yoon,
D.T. Grubb, I. Mita, 1991, ISBN: 1-55899-121-2

Volume 228—Materials for Optical Information Processing, C. Warde,
J. Stamatoff, W. Wang, 1991, ISBN: 1-55899-122-0

Volume 229—Structure/Property Relationships for Metal/Metal Interfaces,
A.D Romig, D.E. Fowler, P.D. Bristowe, 1991, ISBN: 1-55899-123-9

Volume 230—Phase Transformation Kinetics in Thin Films, M. Chen,
M. Thompson, R. Schwarz, M. Libera, 1991, ISBN: 1-55899-124-7

Volume 231—Magnetic Thin Films, Multilayers and Surfaces, H. Hopster,
S.S.P. Parkin, G. Prinz, J.-P. Renard, T. Shinjo, W. Zinn, 1991,
ISBN: 1-55899-125-5

Volume 232—Magnetic Materials: Microstructure and Properties, T. Suzuki,
Y. Sugita, B.M. Clemens, D.E. Laughlin, K. Ouchi, 1991,
ISBN: 1-55899-126-3

Volume 233—Synthesis/Characterization and Novel Applications of Molecular
Sieve Materials, R.L. Bedard, T. Bein, M.E. Davis, J. Garces,
V.A. Maroni, G.D. Stucky, 1991, ISBN: 1-55899-127-1

Volume 234—Modern Perspectives on Thermoelectrics and Related Materials,
D.D. Allred, G. Slack, C. Vining, 1991, ISBN: 1-55899-128-X

*Prior Materials Research Society Symposium Proceedings
available by contacting Materials Research Society.*

PART I

Synthesis and New Materials
for Thin Films

ORGANO-SOLUBLE, SEGMENTED RIGID-ROD POLYIMIDES: SYNTHESIS AND PROPERTIES

F.W. HARRIS, S.L.C. HSU, C.J. LEE, B.S. LEE, F. ARNOLD AND S.Z.D. CHENG
Institute and Department of Polymer Science, The University of Akron, Akron, OH 44325-3909 USA

ABSTRACT

Several segmented, rigid-rod polyimides have been pre-pared that are soluble in organic solvents in their fully imidized form. The polymers were prepared from commercial dianhydrides and 2,2'-bis(trifluoromethyl)-4,4'-diaminobi-phenyl (TFMB). Their intrinsic viscosities ranged from 1.0 to 4.9 dL/g. Tough, colorless films could be cast from m-cresol solutions at 100°C. The polymers had glass tran-sition temperatures (Tgs) above 275°C and displayed out-standing thermal and thermo-oxidative stability. Fibers were prepared from the 3,3',4,4'-tetracarboxybiphenyl dianhydride (BPDA) based polymers that had moduli of 130 GPa and tensile strengths of 3.2 GPa. The thermal expansion coefficients and dielectric constants of thin films (20-25 μm) of the polymers were as low as -2.40x10^{-6} and 2.5, respectively.

INTRODUCTION

This research was part of an ongoing program aimed at the synthesis of soluble, rigid-rod or segmented rigid-rod polyimides[1-3]. Such polymers are sought for use as high modulus fibers and reinforcing components in molecular compos-ites[4]. It is also anticipated that the polyimides that contain rigid-rod segments will prove useful as composite resins and in microelectronic applications.

The approach described in this paper involved the poly-merization of 2,2'-bis(trifluoromethyl)-4,4'-diaminobiphenyl (TFMB). The incorporation of this moiety in aromatic poly-amides has been shown to dramatically affect their proper-ties[5]. The substituents in the 2- and 2'-positions of the aromatic rings force them into a noncoplanar conformation, which decreases the polymers' crystallinity and enhances their solubility. The twisted conformation also interrupts the conjugation along the backbone and, thus, reduces or elimi-nates absorption of visible radiation. It was postulated that similar effects would be observed in polyimide systems. In fact, it was felt that the all-para-linked, rigid segments would enhance the polymers' mechanical properties, in parti-cular the moduli. The high chain rigidity was expected to result in very low thermal expansion coefficients[6]. The perfluoromethyl groups were also expected to enhance other desirable properties such as thermal stability while decreasing water absorption and permittivity.

RESULTS AND DISCUSSION

Monomer Synthesis

The TFMD was synthesized from 2-bromo-5-nitrobenzotri-fluoride by the described procedure[6].

The following dianhydrides were purchased from commercial sources and were heated at 150 to 180°C under reduced pressure overnight prior to use.

Dianhydrides Used in Polymerization

Polymerizations

The polymerizations of TFMB with the dianhydrides were carried out in refluxing m-cresol containing isoquinoline (2%, w/v) with a solids concentration of 10% (w/v). Under these conditions the intermediate poly(amic acids) spontaneously cyclized to the corresponding polyimides. The water of imidization that was generated was removed by distillation.

With the exception of the polymer prepared from PMDA, all of the polymers remained in solution throughout the polymerization. When the viscous solutions of the polymers prepared from BTDA, 6FDA, and BPDA were cooled below 80°C, they set to an opaque gel-like structure. This phenomenon was reversible as the clear, free-flowing solutions could be regenerated by heating to 80 to 100°C. The gellation phenomenon could be delayed or prevented by diluting the reaction mixtures with m-cresol. All of the polymers were isolated by precipitation in methanol.

In an attempt to obtain a more rigid polyimide, an alternating copolymer was prepared from BPDA and pyromellitic dianhydride (PMDA). Thus, BPDA was treated with a 2:1 molar excess of TFMB to afford a soluble oligomer (7), which was polymerized with an equivalent amount of PMDA. In comparing copolymer 8 with the homopolymer prepared from BTDA, one can see that the length of the rigid segment between the biphenyl moieties has been increased. Thus, the copolymer contains longer rigid-rod segments.

Polymer Properties

The polymers' solubility in solvents other than m-cresol depended on their structure (Table 1). Their intrinsic viscosities in m-cresol at 30°C varied from 1.0 to 4.9. Tough, colorless films could be cast from m-cresol solutions at 100°C. The UV-visible spectra of these films showed 85-90% transmittance above 400 nm.

Reaction scheme showing PFMB + BPDA in m-Cresol/Isoquinoline yielding compound 7, then with PMDA yielding compound 8.

Table 1. Polyimide Properties

Polyimide	Dianhydride Used	[η]ᵃ (dL/g)	Tgᵇ (°C)	TGAᶜ (°C)	Solventsᵈ
1	PMDA	INS	ND	555	INS
2	BTDA	1.6	ND	550	MC
3	ODPA	1.1	275	570	MC,NMP,TCE
4	DSDA	1.0	320	540	MC, NMP
5	6FDA	1.9	320	530	MC,NMP,TCE
6	BPDA	4.9	ND	600	MC
8	BPDA/PMDA	3.0	ND	600	MC

a. Intrinsic viscosity determined in m-cresol at 30°C.
b. Mid-point of change in slope on DSC thermogram obtained with a heating rate of 20°C/min.
c. Temperature at which a 5% weight loss occurred on TGA thermogram obtained with a heating rate of 10°C/min.
d. INS=insoluble; MC=m-cresol; NMP=N-methylpyrrolidone; TCE=sym-tetrachloroethane.

   The polymers prepared from ODPA, DSDA and 6FDA displayed glass transition temperatures (Tgs) of 275°C, 320°C, and 320°C, respectively (Table 1). The Tgs of the other polymers could not be detected by Differential Scanning Calorimetry (DSC). The Thermal Gravimetric Analysis (TGA) thermograms of the polymers showed 5% weight losses between 530 and 600°C in air and between 515 and 600°C in nitrogen.

Samples of the polymers prepared from ODPA and 6FDA were compression molded at 420°C and 2500 psi. The flexture moduli of these materials were 3.70 and 4.05 GPa, respectively. Their fracture energies (G_{Ic}) were 0.68 and 0.89 kJ/m^2.

The thermal expansion coefficients (α) of unoriented thin films (20 μm) of the homopolymer (6) and the copolymer (8) prepared from BPDA were determined as a function of stress from 50 to 200°C (Figure 1). Extrapolation of the linear regression to zero stress gave an α of 5.43x10^{-6} for the homopolymer and an α of -2.40x10^{-6} for the copolymer.

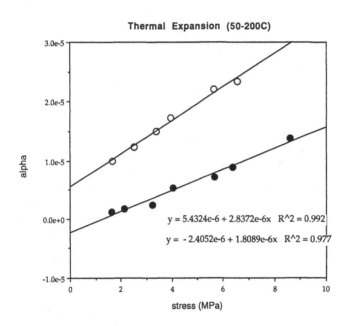

Thermal Expansion (50-200C)

y = 5.4324e-6 + 2.8372e-6x   R^2 = 0.992

y = - 2.4052e-6 + 1.8089e-6x   R^2 = 0.977

Figure 1. Thermal expansion coefficient (50% to 200°C) vs stress for thin (25 μm) films of polymer 6 (-O-) and copolymer 8 (-●-).

The dielectric constants of thin films (25 μm) of the homopolymer and copolymer, which were determined according to ASTM D-150, were 2.6 and 2.5, respectively. The dielectric constant temperature dependence at different frequencies was determined for a thin film of the homopolymer using DEA (Figures 2 and 3). The high permittivity observed in the first heating cycle was due to residual solvent. The subsequent cooling cycles and second heating cycles were almost identical. The permittivities below 200°C were almost constant.

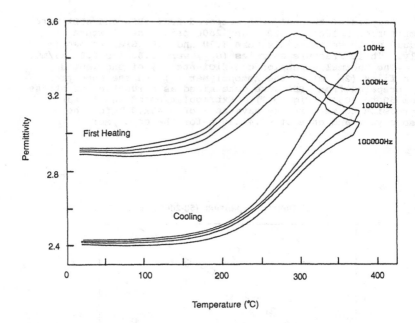

Figure 2. Permittivity at different frequencies vs temperature for a thin (25 μm) film of polymer 6.

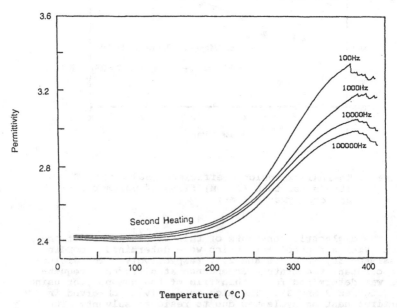

Figure 3. Permittivity at different frequencies vs temperature for a thin (25 μm) film of polymer 6.

Liquid-Crystalline Phenomena

The gel-like structures of the polymers in m-cresol were examined microscopically through a cross-polarizer. The gel of the polymer prepared from 6FDA contained liquid-crystalline spherulites. This behavior was exibited previously by gels of a rigid-rod polyimide[7]. The gels of the polymers prepared from BTDA and BPDA were highly bire-fringent, but did not possess any apparent textures. When the BPDA-based gel was annealed at 90°C, it developed a Schlieren texture typical of a nematic-like phase.

Fiber Spinning Study

A study was carried out to determine the conditions necessary for the preparation of fibers of the polyimide based on BPDA. Fibers were successfully spun from hot m-cresol solutions of the polymer. These fibers were subsequently drawn at elevated temperatures. The tensile strength and tensile moduli of the fibers were 3.2 GPa and 130 GPa, respectively[8].

Acknowledgment

The support of this work by the NASA-Langley Research Center Materials Division under Grant NAG-1-448 and by the Edison Polymer Innovation Center is gratefully acknowledged. We also wish to thank Dr. Z. Wu for the fiber spinning studies.

Literature Cited

[1] F.W. Harris and Y. Sakaguchi, ACS Preprints 60, 187 (1989).
[2] F.W. Harris and S.L.-C. Hsu, ACS Preprints 60, 206 (1989); High Performance Polymers 1, 3 (1989).
[3] F.W. Harris, S.L.-C. Hsu and C.C. Tso, ACS Polym. Preprints 31(1), 342 (1990).
[4] D.R. Wiff, S. Timms, T.E. Helminiak and W.F. Hwang, Polym. Eng. Sci. 27(6), 424 (1987).
[5] H.G. Rogers, R.A. Gaudiana, W.C. Hollinsed, P.S. Kalyanaraman, J.S. Manello, C. McGowan, R.A. Minns and R. Shatjian, Macromolecules 18, 1058 (1985).
[6] S. Numata, S. Ohara, K. Fiyisaki, J. Imaizumi and N. Kinjo, J. Appl. Polym. Sci. 31, 101 (1986).
[7] S.Z.D. Cheng, S.K. Lee, J.S. Barley, S.L.-C. Hsu and F.W. Harris, Macromolecules 24, 1883 (1991).
[8] S.Z.D. Cheng, Z.Q. Wu, M. Eashoo, S.L.-C. Hsu and F.W. Harris, Polymer, in press.

A STUDY OF POLYAMIC-ACID INTERCHANGE REACTIONS

John A. Kreuz and David L. Goff
Du Pont Electronics, E. I. du Pont de Nemours and Company,
Circleville Research Laboratory, Circleville, OH 43113
and Experimental Station Laboratory, Wilmington, DE 19898.

INTRODUCTION

C. C. Walker[1] first demonstrated that size exclusion chromatography (SEC) is an effective analytical technique for observing molecular weight distribution changes (MWD) of polyamic-acids with time. The specific polyamic-acid studied was that based on pyromellitic dianhydride and 4,4'-oxydianiline (PMDA/ODA), and the MWD narrowed on aging. Essentially, a decrease occurred in the weight average molecular weight (M_w), but no change took place in the number average molecular weight (M_n). Redistribution of molecular weight was postulated to be due to an equilibration reaction arising from the reversibility of amic-acids[2]. Additional data on M_w decreases with time of polyamic-acids other than PMDA/ODA have been reported by Miwa and Numata[3] and Kreuz[4], and although trends are similar, rates of equilibrations of various polyamic-acids appeared to be partially due to backbone structure. The dianhydrides of 3,3',4,4'-biphenyl dianhydride (BPDA) and 3,3',4,4'-oxydiphthalic anhydride (ODPA) tended to give more stable polyamic-acid solutions.

If polyamic-acids undergo interchange reactions through reversibility, the concern arises about the fate of polyamic-acids between the time they are preferentially synthesized into ordered sequences and the time their structures are immobilized by polyimide formation. The work of Fukami[5] was the earliest to have addressed this issue. He first utilized specific modes of syntheses to promote either random or sequenced copolyamic-acids. Films fabricated from these synthetic paths had properties that reflected either random or sequenced precursor systems. Closely followed were similar reports by de Visser[6] and then Babu and Samant[7,8]. More recently, Yang and Hsiao[9] showed that regular sequencing of diamines consisting of benzidine, ODA, p-phenylenediamine (PPD), and 4,4'-diaminodiphenylmethane with PMDA would provide different polyimide properties than if the polyamic-acid segments were randomly distributed. Extensive commercial activity in the field of controlling the sequences of polyamic-acid units and subsequent polyimide properties is evidenced by the many reports, patents, and patent applications being generated. Examples of these are the studies and disclosures by Nagano et al[10], Takushi et al[11], and Suzuki et al[12]. Additionally, Mita et al[13] and Yokota et al[14] utilized facile interchange reactions of polyamic-acids to make composite polyimide materials. After blending stiff segment polyamic-acids with flexible segment polyamic-acids, copolyamic-acids were obtained from immiscible polymers that became miscible. The works of Breckner and Feger[15,16], and also Yoon et al[17] examined these exchange reactions. The mechanism cited by Cotts[2] and Feger[16] rationalizes exchange because of the reversible formation of dianhydride and diamine from the polyamic-acid (I).

Based on these aforementioned concerns and results, several intents were pursued in the work reported here. Initial objectives were to examine rates of polyamic-acid interchanges via SEC, and to observe, if possible, bimodal to monomodal SEC changes of blended polyamic-acids. A second objective was to observe changes in polyimide film properties from random vs sequenced copolyamic-acids, as aging proceeded. A third objective was to determine some polyimide film properties from equilibrated blends of polyamic-acids. Two combinations of four different polyamic-acid backbones were used to accomplish the objectives of these studies. The first was the combination of PMDA/ODA with PMDA/PPD. The second was the combination of BPDA/PPD with perfluouroisopropylidene-bis-phthalic anhydride (6FDA) and ODA.

EXPERIMENTAL

Preparations of Polyamic-Acids

Polyamic-acids and random copolyamic-acids were prepared by addition of solid dianhydride to the diamine(s) contained in the solvent. A saturated solution of PMDA in DMAC was added at the end of the polymerizations to adjust viscosity and to attain stoichiometric equivalencies. In the case of the polymers made from PMDA, ODA, and PPD the solvent was DMAC and with polymers made from BPDA, PPD, 6FDA, and ODA the solvent was NMP. All polymers were made in dry, N_2 atmosphere, and stirring was done with a constant speed stirrer. About 2 h elapsed for the polymerization reactions, and afterwards the polymers were frozen over Dry Ice until they were analyzed, aged, or cast into films. All polyamic-acids were made at concentrations of 12-25% solids.

The sequenced copolyamic-acids were made by combining oligomeric polyamic-acids with excess amino functionality, and then coupling the chains by quantitative addition of a 6% solution of PMDA in DMAC to the mixture.

The DMAc and NMP were HPLC grade materials from Burdick & Jackson, and were dried over 4A Molecular Sieves, prior to use. Polymerization grade PMDA, 6FDA, ODA, and PPD were all obtained from within Du Pont, and were used without further purification. The BPDA was obtained from Mitsubishi Chemical and Ube.

Blending and Aging

Polyamic-acid solutions were mixed 50:50 by weight, unless otherwise specified, in a N_2 atmosphere and in sufficient amounts to do SEC analyses (1-2gms) or SEC analyses and film preparations (30-50gms). Vessels were stoppered and aged before analysis/fabrication.

Preparation of Polyimide Films

Films from polymers containing PMDA, PPD, and ODA were made by casting solutions of copolyamic-acids or of homopolyamic-acid blends, and immersing them in a mixture of 1:1 acetic anhydride:β-picoline. Subsequently, the self-supporting films were attached to pin frames and heated for 30 minutes in vacuum under a N_2 bleed at 300°C. A final heat treatment of the films while on the pin frames was carried out at 400°C for 5 minutes.

Films from polymers containing BPDA, PPD, 6FDA, and ODA were made by casting and thermally converting the dried polyamic-acid films at 350°C for 60 minutes (N_2) and finally at 400°C for 30 minutes (N_2).

Viscometric and SEC Analyses

Viscosities were done at 30°C using a Cannon-Ubelohde semimicro dilution viscometer. Concentrations of polyamic-acid were 0.5 gm. per 100mL of solution in Burdick & Jackson DMAC without added electrolyte. The SEC analyses on the polyamic-acids, copolyamic-acids, and the copolyamic-acids blends from PMDA, ODA, and PPD were carried out using the procedure of Walker[1]. The chromatographs were either a Du Pont 8800 or a Hewlett-Packard 1050, each with a column containing Du Pont Zorbax® PSM (60-1000) bimodal mixed bed absorbant. The detection wavelength was set at 254nm in each case.

The SEC analyses on the polyamic-acids, and blends thereof from BPDA/PPD and 6FDA/ODA were also run using the mobile phase of Walker[1]. Three Phenogel® columns were used in series with a differential refractive index detector. The chromatograph was a Waters Associates, Model GPCII. Absolute molecular weights were not obtained.

DISCUSSION OF RESULTS

Aging of Blended Polyamic-Acid Solutions

Blends of high molecular weight (MW) PMDA/ODA with low MW PMDA/PPD, as well as blends of low MW PMDA/ODA with high MW PMDA/PPD were studied as a function of time and temperature with SEC. Examples of the SEC curves are shown in Figure 1, and the MW data are summarized in Table I. The striking feature of the chromatograms is that the "zero" time blends revealed strongly the presence of two distinct molecular domains, which quickly merged into one population peak. The obvious interpretation of these results is that polyamic-acid interchanges occurred rapidly and copolymer species were being generated.

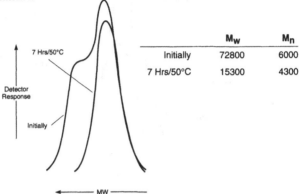

	M_w	M_n
Initially	72800	6000
7 Hrs/50°C	15300	4300

Figure 1 – High MW PMDA/ODA and Low MW PMDA/PPD

Table I – Interchanges of PMDA/ODA with PMDA/PPD

A. High MW PMDA/ODA with Low MW PMDA/PPD[1]:

	40°C			50°C			60°C		
Hrs	M_w	M_n	Inh	M_w	M_n	Inh	M_w	M_n	Inh
0.0	83300	6300	1.25	83300	6300	1.25	83300	6300	1.25
0.5	73200	5900	1.13	38100	5500	0.89	35100	5100	0.86
1.0	63500	5500	1.05	36500	5400	0.85	19400	4300	0.67
3.0	46600	5200	0.91	23200	4800	0.66	11100	3800	0.52
7.0	29100	4800	0.76	15300	4300	0.54	9500	3600	0.49
16.0	18300	4300	0.60	12300	4100	0.48	9000	3500	0.43
24.0	15500	4100	0.52	11700	4200	0.45	9600	3600	0.33

B. High MW PMDA/PPD with Low MW PMDA/ODA[2]:

	40°C			50°C			60°C		
Hrs	M_w	M_n	Inh	M_w	M_n	Inh	M_w	M_n	Inh
0.0	71100	7000	1.24	71100	7000	1.24	71100	7000	1.25
0.5	49600	6800	1.04	33500	6800	0.88	29700	5600	0.90
1.0	43700	6400	0.98	32100	6500	0.85	22700	5300	0.74
3.0	35900	6200	0.90	23800	6100	0.76	14700	4900	0.61
7.0	25600	5900	0.82	18300	5900	0.66	12900	5400	0.59
16.0	20100	5700	0.74	15600	5700	0.60	14800	5400	0.57
24.0	18000	5400	0.66	16400	5600	0.57	14200	5400	0.54

[1] High MW PMDA/ODA; M_w = 185400; M_n = 20700; MWD = 8.96.
Low MW PMDA/PPD; M_w = 13300; M_n = 3600; MWD = 3.69.
[2] High MW PMDA/PPD; M_w = 109000; M_n = 19200; MWD = 12.98.
Low MW PMDA/ODA; M_w = 8400; M_n = 3500; MWD = 2.40.

The effect of phthalic anhydride amine end-capping on the course of these interchanges was also studied. Accordingly, end-capped blends of high MW PMDA/ODA with low MW PMDA/PPD, as well as low MW PMDA/ODA with high MW PMDA/PPD were prepared. Examples of the corresponding SEC curves are shown in Figure 2, and the MW data are summarized in Table II. The same type of bimodal response was observed with the zero time blends, but after short aging, the peaks merge into one. The time required to accomplish development of one peak is similar to non end- capped blends. It suggests that terminal amino groups did not significantly alter the rate of this interchange process, although they might have indeed been operative in the overall interchange reaction.

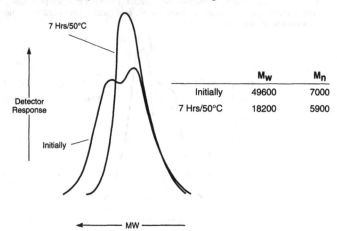

	M_w	M_n
Initially	49600	7000
7 Hrs/50°C	18200	5900

Figure 2 – Phthalic Anhydride End-Capped: High MW PMDA/PPD and Low MW PMDA/ODA

Table II – Interchanges of PMDA/ODA with PMDA/PPD; Both End-Capped

A. High MW PMDA/ODA with Low MW PMDA/PPD[1]:

Hrs	40°C M_w	M_n	Inh	50°C M_w	M_n	Inh	60°C M_w	M_n	Inh
0.0	55400	5100	1.06	55400	5100	1.06	55400	5100	1.06
0.5	51800	5100	1.04	45200	4900	0.99	36100	4900	0.91
1.0	49600	5000	0.98	–	–	0.91	–	–	0.80
3.0	42100	5000	0.95	–	–	0.85	19800	4500	0.67
7.0	33800	4800	0.86	21500	4500	0.73	–	–	0.60
16.0	24800	4700	0.77	16100	4400	0.66	11700	4100	0.48
24.0	23400	4600	0.70	–	–	0.56	–	–	0.41

B. High MW PMDA/PPD with Low MW PMDA/ODA[2]:

Hrs	40°C M_w	M_n	Inh	50°C M_w	M_n	Inh	60°C M_w	M_n	Inh
0.0	42100	5400	0.91	42100	5400	0.91	42100	5400	0.91
0.5	40400	5400	0.89	33900	5300	0.85	29100	5200	0.77
1.0	39000	5300	0.86	32000	5300	0.82	–	–	0.73
3.0	33700	5300	0.85	–	–	0.77	17400	4800	0.67
7.0	28200	5300	0.80	20000	5000	0.66	–	–	0.56
16.0	22600	5100	0.69	15300	4800	0.57	12500	4600	0.49
24.0	20200	5000	0.66	14800	4700	0.54	–	–	0.46

[1] High MW PMDA/ODA; M_w = 100600; M_n = 17100; MWD = 5.88.
Low MW PMDA/PPD; M_w = 23200; M_n = 3200; MWD = 7.25.
[2] High MW PMDA/PPD; M_w = 73100; M_n = 12700; MWD = 5.76.
Low MW PMDA/ODA; M_w = 9500; M_n = 3000; MWD = 3.17.

Temperature dependencies of the interchange reaction rates at 40°C, 50°C, and 60°C are also given by specific molecular weight data in Tables I and II, which are clarified in Figures 3 and 4, where MWD data are plotted vs time. Although there was scatter in the data, the reaction rates roughly doubled for each ten degrees rise in temperature, and levelling of the MWD was obtained consistently in about 7 to 16 hours at 50°C from the onset of aging. Beyond these times and temperature, decreases in both M_n and M_w suggest that degradation of the chains is occurring by unknown pathways.

A. High MW PMDA/ODA with Low MW PMDA/PPD

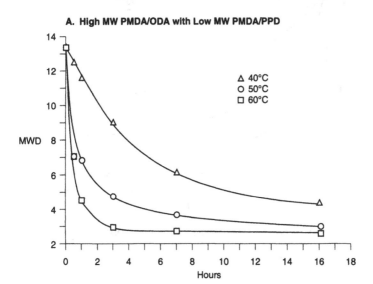

B. High MW PMDA/PPD with Low MW PMDA/ODA

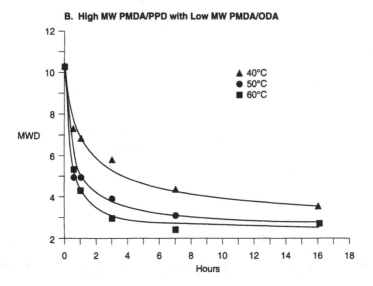

Figure 3 – Interchanges of PMDA/ODA with PMDA/PPD

A. High MW PMDA/ODA with Low MW PMDA/PPD

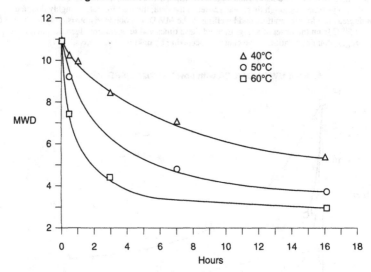

B. High MW PMDA/PPD with Low MW PMDA/ODA

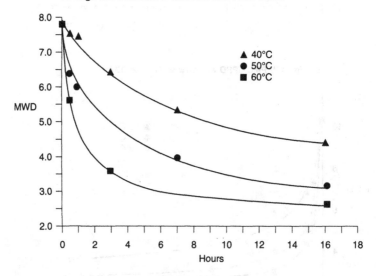

Figure 4 – Interchanges of PMDA/ODA with PMDA/PPD; both Polyamic-Acids End-Capped

Blending studies were then extended to phthalic anhydride end-capped polyamic-acids of BPDA/PPD and 6FDA/ODA (Table III). Low and high MW polymer solutions of each backbone type were mixed, allowed to stand at room temperature for specified time periods, and then monitored by SEC. The low MW BPDA/PPD and high MW 6FDA/ODA immediately formed a homogeneous solution and an SEC analysis of an immediately frozen sample revealed a MWD intermediate between those of the starting polymers and it was also somewhat broader in polydispersity. The blend was followed by SEC for 16 days and was observed to shift slightly to lower M_W and narrower MWD.

The blend of high MW BPDA/PPD and low MW 6FDA/ODA initially gave a non-homogeneous blend and the corresponding SEC in Figure 5 showed two distinct peaks which were essentially the same as the homopolymers. Subsequently, results were similar to the blend of low MW BPDA/PPD and high MW 6FDA/ODA, but at a much slower rate. Additional aging produced a monomodal distribution after 5 days of bimodal·behavior. As aging ensued up to 16 days, the M_W lowered and the MWD narrowed. The seemingly important aspect of these blending phenomena with the BPDA/PPD and 6FDA/ODA is that the polyamic-acids appear to transamidate, albeit rates are different and depend on the backbone structure of the polyamic-acid.

Table III – Interchanges of BPDA/PPD with 6FDA/ODA – Both Polyamic-Acids End-Capped

A. High MW 6FDA/ODA with Low MW BPDA/PPD[1] at 25°C:

Days	M_W	M_n	MWD
0.0	52800	19500	2.71
5.0	48300	23100	2.09
9.0	45900	21600	2.12
16.0	45700	24400	1.87

B. High MW BPDA/PPD with Low MW 6FDA/ODA[2] at 25°C:

Days	M_W	M_n	MWD
0.0	148000	32800	4.52
1.0	120000	31600	3.80
3.0	95200	32600	2.91
5.0	80600	32700	2.46
7.0	74500	28900	2.57
9.0	71700	30600	2.34
16.0	66600	30800	2.16

[1] High MW 6FDA/ODA; M_W = 85800; M_n = 29100; MWD = 2.94.
Low MW BPDA/PPD; M_W = 30200; M_n = 16400; MWD = 1.84.
[2] High MW BPDA/PPD; M_W = 236000; M_n = 84400; MWD = 2.81.
Low MW 6FDA/ODA; M_W = 28500; M_n = 14900; MWD = 1.91.

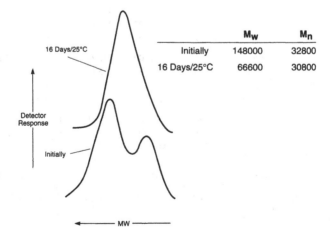

	M_W	M_n
Initially	148000	32800
16 Days/25°C	66600	30800

Figure 5 – Phthalic Anhydride End-Capped: High MW BPDA/PPD and Low MW 6FDA/ODA

Copolyimide Films From Aged Copolyamic-Acids

Copolyamic- acid equilibrations were carried out with randomly and sequentially synthesized copolyamic-acids based on PMDA/PPD/ODA. The mole ratio of PPD to ODA was maintained at 2:3, so that enough of the stiff codiamine segment, PPD, was present to observe its potential effects on polyimide film properties. The random copolymers were made by addition of PMDA to a solution of the diamines and polymerization was allowed to occur so that the only sequential ordering was due to the preferential polymerization of any given amino group. The sequenced copolymers were made by separately synthesizing oligomeric PMDA/PPD and PMDA/ODA, mixing them, and finally adding more PMDA to couple all of the smaller oligomers. The stoichiometry was chosen so that the average degree of polymerization (DP) of PMDA/PPD segments was 20, and the average DP of the PMDA/ ODA segment was 50.

The copolyamic-acids from both the random and sequenced synthetic schemes were aged for periods of 7 and 16 hours at 50°C, and in addition to SEC molecular weights being determined, chemically converted copolyimide films were made. Data in Table IV demonstrate that equilibrations occurred. Table IV also contains the film physical property of prime interest, viz., the coefficients of thermal expansion (CTE). This measurement was chosen, because of the success that was achieved in the reports of Numata et al[18,19,20,21], which showed that introduction of rod-like segments in the polyimide backbone resulted in reduction of the in-plane CTE. It was felt that the synthetic introduction of PPD into PMDA/ODA would not only stiffen the chain, but if it could be extended as an oligomeric block, then the rod-like effect on the CTE could be enhanced.

Table IV – Equilibrations of Copolyamic-Acids from PMDA/0.4PPD/0.6ODA and Corresponding Copolyimide Films

Copolyamic-Acid	Random/ Sequenced	Hours @ 50°C	M_w	M_n	MWD	Inh	CTE* x 10^6
A	Rand.	0.0	248800	45300	5.49	2.24	19
A	Rand.	7.0	170000	41100	4.14	1.96	20
B	Seqd.	0.0	249900	47000	5.32	2.44	13
B	Seqd.	7.0	172900	42400	4.08	2.13	12
C	Rand.	0.0	238900	47300	5.05	2.24	–
C	Rand.	16.0	165500	37300	4.44	2.01	21
D	Seqd.	0.0	191200	30100	6.35	2.35	–
D	Seqd.	16.0	121500	29900	4.06	1.77	11
E	Rand.	0.0	258000	42800	6.03	2.64	21
E	Rand.	7.0	171900	39700	4.33	2.35	19
E	Rand.	16.0	171500	40000	4.29	1.98	19
F	Seqd.	0.0	169800	29100	5.84	2.14	12
F	Seqd.	7.0	114400	28500	4.01	1.87	14
F	Seqd.	16.0	103100	25800	4.00	1.81	13
G	Seqd.	0.0	253200	39500	6.41	2.76	11
G	Seqd.	7.0	206300	41700	4.95	2.54	10
G	Seqd.	16.0	211900	41100	5.16	2.45	11

* Measured on a Mettler TMA 50 using 0.05N force at a heating rate of 10°C/min. in N₂. Films nominally 50µm thick. Second pass results reported between 40 and 260°C.

Several examples from separately synthesized samples clearly showed two distinctive features in the film physical properties. The first of these was that copolyimides made from preferred synthetic pathways displayed a CTE in line with those syntheses. Whereas the initial average CTE of the randomly made copolyimide films was 20PPM, the initial average CTE of the sequentially synthesized copolyimides was 12PPM.

The second distinctive result is that despite aging and equilibration, the CTE's of the copolyimide films, were not only representative of the initial synthetic paths, but they essentially did not change under the experimental conditions imposed. Even after aging the randomly synthesized copolyamic-acid for 16 hours at 50°C, the corresponding copolyimide films indicated an average CTE of 20PPM. A similar phenomenon was observed for the copolyimide films from the sequentially synthesized copolyamic-acids, which were aged for 16 hours at 50°C. The average CTE was 12PPM.

In addition to measuring the CTE, moduli of the copolyimide films were determined. As expected, there is also a trend for films from the random synthetic approach to give moduli that were lower than moduli of films from the sequential approach. Of all the samples tested, the "random" films had 410Kpsi tensile modulus and the "sequential" films had 490Kpsi tensile modulus. After 16 hours at 50°C, the results from the "random" polymers were 475Kpsi and those from the "sequential" polymers were 479Kpsi. The narrow moduli differences, together with the erratic data obtained precluded any conclusions based on moduli, even though it would be tempting to suggest that random polyamic-acids might tend to become ordered on aging. The reasons for erratic moduli are thought to arise in part from the varied thicknesses and bulk defects, which are prevalent in laboratory films, as well as sporadically different orientations.

In light of the convincing CTE data, in which the measurements are relatively insensitive to the film nonuniformities of laboratory samples, the cause(s) for the CTE differences based on initial copolyamic-acid syntheses is intriguing. First of all, it appears that these conditions of aging along with the attendant equilibrations that occurred, did not alter the chain segmental distributions in a way that affected the final film properties. Perhaps a rationalization as to how this can occur is to view equilibration proceeding through portions of chains that are more reactive to interchange. In this case, it would be the PMDA/ODA segmental lengths of any given chain. It seems, however, that other techniques to measure the integrity and changes of individual chain lengths are needed to facilitate an understanding of these transamidation interchanges.

Films from Blends of PMDA/ODA and PMDA/PPD Polyamic-Acids

In another experiment, a blend of polyamic-acids, one from PMDA/ODA, and another from PMDA/PPD, was prepared so that the mole ratio of the PPD to ODA was 2:3. The calculated DP of each polyamic-acid based on stoichiometry was 190 for PMDA/ODA and 33 for PMDA/PPD. Films were cast and chemically converted immediately after mixing and then again after 16 hours at 50°C. It was observed that the initial mixtures were immiscible, but after aging, miscibility developed. In contrast, the low molecular weight mixtures of polyamic-acids that had been prepared earlier for the sequential copolymers were immediately miscible. Also, as shown in Table V, the molecular weight of the blend was initially low, but after aging, the drop was severe and is probably due to extensive interchange, as well as unknown degradation paths.

The films reflected the lower molecular weights, since they often cracked on crease testing, but the CTE's were in the vicinity of 14PPM (Table V). This might suggest that the integrity of the PMDA/PPD chains maintained more of their initial character than did the PMDA/ODA chains.

Table V – Polyamic-Acid Interchanges from PMDA/ODA and PMDA/PPD and Corresponding Composite Polymide Films

Polyamic-Acid	Hours @ 50°C	M_w	M_n	MWD	Inh	CTE* x 10^6
PMDA/ODA(1)	0.0	173100	32200	5.38	2.27	–
PMDA/PPD(2)	0.0	16200	17300	6.72	1.78	–
Blend						
60%(1)+40%(2)	0.0	159500	24400	6.53	1.91	18
60%(1)+40%(2)	16.0	48000	16400	2.93	1.08	15

*CTE's measured by the same method as reported in Table IV.

Films from Blends of BPDA/PPD and 6FDA/ODA Polyamic-Acids

Blends of polyamic-acids were also prepared from BPDA/PPD and 6FDA/ODA homopolymers over a range of weight compositions from 95/5 to 80/20. The calculated DP of each polyamic-acid, based on stoichiometry, was 40 for BPDA/PPD and 60 for 6FDA/ODA. Initial mixtures were immiscible, but they gradually became completely miscible after 5-7 days at 25°C. Analysis of these blends by SEC indicated that some degree of polymer chain redistribution had occurred as evidenced by a shift in M_W relative to starting homopolymers (Table VI). Extent of redistribution and whether or not an equilibrium state had been achieved was not investigated. The BPDA/PPD and 6FDA/ODA homopolymers demonstrated only minimal M_W variance upon aging.

The aged blends were cast and thermally converted to provide tough, flexible films. A trend of diminishing mechanical performance and increasing in-plane CTE was noted which paralleled the weight percent increase of 6FDA/ODA. The blend CTE's ranged from 19 to 28PPM. Compared to the starting homopolymers and in view of the previous low/high MW blending, these results support an inter-chain polymer reorganization. This strongly suggests the formation of new copolymer species that contain extended BPDA/PPD segments, which maintain the rod-like impact on the CTE.

Table VI – Polyamic-Acid Interchanges from BPDA/PPD and 6FDA/ODA and Corresponding Composite Polyimide Films

Polyamic-Acid	Days @ 25°C	M_W	M_n	MWD	CTE* x 10^6
BPDA/PPD(1)	0.0	166000	62300	2.68	12
BPDA/PPD(1)	7.0	164800	60200	2.72	–
6FDA/ODA(2)	0.0	107000	47300	2.27	75
6FDA/ODA(2)	7.0	109200	48600	2.25	–
Blends					
95%(1)+5%(2)	7.0	128000	53400	2.41	19
90%(1)+10%(2)	7.0	127000	56000	2.26	23
80%(1)+20%(2)	7.0	117000	45600	2.58	28

* Measured on a Perkin-Elmer model TMA 7 in extension mode using 0.05N force at a heating rate of 10°C/min. in N_2. Films nominally 35μm thick. Second pass results reported between 50°C and 250°C.

CONCLUSIONS

- Evidence for interchange is given by bimodal to monomodal transitions occurring in SEC's of mixtures of solutions of PMDA/ODA // PMDA/PPD and of BPDA/PPD // 6FDA/ODA. Equilibria conditions prevail, after monomodal peaks develop.
- Interchanges through transamidations are also suspected in equilibrations of copolyamic-acids of PMDA/PPD/ODA that are either randomly or sequentially synthesized.
- Rates of transamidation appear to depend not only on temperature, but also backbone structure.
- Films of equilibrated random and sequenced copolyamic-acids have properties consistent with the notion that ordered, alternating sequences of homopolyamic-acid segments do form from preferred synthetic routes, and the sequences tend to maintain original order after some equilibration and interchange has occurred.
- Film properties of blends are believed to reflect copolymers properties synthesized by interchange reactions.

ACKNOWLEDGMENTS

We are indebted to William T. Hatzell, Robert T. Lutz, Richard L. Morgan, Richard H. Sill, and Arturo Vatvars who made this paper possible with experimental support. We also acknowledge John D. Craig for experimental suggestions regarding these studies.

REFERENCES

1. C. C. Walker, J. Polym. Sci. A, 26, 1649 (1988).
2. P. M. Cotts and W. Volksen, ACS Symp. Ser., 242, 227-237 (1984).
3. T. Miwa and S. Numata, Polymer, 30, 893 (1989).
4. J. Kreuz, J. Polym. Sci. A, 28, 3787 (1990).
5. A. Fukami, J. Polym. Sci., Polym. Chem. Ed., 15, 1535 (1977).
6. A. C. de Visser, D. E. Gregonis, and A. A. Driessen, Makromol. Chem., 179, 1855 (1978).
7. G. N. Babu and S. Samant, Eur. Polym. J., 17, 421 (1981).
8. G. N. Babu and S. Samant, Makromol. Chem., 183, 1129 (1982).
9. Chin-Ping Yang and Sheng-Huei Hsiao, J. Appl. Polymer Sci., 31, 979 (1986).
10. H. Nagano, H. Kawai, K. Yonezawa, Symposium On Recent Advances In Polyimides And Other High Performance Polymers, Reno, Nevada, July 13-16, 1987.
11. Sato Takushi, Kawamata Motoo, Takahashi Shigeru, Ibl Akira, Naito Mituyuki, Shindo Kazumi, Shishido Shigeyuki, European Patent Application, #0125599, c/o Mitsui Toatsu Chemicals, Inc., November 21, 1984.
12. Atsushi Suzuki and Yoriko Fujimura, Japanese Patent Application, #62-161831, c/o Toray Industries, Inc.
13. I. Mita, M. Kochi, M. Hasewaga, T. Iizuka, H. Soma, R. Yokota, and R. Horiuchi, Polyimides: Materials, Chemistry, and Characterization, p.1, Elsevier, Amsterdam (1989).
14. R. Yokota, R. Horiuchi, M. Kochi, C. Takahashi, H. Soma, and I. Mita, in Polyimides: Materials, Chemistry, and Characterization, Elsevier, Amsterdam (1989), p. 13.
15. M. Brekner and C. Feger, J. Polym. Sci., 25, 2479 (1987).
16. C. Feger, Symposium on Recent Advances In Polymides And Other High Performance Polymers , San Diego, CA, January 22-25, 1990.
17. D. Y. Yoon, M. Ree, W. Volksen, D. Hofer, L. Depero, and W. Parrish, The Third International Conference on Polyimides, Mid-Hudson Section of SPE, November 2-4, 1988, Ellenville, New York.
18. S. Numata, S. Oohara, J. Imaizumi, Polymer J., 17, #8, 981 (1985).
19. S. Numata, S. Oohara, K. Fujisaki, J. Imaizumi, and N. Kinjo, J. Appl. Polymer Sci., 31, 101 (1986).
20. S. Numata, K. Fujisaki, and N. Kinjo, Polymer, 28, 2282 (1987).
21. S. Numata, K. Kinjo, and D. Makin, Polym. Eng. & Sci., 28, #14, 906 (1988).

CHEMISTRY AND CHARACTERIZATION OF POLYIMIDES DERIVED FROM POLY(AMIC ALKYL ESTERS)

WILLI VOLKSEN, D. Y. YOON, J. L. HEDRICK AND D. HOFER
IBM Research Division, Almaden Research Center, Dept. K92/801
650 Harry Road, San Jose, CA 95120-6099

ABSTRACT: The modification of the pendant acid groups along the poly(amic acid) backbone in the form of alkyl ester groups leads to greatly improved polyimide precursors. These poly(amic alkyl esters) are characterized by the absence of hydrolytic instability due to elimination of the "monomer-polymer" equilibrium associated with poly(amic acids), a broad imidization temperature regime, improved solubility characteristics, and enhanced mechanical properties. In the absence of hydrolytic instability, it is now possible to use an aqueous work-up of the polyimide precursor. This presents an attractive synthetic pathway for the preparation of well-defined, amine-terminated oligomers. Such oligomers can then be utilized both in the preparation of low molecular weight, chain-extendable polyimide precursors as well as polyimide block copolymers. The higher imidization temperatures offered by the "amic ester" chemistry allows for more efficient chain extension prior to imidization. Alternatively, the lack of the "monomer-polymer" equilibrium and accompanying propensity for monomer randomization reactions presents a potential pathway for the preparation of polyimide blends.

INTRODUCTION

The utility of high temperature polymers as functional dielectric insulators for microelectronic applications has been recognized for many years and is well documented in the literature [1,2]. In this respect, polyimides have emerged as the primary class of materials with the desired properties necessary to realize the successful fabrication of microelectronic devices.

SCHEME 1

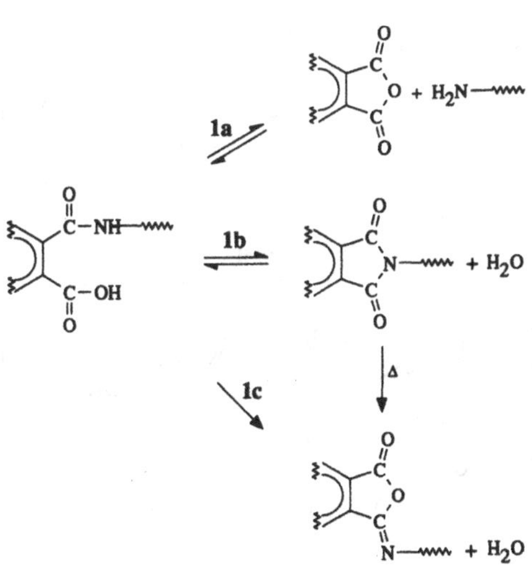

However, in contrast to the excellent properties associated with the final polyimide, the processable polyimide precursor, i.e. poly(amic acid), exhibits a number of undesirable characteristics, which severly limit the potential utility of this preparative approach.

The majority of these problems originate from the complex behavior of the polyimide precursor in solution; see Scheme 1. This scheme was first proposed over 20 years ago [3] and makes it possible to understand the unusual solution behavior of poly(amic acids). Pathway *1a*, which leads to anhydride endgroups, illustrates both the hydrolytic instability of the system as well as a potential mechanism for polymer fragmentation and recombination of different fragments, i.e. monomer randomization. Pathways *1b* and *1c* demonstrate the mechanism for potential gelation due to chemical or physical crosslinking as a result of diamide or imide formation, respectively.

Alternatively, chemical modification of the poly(amic acid) backbone in the form of its related poly(amic alkyl ester) not only appears to eliminate the various equilibria illustrated in Scheme 1, but also provides additional benefits in terms of synthetic flexibility and higher imidization temperatures [4,5,6]. Although the chemical modification of the poly(amic acid) backbone via conversion to the isoimide and subsequent reaction with alcohol to the poly(amic alkyl ester) has been reported [7], well-defined poly(amic alkyl esters) are best prepared as illustrated in Scheme 2. First, the aromatic dianhydride is reacted with a suitable alcohol to yield the corresponding diester diacid, Next, the diester diacid is converted into a diester diacyl chloride to enable the monomer to enter a polymer forming reaction. Reaction of the diester

SCHEME 2

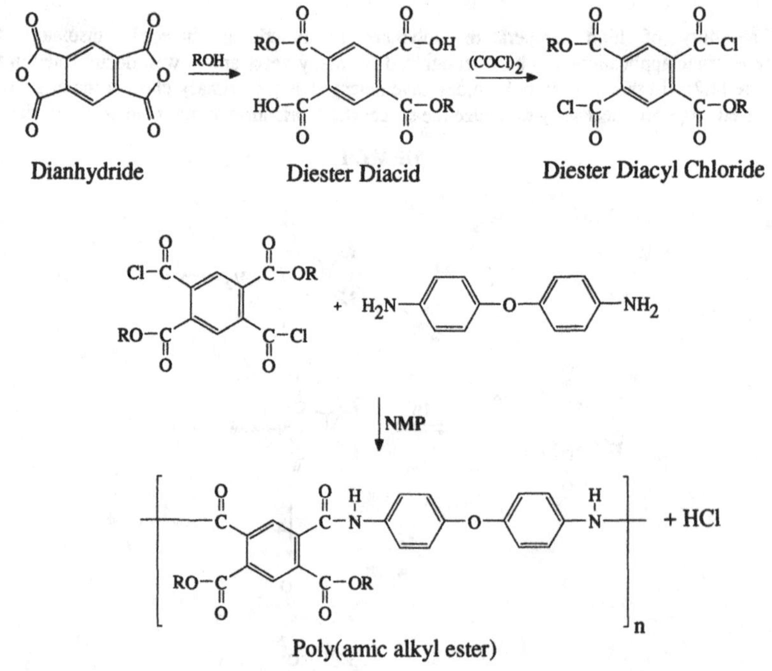

| Dianhydride | Diester Diacid | Diester Diacyl Chloride |

Poly(amic alkyl ester)

diacyl chloride with an aromatic diamine provides the desired poly(amic alkyl ester) via low temperature solution polycondensation in a polar, aprotic solvent such as N-methylpyrrolidone (NMP). The polymer is then precipitated in water and thoroughly washed to remove the by-product HCl, followed by an alcohol extraction to promote vacuum drying of the polymer.

EXPERIMENTAL

Starting Materials

Aromatic dianhydrides and diamines were commercially available materials, which were purified by one sublimation thru 30 mesh, neutral alumina to remove colored impurities, followed by a second, neat sublimation to insure dryness. N-methylpyrrolidone was vacuum distilled from P_2O_5 and stored under nitrogen.

Dialkyl dihydrogen pyromellitates were prepared similar to the procedure of Bell and Jewell [8], whereas dialkyl dihydrogen pyromellitates based on acidic alcohols, such as ethyl glycolate and trifluoroethanol, as employed for chain-extension experiments were prepared according to the procedure previously described [9].

PMDA/ODA based Poly(amic ethyl ester)

Diethyl dihydrogen pyromellitate was converted to the diacyl chloride by reaction with an excess of oxalyl chloride in ethyl acetate at 60 ^{o}C. Once gas evolution subsided and the reaction mixture had turned homogeneous, the solvent was stripped under vacuum and the crystalline residue crystallized from hexane. Polymerizations were then conducted at 0 - 10 ^{o}C by adding a solution of the diacyl chloride in dry tetrahydrofuran to a solution of p,p-oxydianiline (ODA) in dry NMP with vigorous mechanical stirring. Upon completion of acyl chloride addition, the temperature was allowed to gradually return to ambient temperature and stirring was continued for additional 12 hours before precipitating the polymer in water using a Waring blender. For high molecular weight samples the stoichiometric imbalance, r, was ca. 0.980, whereas for low molecular weight, amine-terminated oligomers, DP=10, r was 0.8181 using the aromatic diamine in excess.

High Solids, Chain-extendable Polyimide Precursors

The amino-terminated PMDA/ODA based oligo(amic alkyl esters) were formulated in NMP with a suitable chain-extender as previously described for (amic acid) based systems [9]. Formulation ranging from 35-50 weight% solids were then doctor-bladed onto glass slides and ramp-cured to 350 ^{o}C at 10 ^{o}C/min on nitrogen blanketed hot plates. Specimens were lifted off the glass by scoring the edges and immersing the samples in water. Polyimide strips 0.25 inches in width were cut from the larger films on a JDC® precision cutter for mechanical measurements.

Measurements

Thermal analysis measurements, both isothermal and variable temperature, were obtained from a Perkin-Elmer, Model TGA-7, utilizing a heating rate of 10 ^{o}C/minute in a nitrogen atmosphere. Sample weights of at least 20 mg were employed.

Measurements of tensile properties and adhesion strength were performed on an Instron 1122 at ambient temperature and humdity. A gauge length of 2.0 inches and a strain rate of 0.2mm/minute were typically employed.

Dynamic mechanical thermal analyses wer obtained from a Polymer Laboratories Dynamic Mechanical Thermal Analyzer (DMTA) in the tensile mode at a frequency of 10 Hz, using a heating rate of 10 ^{o}C/minute. All measurements were done in air.

RESULTS AND DISCUSSION

PMDA/ODA Based Poly(amic alkyl esters)

Utilizing the general approach illustrated in Scheme 2, we were able to prepare a variety of different poly(amic alkyl esters) with varying pendant ester groups. In addition, for each particular poly(amic alkyl ester), pure structural isomers could be prepared by starting with either the meta- or para-dialkyl dihydrogen pyromellitate monomer. These monomers were readily accessible via fractional crystallization or solvent extraction of the initial isomer mixture. The corresponding dialkyl diacyl chlorides are highly crystalline solids, which could be easily purified by crystallization from hexane enabling the high yield preparation of the

TABLE I. Viscosity Data of PMDA/ODA Based Poly(amic ethyl ester)

Stoichiometry[a]	*meta*-Diester [η] (dl/g)	*para*-Diester [η] (dl/g)
0.926	0.37	0.74
0.962	0.44	1.18
0.980	0.51	1.58
1.000	1.10	2.67

[a] dianhydride component/diamine, using diamine in excess

desired polyimide precursor with controlable molecular weights. As shown in Table I, a wide range of molecular weights is achievable, ranging from weight average molecular weights of ca. 20,000 to 90,000 [10].

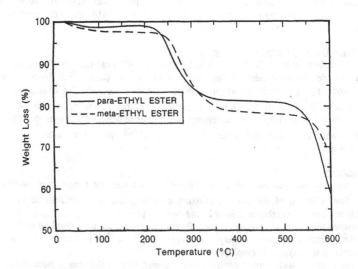

Figure 1. Thermogravimetric analysis (TGA) of PMDA/ODA based poly(amic ethyl ester) isomers; heating rate = 10 °C/minute.

Characterization with respect to the imidization behavior of the various PMDA/ODA based poly(amic alkyl esters) points toward an interesting trend with respect to the chemical nature of the pendant substituent group as well as its corresponding isomer composition. As shown in Figure 1, the para-isomer exhibits onset of imidization ca. 10-20 °C lower than the corresponding meta-isomer. This behavior appears to quite general for all the materials examined so far and suggests a more favorable conformation of the para-isomer toward thermal ring closure. In addition, within each isomer group, imidization temperatures exhibit an inverse correlation with the relative acidity of the parent alcohol used to prepare the corresponding ester derivative; see Figure 2. In retrospect, this is not too surprising since the relative alcohol acidity relates to the electron-withdrawing power of the alkyl fragment, which in the ester form translates to relative susceptibility of the carbonyl group toward nucleophilic attack by the nitrogen in the ortho-amide linkage.

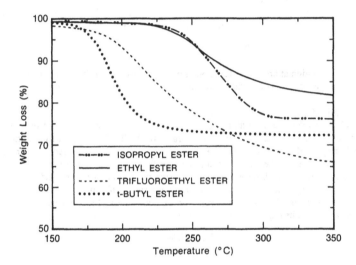

Figure 2. Thermogravimetric analysis (TGA) of various PMDA/ODA based poly(amic alkyl esters). Samples are > 95% para-isomers, heating rate = 10 °C/minute.

Characterization of the mechanical properties of polyimides derived from the thermal cyclization of poly(amic alkyl esters) correlate quite well with those obtained from poly(amic acid) derived polyimides. However, the elongation at break, ε_b, which is quite sensitive to the molecular weight of the polyimide [11], is substantially higher for poly(amic alkyl ester) derived polyimides; see Figure 3. Whereas polyimides derived from poly(amic acids) exhibit a plateau at ca. 50-60% elongation for molecular weights in excess of 10,000, poly(amic alkyl ester) precursor based polyimides increase past this value, reaching values of ca. 120% elongation. This suggests that poly(amic acid) derivatives are much less prone to molecular weight breakdown upon thermal imidization due to the absence of the by-product water, in analogy with the mechanical behavior observed for chemically imidized poly(amic acid) samples [12].

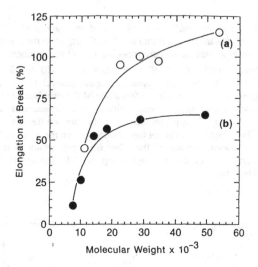

Figure 3. Elongation at break as a function of molecular weight for PMDA/ODA based
polyimides; a) poly(amic ethyl ester) precursor; b) poly(amic acid) precursor.

Chain-Extendable Oligo(amic alkyl esters)

Polyimide resins derived from monomeric and/or oligomeric systems capable of undergoing
molecular weight build-up during curing are desirable for high solids content formulations with
significantly improved global planarization characteristics. Although a variety of such systems
have gained popularity in the aerospace industry for fabrication of lightweight, high strength
composites, approaches which provide chain-extension without cross-linking and change in the
overall chemical composition of the final polyimide are rare. Of course, such an approach has
the distinct advantage of yielding a material with optimum thermal and mechanical properties,
indistinguishable from its high molecular weight counterpart.

SCHEME 3

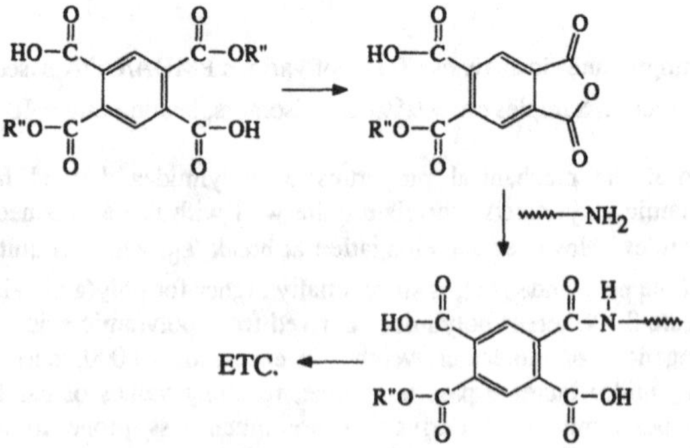

where R" = -CH$_2$CF$_3$, -CH$_2$CO$_2$C$_2$H$_5$, etc.

However, semiflexible and rigid polyimides, with high glass transition temperatures ($T_g > 350$ °C), do not exhibit the expected tracking of the T_g with curing temperature. Upon imidization, the T_g jumps very quickly to its maximum value, thus effectively vitrifying the system at relative low temperatures. For this reason, a chain-extension scheme which occurs at temperatures preceeding the maximum imidization temperature is required. Along these lines we have previously reported a chain-extension approach, which utilized oligo(amic acids) in conjunction with "activated" dianhydride derived diester diacids as chain-extenders [9]. These "activated" diacid diesters revert back to an anhydride intermediate at elevated temperatures, allowing for reaction with amine functionalities, see Scheme 3. This chain-extension approach provides for linear chain-extension as well as a final polyimide identical in chemical composition to polyimides derived from high molecular weight precursors.

It is not too surprising that both the tendency for anhydride formation of the chain-extender and the amic-acid cyclization reaction to yield imide occur in a similar temperature regime due

SCHEME 4

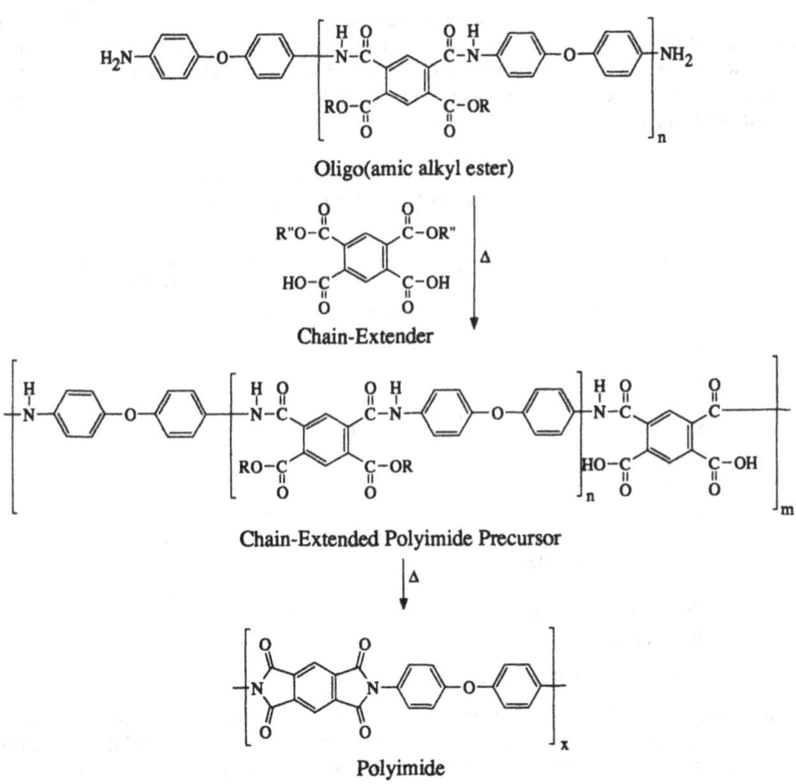

Oligo(amic alkyl ester)

Chain-Extender

Chain-Extended Polyimide Precursor

Polyimide

to related chemistries. This overlap in the temperature requirements for anhydride formation and imidization necessitates the use of highly "activated" chain-extenders, i.e. anhydride formation significantly lower than 150 °C, to facilitate molecular weight build-up prior to imidization.   Of course, the enhanced reactivity of the chain-extender at lower temperatures translates directly into less stable polyimide precursor formulations. Alternatively, It would be highly advantageous to push the imidization reaction toward higher temperatures allowing for a broader process window for chain-extension and, thus, the use of less active (more stable) chain-extenders. In light of this discussion, it seems only appropriate to consider amic-acid derivatives for use as oligomers with high associated imidization temperatures. As shown in Scheme 4, an oligo(amic alkyl ester) is combined with a suitable chain-extender as dictated by the imidization characteristics of that particular oligomer. The formulation is then slowly cured via a ramp-soak type temperature profile to promote chain-extension and finally heated to a temperature in excess of 300 °C to fully imidize the system.   In addition to imidization characteristics, solubility considerations as offered by the respective structural isomers and the chemical nature of the ester group further help to optimize the design of high solid content formulations. Table II tabulates the mechanical properties, such as elongation and tensile strength, of PMDA/ODA based polyimides derived from a high molecular weight poly(amic acid) precursor as well as the identical polyimides as derived from an oligo(amic acid) and various oligo(amic alkyl ester) based chain-extendable formulations.   As reported in the literature for PMDA/ODA based polyimides, the elongation at break is a good indicator as to the relative molecular weight of the polymer, dropping off rather sharply at weight average molecular weights below 10,000 [11]. Comparison of the experimental data in Table II clearly demonstrates the effectiveness of most oligo(amic alkyl ester) based chain-extendable formulations, particularly when compared to the mechanical properties of poly(amic acid) and oligo(amic acid)/diethylglycolyl pyromellitate (EGX) derived polyimides.

TABLE II. Mechanical Properties of PMDA/ODA Based Polyimides Derived From
Low Molecular Weight, Chain-Extendable Precursors

Formulation	Tensile Strength (psi)	Elongation (%)
Poly(amic acid) DP = 70	19,900	52
Oligo(amic acid) DP=5, EGX	16,900	30
Oligo(amic ethyl ester) DP=12, EGX	19,600	81
Oligo(amic isopropyl ester) DP=11, EGX	16,800	77
Oligo(amic ethyl ester) DP=12, MEX	-	brittle
Oligo(amic isopropyl ester) DP=11, MEX	16,600	56

Comparison of the EGX based chain-extender formulations reveals exceptional elongation values, reflecting efficient molecular weight build-up during cure, in support of the mechanistic aspects discussed earlier.   On the other hand, dimethyl pyromellitate, MEX, based chain-extenders, which are expected to exhibit considerable less reactivity than their EGX counterparts, still provided one formulation, based on the oligo(amic isopropyl ester), which yielded a polyimide with respectable mechanical properties. Although their is only a 15-20 °C difference in the observed maximum imidization temperature of the oligo(amic ethyl ester) and oligo(amic isopropyl ester), this relative small imidization temperature difference was sufficient to prevent efficient chain-extension in case of the former oligomer. Although these are just a few of the many possible oligomer/chain-extender combinations, they demonstrate the possibility of designing a variety of low molecular weight, high solids content polyimide precursors by the proper selection of chain-extender and olig(amic alkyl ester).

Imide Containing Multiphase Block Copolymers and Blends

Another area amenable to poly(amic alkyl ester) chemistry is in the preparation of polyimide block copolymers. The use of block copolymers allows for the design of materials with varying molecular architecture, block lengths and composition, yielding a wide range of properties and morphologies. With the proper choice of polymer morphology it is then possible to taylor block copolymers which display the desirable features of each homopolymer constituent. Polyimides, as a class of materials, have received relatively little attention as a component in the synthesis of block and segmented copolymers [13,14]; among these the imide-siloxane copolymers [14] have been most widely studied. However, even in these cases only flexible imide components, i.e., those exhibiting low T_g and good solubility in the imide form, have been reported.

Although rigid and semi-rigid polyimides meet many of the material requirements for microelectronic applications, the presence of an ordered morphology coupled with the lack of a well-defined T_g, result in extremely poor self-adhesion, i.e., the adhesion of polyimide to itself.

SCHEME 5

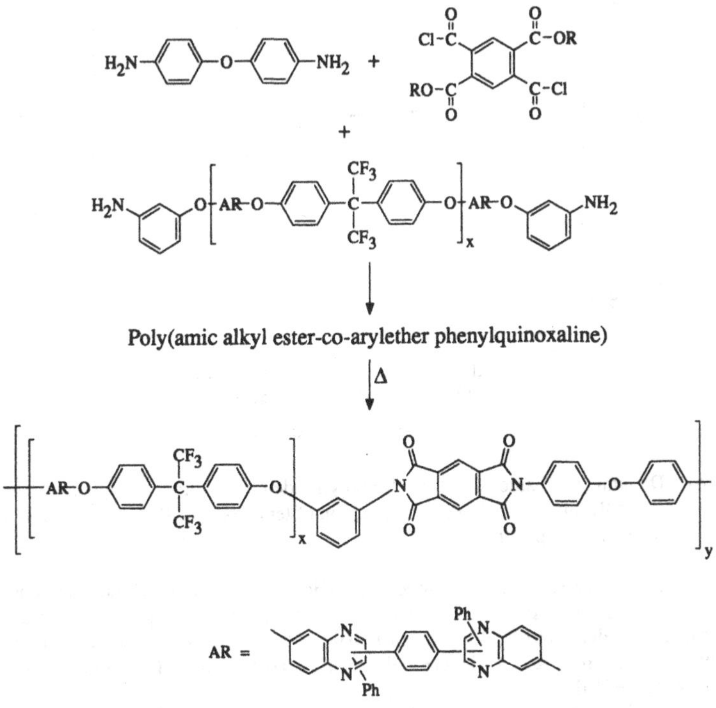

Poly(amic alkyl ester-co-arylether phenylquinoxaline)

Alternatively, thermally stable engineering thermoplastics, i.e., poly(phenylquinoxalines) poly(aryl ether-sulfones), etc., exhibit excellent self-adhesion, but lack sufficient high temperature dimensional stability and/or mechanical properties. Consequently, it would be desirable to combine the self-adhesion of an engineering thermoplastic with the ordered morphology and excellent mechanical properties of a rigid polyimide. Utilizing a poly(amic alkyl ester) based chemistry it is possible to design a synthetic pathway amenable to block copolymer preparation. As illustrated in Scheme 5, a suitable amine-terminated oligomer, i.e. aryl ether phenylquinoxaline (PQE), is co-reacted with a monomeric diamine and a dianhydride derived dialkyl ester diacyl chloride. This approach offers enhanced synthetic flexibility over traditional poly(amic acid) precursor route due to the compatibility of the starting materials and resulting intermediate block copolymer in a variety of mixed solvents. The intermediate block copolymer is then precipitated and thoroughly washed to remove any homopolymer contamination, a feature presented solely due to the hydrolytic stability of the amic alkyl ester linkages. Analogous to traditional polyimides, this precursor is then formulated in a suitable solvent and thermally cured to yield the desired polyimide block copolymer. The salient features of such a system are clearly shown by the dynamic mechanical behavior as a function of temperature in Figure 4. Two transitions, as reflected by the drop in the loss modulus, E' are apparent, indicating a phase-separated morphology. The first transition in the 250 °C range, resulting from the PQE glass transition, and the second transition in the 360 °C range, identical

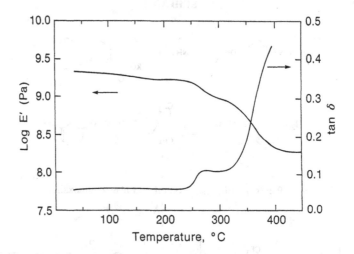

Figure 4. Dynamic mechanical behavior for poly(imide-co-aryl ether phenyl quinoxaline) (PI/PQE) block copolymer as a function of temperature. PQE content is 29 wt% with a block size of 15,500.

to the transition observed in the dynamic mechanical spectra of polyimide homopolymers, clearly support this conclusion. The minimal drop in modulus at high temperatures further indicates that the ordered morphology, characteristic of the parent polyimide, has been retained. Accompanying properties such as the thermal expansion coefficient and self-adhesion have been reported elsewhere [15].

A more practical approach to the modification of polymer properties is through the use of polymer blends. However, in the case of polyimides the blending of the respective homopolymers has to occur in the precursor state since the majority of aromatic polyimides are insoluble. Since the polymer molecules tend to be more flexible in the precursor form, large scale phase-separation of the otherwise rigid polymer molecules can be prevented. In the case of poly(amic acid) mixtures, the respective homopolymers do not form a well-defined molecularly mixed blend. Instead, the "monomer-polymer" equilibrium causes fragmentation of the respective homopoly(amic acids) with recombination of different fragments leading to the initial formation of segmented, block copoly(amic acids) and finally extensive monomer randomization [16]. As a result, such solution mixtures age with time and ultimately result in polymer properties which are comparable to a random copolymer of similar composition. If one of the blend components is unable to display this fragmentation-recombination behavior, as in the case of poly(amic alkyl esters), then mixing is possible. By carefully controlling the composition, molecular weight of the components, curing temperature profile and film thickness, transparent polyimide films can be obtained. This multiphase behavior is demonstrated by the high temperature dynamic mechanical properties. Figure 5 shows the storage modulus, E', and loss modulus, E", of the PMDA-PDA poly(amic ethyl ester)/6F-BDAF poly(amic acid) derived polyimide blend as a function of temperature and composition. The rigid PMDA-PDA homopolyimide exhibits only a small decrease in E' up to 500 °C .In contrast, 6F-BDAF polyimide shows a distinct T_g at ca. 260 °C. All of the mixtures studied reflect a distinct softening as indicated by a drop in E' at a temperature associated with the glass transition of the 6F-BDAF component. This constant Tg indicates almost total exclusion of the two components irrespective of the relative composition.

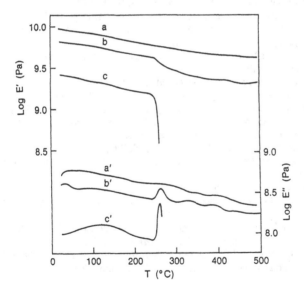

Figure 5. Temperature dependence of the storage (E') and loss (E") modulus at 10 Hz for a PMDA/ODA-6FDA/BDAF (10K) polyimide blend. a) PMDA/PDA homopolymer; b) 70/30 PMDA/PDA-6FDA/BDAF blend; c) 6FDA/BDAF homopolymer.

CONCLUSIONS

Although poly(amic alkyl esters) have been known for many years, their potential utility has not been fully recognized until recently. The fact that poly(amic alkyl ester) derived polyimides exhibit enhanced mechanical properties, particularly with respect to elongation, holds out hope that otherwise brittle polyimides can be similarly improved. In addition, several new approaches leading to high solids content polyimide precursor formulations as well as polyimides with multiphase morphologies have been successfully explored. These approaches were made possible by utilizing derivatized amic acid materials which are polyimide precursors devoid of hydrolytic instability and which can be designed to imidize at signficantly higher temperatures than their amic acid counterparts. The combination of enhanced solubility and higher imidization temperature enabled the design of a condensation chemistry based, linear chain-extension scheme.

As a result, traditional shortcomings of aromatic polyimides, such as shelf-life, processability, and adhesion can be effectively addressed and optimized without sacrificing any of the other desirable properties associated with polyimides.

REFERENCES

1. C.E. Sroog, *J. Polym. Sci.: Macromol. Rev.*, **11**, 161 (1976).
2. A.M. Wilson, *Thin Solid Films*, **83**, 145 (1981).
3. L.W. Frost and I. Kesse, *J. Appl. Polym. Sci.*, **8**, 1039 (1964).
4. S. Nishizaki and T. Moriwaki, *J. Chem. Soc. Japan*, **71**, 1559 (1967).
5. S.N. Kharkov, Ye.P. Krasnov, S.N. Lavrona, S.A. Baranova, V.P. Akesovova and A.S. Chengolya, *Vysokomol. Soyed.*, **13**, 8233 (1971).
6. V.V. Khorshak, s.V. Vinogradova, Ya.S. Vygodskii and Z.V. Gerashenko, *Vysokomol. Soyed.*, **13**, 1197 (1971).
7. M. Fryd et. al., US Patents 4,533,574 and 4,562,100 (1985).
8. V.L. Bell and R.A. Jewell, *J. Polym. Sci., Part A-1*, **5**, 3043 (1967).
9. W. Volksen, R. Diller and D.Y. Yoon, <u>Recent Advances In Polyimide Science and Technology</u>, Proceedings of 2nd Technical Conference on Polyimides, Society of Plastics Engineers, Ellenville, New York, October 1985, page 102.
10. P.M. Cotts and W. Volksen, *Polymer News*, **15**, 106 (1990).
11. W. Volksen, P.M. Cotts and D.Y. Yoon, *J. Polym. Sci., Part B:Polymer Physics*, **25**, 2487 (1989).
12. M.L. Wallach, *J. Polym. Sci.:Part A-2*, **6**, 953 (1968).
13. B.J. Jensen, P.M. Hergenrother and R.G. Bass, *Proc. Polym. Mat. Sci. & Eng.*, **60**, 294 (1989).
14. B.C. Jonson, I. Yilgor and J.E. McGrath, *Polym. Preprints*, **25(2)**, 54 (1984).
15. J. Hedrick, J. Labadie and W. Volksen, *SAMPE Proc.*, **4**, 214 (1990).
16. M. Ree, D.Y. Yoon and W. Volksen, *Polym. Mat. Sci. & Eng.*, **60**, 179 (1989).
17. S. Rojstaczer, D.Y. Yoon, W. Volksen and B.A. Smith, *MRS Symp. Proc.*, **171**, 171 (1990).

POLYIMIDE COPOLYMERS CONTAINING VARIOUS LEVELS OF THE 6F MOIETY FOR HIGH TEMPERATURE AND MICROELECTRONIC APPLICATIONS

M. HAIDER, E. CHENEVEY, R. H. VORA, W. COOPER, M. GLICK and M. JAFFE; Hoechst Celanese Research Division, 86 Morris Avenue, Summit, New Jersey 07901

ABSTRACT

Trifluoromethyl group-containing polyimides not only show extraordinary electrical properties, but they also exhibit excellent long-term thermo-oxidative stability. Among the most thermomechanically stable structural polyimides are those from 6F dianhydride (6FDA) and 6F diamines. The effects of substituting non-fluorine containing monomers such as BTDA, mPDA and 4,4'-DADPS for the hexafluoroisopropylidene monomers on the dielectric, thermo-oxidative, thermal and mechanical properties of the copolymers were studied.

INTRODUCTION

Polyimide polymers are generally prepared by the condensation polymerization of dianhydrides with aromatic diamines. These polymers usually have excellent high temperature properties with the glass transition temperatures (Tg) typically ranging from 200°C for polysiloxane imide to 400°C for duPont Kapton. The continuous service temperature for wholly aromatic polyimides range from 220-300°C [1]. Due to their high thermo-oxidative stability, polyimides can withstand exposure to 300°C for periods of 100 hours or longer.

Linear aromatic condensation polyimides are being used increasingly as high performance film and coating materials for the electronic applications. The primary applications are in the areas of microelectronics as (1) fabrication media such as photo resists, planarization layers and ion implant masks; (2) passivation overcoat and interlayer dielectrics; (3) adhesives; and (4) substrates for electronic devices. Polyimides are also used in the aerospace structural composites and automotive industries where thermal and/or chemical stability is required.

The dielectric constants of commercially available polyimides range from 3.0 to 4.0 MHz. Unfortunately, as the polymer absorbs moisture, the dielectric constant rises, making measurement and operation of electronic devices more complicated. Recently, at Hoechst Celanese, several novel fluorinated polyimides have been synthesized that have excellent electrical, thermal, mechanical and moisture uptake properties. The dielectric constants of those fluorinated polyimides range from 2.4 to 2.8.

The purpose of this investigation is to synthesize and characterize a series of polyimide polymers with various compositions of dianhydrides and diamines and compare them with 6F fluorinated polyimides. To study the structure-property relationships, two approaches were used. The first approach was to reduce or replace 6F dianhydride with BPDA, BTDA and ODPA. The second approach was to reduce or replace 6F diamine with other diamines.

EXPERIMENTAL

Materials

The chemical structure and nomenclatures of the monomers, aromatic diamines and dianhydrides, used in this study, are shown in Figure 1. High purity polymer-grade monomers of PMDA, BPDA, BTDA, ODPA and diamine monomers 4,4' DDSO2 and mPDA were commercially available. Electronic grade 6FDA and 6F-diamines were obtained in-house. All dianhydrides and diamines were purified by vacuum sublimation prior to use. N-methyl pyrrolidone (NMP), used as polymer solvent, was vacuum distilled. The acetic anhydride and beta-picoline were used as received.

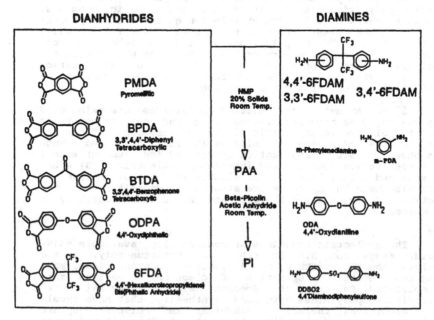

Figure 1. MONOMER STRUCTURE, NOMENCLATURES AND POLYMERIZATION CONDITIONS

Polymerization

Preparation of polyimides is reported in the literature by various methods [2,3,4]. The most common procedure, also used in this investigation, is a two-step synthesis process wherein a dianhydride such as pyromellitic dianhydride (PMDA) is reacted with a diamine to form a soluble polyamic acid (PA) which is then cyclized to form a polyimide (PI). Cyclization can occur by either thermal or by chemical means [5,6,7]. The solid dianhydride monomer was added to an equimolar amount of diamine dissolved in NMP to make a 20% solid concentration. The reaction mixture was stirred for about 15 to 20 hours at room temperature under nitrogen to make a polyamic acid solution. Chemical imidization was carried out by the addition of beta-picoline and acetic anhydride (dehydrating agent) at room temperature while being stirred for an additional 15 to 20 hours. The chemically imidized polymer was then precipitated with methanol. Synthesis of copolyimides was accomplished by similar reaction conditions using multiple dianhydride and diamine monomers. The monomer ratios for copolyimides 6FDA+BTDA+4,4'DDSO2 and 6FDA+BTDA+4,4'DDSO2+ mPDA were 37.5/12.5/50.0 and 37.5/12.5/25.0/25.0, respectively. The monomers used in this study are shown in Figure 1.

RESULTS AND DISCUSSION

Properties of polyimides prepared from the monomers shown in Figure 1 are listed in Tables 1 through IV. Polymer molecular weight, inherent viscosity, and solubility are listed in Table 1. The glass transition temperatures of these materials (Table II) ranged from 252-406°C. The wide range in Tg measurements reflects the large variation in molecular structures. Consistent with observations made by St. Clair and coworkers [8], and with the explanation presented by Fryd [11], the polymer Tg increased with the rigidity of the dianhydride structure. Rigidity of the dianhydride structure was controlled via incorporation of various flexibilizing linkages or "separator" groups such as -CO, -O, -CF3, -SO2. The observed Tg increase based on the dianhydride structure in this study was in the following order: PMDA>BPDA>6FDA>ODPA,BTDA. There was a significant reduction in Tg of the polymers prepared using the meta-oriented 3,3'-6F diamine with 6FDA as compared to the 4,4'-6F or 3,4'-6F diamine isomers. This was due to the distortion of the linearity of the polyimide chain thereby reducing rotational energy [9].

Precipitated polymer solubility was tested in NMP and DMAc at 20-50°C. The data (Table I) shows that polymers containing dianhydride PMDA are insoluble. The remaining polymers used in this study proved to be soluble. PMDA-based polyimide was also solubilized by copolymerizing PMDA and 6FDA (two dianhydrides) with two aromatic diamines, ODA and 4,4'-6F [10,11].

TABLE I. Polyimides – Solubility and Intrinsic Properties

POLYMER COMPOSITION	IV(dl/g)[1] PA	IV(dl/g)[2] PI	Mw(PI)	Mn(PI)	Mw/Mn	SOLUBILITY[3]
6FDA + 3,3' 6FDAM	0.73	0.53	131,000	52,100	2.5	Sol
6FDA + 4,4' 6FDAM	1.14	1.00	243,000	135,000	1.8	Sol
6FDA + 3,4' 6FDAM	0.42	0.31	42,130	13,900	3.0	Sol
PMDA + 4,4' 6FDAM	0.50	Ins	N/A	N/A	N/A	Ins
BPDA + 4,4' 6FDAM	1.07	0.97	140,000	80,000	1.7	Sol
BTDA + 4,4' 6FDAM	1.15	0.73	95,000	32,700	1.7	Sol
ODPA + 4,4' 6FDAM	1.35	1.10	117,000	47,000	2.4	Sol
6FDA+BTDA+4,4'DDSO2	0.78	0.50	42,500	14,600	3.0	Sol
6FDA+BTDA+4,4'DDSO2+mPDA	0.50	0.43	39,800	16,000	2.5	Sol
PMDA + ODA	1.50	Ins	N/A	N/A	N/A	Ins
PMDA+6FDA+ODA+4,4'6FDAM	0.96	0.59	47,300	24,862	1.9	Sol

(1) Inherent Viscosity of polyamic acid in NMP at 25°C.

(2) Inherent Viscosity of precipitated polyimide in NMP at 25°C.

(3) NMP and DMAc solvents at 20°-50°C.

The thermal stability of the polymers was determined (Table II) by obtaining the temperature of 5% weight loss using dynamic TGA in air at 20°C/min. Since the 5 wt. % decomposition temperatures could not fully indicate the thermal performance of a polymer at a given temperature, isothermal thermo-oxidative (TOS) measurements via TGA at 343°C (650°F) for 300 hours in air were performed. The TGA weight loss of polymers as compared to that of duPont's Kapton is listed in Table II. As expected, the thermal stability of polyimides increased with the presence of -CF3 groups [8]. Also observed was significant degradation for BTDA+4,4'-6FDAM after 300 hours of aging, in air, at 343°C. This was probably due to the carbonyl group oxidation. For 6F polyimides, TOS, in air, generally increased with the following order of dianhydride moieties present in the polymer backbone: 6FDA>ODPA>BPDA>BTDA.

Long-term isothermal, thermo-oxidative studies of poly-imides, including the commercially-available duPont Kapton H, were also performed on film samples air aged for 500 hours at 316°C (600°F) in a forced air oven. Test results are shown in Figure 2. All samples were predried at 150°C for one hour and their weight at this point taken as the reference, or 100% weight value. During the test, crucibles containing the samples were removed from the oven at appropriate times, immediately sealed for cooling, weighed and then returned to the oven for further aging. Neglecting the initial weight loss of all the samples tested (except Kapton) which is thought to be associated with solvent removal, an approxi-mate weight loss of only 1.0 to 3.0% occurred over 450 hours. The weight retention of 6FDA+4,4'6FDAM and 6FDA+3,3'6FDAM over 500 hours of isothermal aging at 600°F was the best observed among the materials tested.

TABLE II. Polyimides - Thermal and Moisture Properties

POLYMER COMPOSITION	DSC Tg(C)	TGA (1) 5%WL at C	TOS (%WL) at (2) 343C/300Hrs	WATER (3) UPTAKE(%)
6FDA + 3,3' 6FDAM	252	525	4.16	1.10
6FDA + 4,4' 6FDAM	320	520	4.16	1.15
6FDA + 3,4' 6FDAM	285	520	N/A	N/A
PMDA +4,4' 6FDAM	405	510	N/A	N/A
BPDA + 4,4' 6FDAM	355	540	5.65	N/A
BTDA + 4,4' 6FDAM	307	535	12.66	N/A
ODPA + 4,4' 6FDAM	307	540	4.27	N/A
6FDA+BTDA+4,4'DDSO2	333	530	N/A	2.95
6FDA+BTDA+4,4'DDSO2+mPDA	321	540	N/A	3.05
PMDA + ODA	406	600	16	3.15
PMDA+6FDA+ODA+4,4'6FDAM	367	538	N/A	N/A

(1) In air, at 20°C/min.

(2) Isothermal TGA in air.

(3) Water vapor at 35°C and 30 mm Hg.

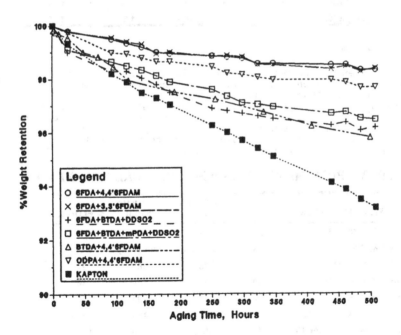

Figure 2.   THERMO-OXIDATIVE STABILITY OF POLYIMIDES AGED IN AIR AT 315°C (600°F)

Electric properties, measured on dried film samples ranging in thickness from 1.0 to 3.0 mils, are reported in Table III. The dielectric constants at 10 MHz ranged from 2.36 for ODPA+4,4'-6FDAM to 3,2 for the commercial duPont polyimide, Kapton H. Overall, the incorporation of fluorine atoms into the backbone of the polymers via addition of -CF3 groups has produced moisture resistant polyimides (Table II) with dielectric constants below 2.6.

TABLE III. POLYIMIDE FILMS – ELECTRICAL PROPERTIES

POLYMER COMPOSITION	F.T. (mil)	D.Const. (10MHz)	D.Stren. (V/mil)	Dis.Fac./10-3 (10MHz)	Vol.Res./10-16 (Ohm/cm)
6FDA + 3,3' 6FDAM	3.00	2.55	1933	1.5	1.97
6FDA + 4,4' 6FDAM	3.00	2.58	1500	1.8	1.92
6FDA + 3,4' 6FDAM	1.25	2.59	3280	4.3	1.98
BPDA + 4,4' 6FDAM	3.00	2.55	1900	1.7	2.12
BTDA + 4,4' 6FDAM	2.00	3.17	3110	2.6	1.74
ODPA + 4,4' 6FDAM	3.00	2.36	1660	3.1	1.46
6FDA+BTDA+DDSO2	2.00	2.56	2050	2.7	1.89
PMDA+ODA (KAPTON)	1.00	3.20	3900	1.6	1.66
PMDA+6FDA+ODA+6FDAM	1.25	2.44	4080	3.9	1.94
BPADA+mPDA (ULTEM)	2.00	3.15*	N/A	1.3*	6.70

*Measured at 1 MHz.

The tensile properties of 6F polyimide films were evaluated and are shown in Table IV. The film samples were solution cast (DMAc or methylene chloride) and vacuum dried, giving essentially no orientation. As a reference, the film properties of the commercial polyimide, Kapton H, are also presented. Compression molded bar properties which are provided for some of the samples correlate very well with the corresponding film properties. The meta-oriented 3,3'-6F

TABLE IV. POLYIMIDE FILMS – MECHANICAL PROPERTIES

POLYMER COMPOSITION	F.T. (mil)	TEN.STREN. (KPSI)	MODULUS (KPSI)	% ELONG.
6FDA + 3,3' 6FDAM	2.0	14	470	4
6FDA + 4,4' 6FDAM	2.3	15 (13.5)	400 (434)	8 (6)
BPDA + 4,4' 6FDAM	3.0	18	390	22
BTDA + 4,4' 6FDAM	2.0	15	380	10
ODPA + 4,4' 6FDAM	3.0	14	294	10
6FDA+BTDA+4,4'DDSO2	*	(9.1)	(499)	(2.1)
6FDA+BTDA+DDSO2+mPDA	*	12.5	(450)	(3.6)
PMDA + ODA	1.0	33	370	72

*Molded bars; Molded properties in parenthesis.

diamine had a higher modulus than the corresponding para material. However, addition of mPDA decreased the modulus of the 6FDA+BTDA+ 4,4'DDSO2 system. Overall, there was no apparent correlation between the tensile properties of the synthesized polyimides and the rigidity of the dianhydride structures.

Melt processing stability of three polyimide samples, namely 6FDA+4,4'-6FDAM (6F44), 6FDA+3,3'-6FDAM (6F33) and 6FDA+BTDA+4,4'-DDSO2, was compared against General Electric Ultem 1000 polyetherimide (PEI) by measuring the complex viscosities as a function of time [12]. The properties of the polymers were measured with a Rheometrics Mechanical Spectrometer. Polymer samples were first molded into disks and melted in the parallel plate test fixture of the rheometer with minimum exposure to oxygen. Polymer viscosities were measured as a function of time for 30 minutes at 50°C above Tg and a frequency of 1 radian per second (rad/s). The viscosities of the samples, plotted as a function of time in Figure 3, indicate that the 6F polyimide resins should behave similarly to Ultem PEI in melt processing. The viscosity of the copolyimide 6FDA+BTDA+DDSO2, although adequate for melt processing, appeared to be the least stable among the samples tested and increased with time.

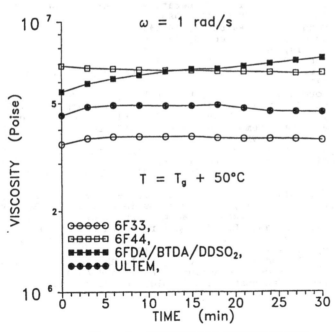

Figure 3. VISCOSITY AS A FUNCTION OT TIME

CONCLUSIONS

A series of stable, high temperature polyimides has been successfully synthesized by the solution polymerization of fluoro-containing dianhydrides and aromatic diamines. These polymers, with glass transition temperatures in excess of 250°C, have been readily processed into thin film and molded articles via conventional techniques. These materials exhibited exceptionally low dielectric constants ranging in value from 2.4 to 2.8, significantly lower than Kapton. Co-polymerization allowed optimization of dielectric properties, solubility and cost for specific end uses.

The Tg of polyimides used in this study could be tailored by controlling the rigidity of the chain through the introduction of rigid or flexibilizing "separator groups."

Polymers such as these are applicable for aerospace, electronic, automotive and other industries where high electrical and moisture resistance, high mechanical properties, chemical stability, thermal stability and ease of processibility are required.

REFERENCES

1. W. B. Alston, "Structure-To-Glass Transition Temperature Relationships in High Temperature Stable Condensation Polyimides," NASA TM-87113, 1985.
2. F. E. Rogers, U.S. Patent 3,356,648.
3. F. E. Rogers, U.S. Patent 3,959,350.
4. A. L. Landis, et al., U.S. Patent 4,645,824.
5. R. H. Vora, U.S. Patent 4,978,737.
6. W. Mueller, R. H. Vora and D. Khanna, U.S. Patent 4,978,738.
7. R. H. Vora and W. Mueller, U.S. Patent 4,978,742.
8. A. K. St. Clair, T. L. St. Clair, W. S. Slump and K. S. Ezzell, "Optically Transparent/Colorless Polyimides," NASA Technical Memorandum 87650, NASA Langley Research Center, Hampton, VA, 1985.
9. M. Fryd, "Structure-Tg Relationships in Polyimides," Proceedings from the First Technical Conference on Polyimides, Ed. by K. L. Mittal, Plenum Press, New York, 1984.
10. A. K. St. Clair, T. L. St. Clair, and E. N. Smith, "Structure-Solubility Relationships in Polymers," Ed. by F. Harris and R. Seymour, Academic Press, New York, 1977.
11. F. W. Harris and L. H. Lanier, "Structure-Solubility Relationships in Polyimides," Ed. by F. Harris and R. Seymour, Academic Press, New York 1977.
12. J. M. Dealy and K. F. Wissbrun, "Melt Rheology and Its Role in Plastics Processing - Theory and Applications," pp. 577-585, Van Nostrand Reinhold, New York, 1990.

FLUORINATED POLYIMIDE BLOCK COPOLYMERS

J.W. Labadie, M.I. Sanchez, Y.Y. Cheng and James L. Hedrick
IBM Research, Almaden Research Center, 650 Harry Road, San Jose, CA
95120-6099

ABSTRACT

Block copolymers based on the poly(amic ester) of poly(4,4'-oxydiphenylene-pyromellitide) and perfluoroalkylene aryl ethers were synthesized from preformed amine terminated perfluoroalkylene aryl ether oligomers. Fluoropolymer structure, wt% incorporation, and molecular weight were varied. Reduction in the dielectric constant to 2.8-2.9 and reduced moisture sensitivity were observed for block copolymers with 55 wt.% poly(perfluoroalkylene aryl ether) content without significant compromise of mechanical or thermal properties.

INTRODUCTION

Polymers are becoming increasingly important as interlayer dielectric materials in electronic components. Semi-rigid aromatic polyimides, e.g. poly(4,4'-oxydiphenylenepyromellitimide) (PMDA-ODA), have proved to be the most suitable materials for this application. Polyimides with lower dielectric constants (2.7-3.0) have been prepared through incorporation of perfluoroalkyl groups in the polyimide structure. Examples include the incorporation of hexafluoroisopropylidine linkages (Hoechst Sixef), main chain perfluoroalkylene groups, and pendent trifluoromethyl groups [1]. Modifying the polymer by introducing flexible perfluoroalkyl groups in the polymer backbone affords a material with diminished final properties, while the addition of pendent trifluoromethyl groups appears to have a less deleterious effect. Alternatively, reduction of the dielectric constant can be achieved through introduction of a fluorinated polymer as a distinct phase in a rigid polyimide matrix. These molecular composites can be prepared by either blending fluorinated poly(amic acids) with rigid poly(amic esters) [2], or by a block copolymer approach [3]. In the case of block copolymers the dissimilar coblocks are covalently bonded, which affords microphase separation for even large differences in the solubility parameter of the component coblocks. In this paper we will describe our continuing investigation of block copolymers based on PMDA-ODA and perfluoroalkylene aryl ethers.

RESULTS AND DISCUSSION

Preparation of the block copolymers required the synthesis of highly fluorinated amine functional oligomers which were soluble in N-methylpyrrolidone (NMP) or NMP solvent mixtures. This led to our investigation of new semifluorinated polymers derived from poly(aryl ethers). Incorporation of aryl ether groups should give better NMP processability while maintaining the thermal stability, and the techniques known to prepare amine functional poly(aryl ethers) can be applied [4]. This lead us to study the synthesis of poly(perfluoroalkylene aryl ethers), or PFAAE, from bisphenols and 1,6-(4-fluorophenyl)perfluorohexane under conventional polyether polymerization conditions [5]. The resulting polymers displayed good processability in a variety of solvents, excellent thermal stability, Tg = 40 - 100 °C., and dielectric constants in the 2.6 range (1 MHz). Amine functional oligomers of controlled molecular weight were prepared by proper adjustment of the

Scheme I

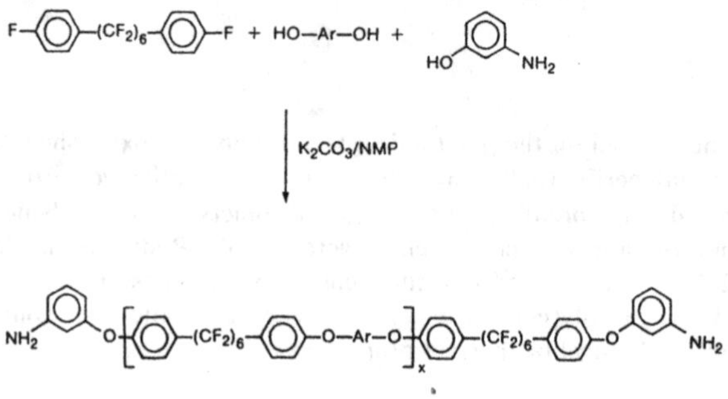

PFAAE

$$Ar = -PhC(CH_3)_2Ph- (6F_2-BisA)$$
$$-PhC(CF_3)_2Ph- (6F_2-BisAF)$$
$$1,3-Ph (6F_2-Res)$$

Scheme II

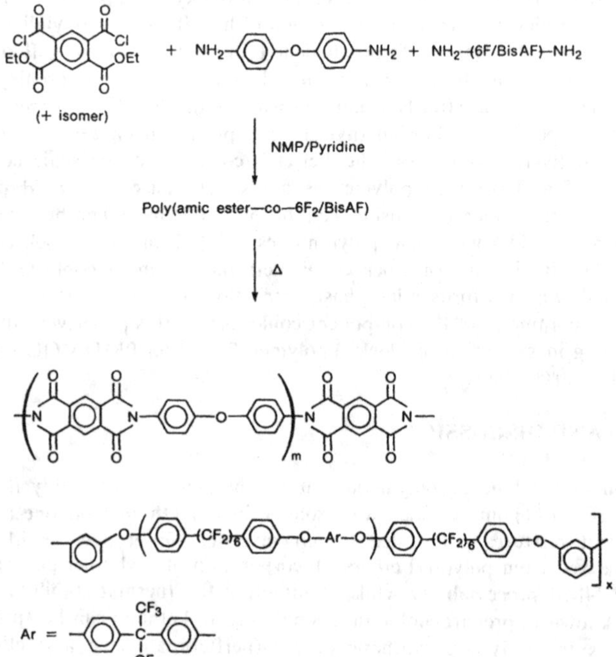

(+ isomer)

NMP/Pyridine

Poly(amic ester—co—6F$_2$/BisAF)

Δ

TABLE 1

IMIDE-PFAAE COMPOSITION

PFAAE Oligomer	Mn	PAE/PFAAE[a]		PI/PFAAE[b]
		Theo	Exptl	
$6F_2$-BisA	6,500	70/30	71/29	67/33
$6F_2$-BisA	6,500	50/50	59/41	54/46
$6F_2$-BisA	6,500	25/75	34/66	29/71
$6F_2$-BisAF	11,900	50/50	49/51	45/55
$6F_2$-BisAF	7,700	50/50	50/50	45/55
$6F_2$-Res	9,000	70/30	64/36	58/42
$6F_2$-Res	9,000	50/50	49/51	44/56

a) poly(amic ester)/PFAAE. b) polyimide/PFAAE

stoichiometry and addition of 1,3-aminophenol as an end-capping agent (Scheme I). This procedure was used to synthesize PFAAE oligomers derived from bisphenol-A ($6F_2$-BisA), bisphenol-AF ($6F_2$-BisAF), and resorcinol ($6F_2$-Res). Block copolymers were prepared by adding a solution of the diester diacyl chloride of pyromellitic dianhydride (PMDA) to a mixture of oxydianiline (ODA) and the PFAAE oligomer, analogous to the preparation of poly(amic ester-co-benzoxazole) [6] (Scheme II). Formation of the polyimide precursor via poly(amic ester) chemistry, rather than a conventional poly(amic acid) preparation, allowed for the use of toluene as a cosolvent with NMP to improve the solubility of the fluorinated coblock and the resulting block copolymer could be isolated by precipitation and washed with toluene to remove unreacted PFAAE homopolymer. Good agreement between charged PFAAE oligomer and % incorporation was observed by H NMR (Table 1). It should be noted that the higher level of incorporation observed for the 50/50 PAE/$6F_2$-BisAF and $6F_2$-Res may be due to error in the H NMR determination, since there is not a distinct resonance for integration as in the case of the isopropylidine linkage in $6F_2$-BisA. The 36% $6F_2$-Res incorporation determined for the 70/30 PAE/$6F_2$-Res is also suspect since it is unlikely poly(amic ester) would be selectively lost in the polymer workup. Thermal cure leads to conversion of the poly(amic ester) to polyimide and a concomitant increase in PFAAE content. The thermal stability of the cured copolymers was very good at 400 °C (0.20 wt%/h for the 50/50 compositions), albeit lower than the polyimide homopolymer.

The dynamic mechanical analysis of copolymers with $6F_2$-BisA compositions of 35% and 50% are shown in Figure 1. The copolymers displayed a transition corresponding to the PFAAE Tg (100 °C) and a transition at 350 °C, which is identical to that observed for PMDA/ODA. The presence of two transitions at temperatures close to that of the pure homopolymers is indicative of a heterogeneous morphology with good phase purity for the PFAAE domains as well as the PMDA/ODA matrix. The retention of the modulus at high temperature shows the polyimide is the major component and acting as the matrix, despite the high PFAAE content (46%). This can be accounted for by the higher density of PFAAE, which translates into a lower Vol.% of $6F_2$-BisA. This is also borne out in

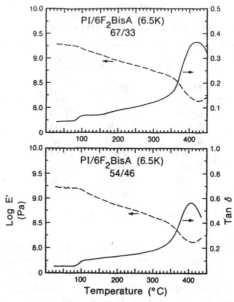

Figure 1. DMTA for Imide-PFAAE block copolymers.

TABLE 2

IMIDE-PFAAE MECHANICAL PROPERTIES

| Composition | | PFAAE | Modulus | σ |
Imide	PFAAE	Mn	(MPa)	(%)
100	0	-	2000	50
6F2-BisAF				
45	55	7,700	2000	72
45	55	11,900	1700	62
0	100	-	1350	35
6F2-BisA				
54	46	6,500	2030	102
53	47	9,900	2000	107
0	100	-	1400	100

the thermal expansion coefficient for the 45/55 composition of $PI/6F_2$-BisAF (11.9 K) which was 40 ppm, equivalent to PMDA-ODA. The mechanical properties of the imide-PFAAE are excellent, with moduli comparable to PMDA/ODA and elongations as high as 100% (Table 2). The dielectric properties of the copolymers showed a reduction relative to PMDA-ODA with the incorporation of PFAAE (Table 3). The lowest values (2.8-2.9) were obtained for the 55% PFAAE compositions with $6F_2$-Res and $6F_2$-BisAF. The PFAAE oligomers with higher DPs showed the best dielectric behavior, indicating phase purity may play an important role. The copolymers showed less sensitivity to ambient moisture relative to

PMDA-ODA or PMDA-6F diamine. The effect of a nitrogen atmosphere saturated with water was followed with time for the $6F_2$-Res copolymers (Figure 2). The magnitude in the increase in dielectric constant was inversely proportional to the $6F_2$-Res content and suggests that the polyimide matrix is absorbing water, with the reduction in total water absorption due to decreasing polyimide content in the film.

TABLE 3

IMIDE-PFAAE DIELECTRIC PROPERTIES

| Composition | | PFAAE | | ε^a |
Imide	PFAAE	Mn(DP)	Wt.%F	Dry/Ambient
100	0	-	0	3.1/3.4
6F2-BisAF				
45	55	7,700(9.7)	24	3.0/3.2
45	55	11,900(15)	24	2.9/3.0
0	100	-	43.5	2.6/2.6
6F2-Res				
58	42	9,000(16)	17	3.0/3.3
44	56	9,000(16)	23	2.8/2.9
0	100	-	40.7	2.6/2.6
Fluorinated Polyimides				
PMDA-6F[b]				2.9/3.3
Hoechst 6F				2.65[c]

a) 1 MHz (b) 6F = 6F-diamine (c) 10 MHz

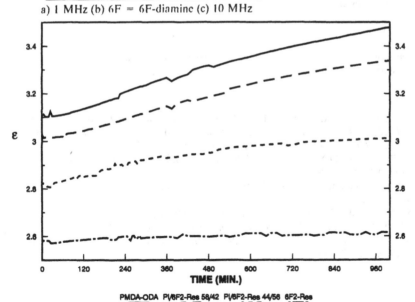

PMDA-ODA PI/6F2-Res 58/42 PI/6F2-Res 44/56 6F2-Res

Figure 2. Dielectric constant as a function of time in water saturated nitrogen.

48

EXPERIMENTAL

The poly(amic ester) block copolymers were prepared by analogy to poly(amic ester-co-aryl ether benzoxazoles) [6]. Films were cast from NMP/xylene (2/1) and heated to 350 °C (5 °C/min) and held for one hour. Dynamic mechanical analysis were performed at 10 Hz and a heating rate of 10 °C/min in the tension mode. Mechanical property measurements were carried out on an Instron tensile tester at a strain rate of 10 mm/min. Samples for dielectric measurements were prepared by evaporating 5 mm gold electrodes on both sides of a free-standing film. Measurements were made at 5 frequencies between 10 KHz and 1 MHz on a HP-4275A multifrequency analyzer, and are reported at 1 MHz. Standard measurement procedure involved equilibration of the sample for three days at ambient conditions, followed by measurement of the dielectric constant through the following temperature profile: -150 °C to 150 °C (10 °C/min)/ hold one hour at 150 °C/ cool to -150 °C/ -150 °C to 150 °C (10 °/min)/ cool to room temperature/ hold 20 h under a sparge of nitrogen saturated with water.

CONCLUSIONS

Block imide-perfluoroalkylene aryl ether copolymers were synthesized and displayed a heterogeneous morphology which allowed the incorporation of approximately 50 wt.% of the perfluoroalkylene aryl ether as a coblock without degrading dimensional stability and mechanical properties. A lower dielectric constant (ε = 2.8-2.9) was observed. The effect of moisture on the dielectric constant was reduced, with the level of reduction in dielectric constant under high humidity primarily to the lower polyimide content in the film.

ACKNOWLEDGEMENT

The author's acknowledge Bruce Fuller for experimental design of the dielectric measurements and Tad Palmer for mechanical measurements.

REFERENCES

1. J.S. Critchley, P.A. Gratan, M.A. White, and J.S. Pippett *J. Polym. Sci.: Part A-1*, **10**, 1789 (1972), T. Matsuura, S. Nishi, M. Ishizawa, Y. Yamada, and Y. Hasuda, *Pacific Polymer Preprints*, **1**, 87 (1989).
2. S. Rojstaczer, D.Y. Yoon, W. Volksen, and B.A. Smith, *Matl. Res. Symp. Proc.*, **171**, 171 (1990).
3. J.W. Labadie and J.L. Hedrick *Proc. of the 4th Intl. SAMPE Elec. Conf.*, **4**, 495 (1990).
4. M. Jurek and J.E. McGrath *Polymer*, **30**, 1552 (1989).
5. J.W. Labadie and J.L. Hedrick, *Macromolecules*, **23**, 5371 (1990).
6. (a) J.L. Hedrick, J.G. Hilborn, J.W. Labadie, and W. Volksen, *Polymer Bulletin*, **22**, 47 (1989).

PROPERTY CONTROL OF NEW FLUORINATED COPOLYIMIDES AND THEIR APPLICATIONS

S. SASAKI, T. MATSUURA, S. NISHI, AND S. ANDO
NTT Applied Electronics Laboratories, 3-9-11, Midori-cho, Musashino-shi, Tokyo, 180, Japan

ABSTRACT

New fluorinated copolyimides are synthesized with 2,2'-bis(trifluoromethyl)-4,4'-diaminobiphenyl (TFDB) and a mixture of 2,2-bis(3,4-dicarboxyphenyl)hexafluoropropane dianhydride (6FDA) and pyromellitic dianhydride (PMDA). The thermal expansion coefficient decreases with decreasing 6FDA content, because the hexafluoroisopropylidene group makes the main chain flexible. The refractive index decreases with increasing 6FDA content, because the fluorine content increases. The thermal expansion coefficient can be controlled between -0.5 x 10^{-5} and 8 x 10^{-5} / $^{\circ}C$, and the refractive index can be controlled between 1.556 and 1.647 at 589.3 nm, by changing the 6FDA content. The polyimides and copolyimides from TFDB can be used in optical filters because of their high optical transparency and low thermal expansion coefficient. They are expected to be used in optical waveguides because of their high optical transparency and refractive index control.

INTRODUCTION

Polyimides have been applied in the microelectronic device[1,2] and aerospace fields[3,4] because of their thermal durability. Especially polyimides have become more important in the microelectronic industry. Applied to interlayer dielectrics, low dielectric constant polyimides are vital for high signal propagation speed[5]. Low thermal expansion polyimides are also needed because they are used with low thermal expansion substrates such as copper, silicon, or silicon dioxide[6]. In applying polyimides to optical uses, optical transparency and refractive index control of the polyimides are required. However, conventional polyimides exhibit low optical transparency. Less colored polyimide films have been reported by St. Clair and co-workers [7,8].

We have been investigating the synthesis and characterization of fluorinated polyimides for application to electrical and optical components. We have already reported a new series of polyimides containing fluorinated alkoxy side chains[9] and two new fluorinated polyimides from 2,2'-bis(trifluoromethyl)-4,4'-diaminobiphenyl (TFDB) [10-13]. One polyimide (6FDA/TFDB), synthesized with TFDB and 2,2-bis(3,4-dicarboxyphenyl)hexafluoropropane dianhydride (6FDA), has high optical transparency and a low refractive index. The other polyimide (PMDA/TFDB), synthesized with TFDB and pyromellitic dianhydride (PMDA), has a low thermal expansion coefficient. Harris and co-workers also reported on the synthesis and properties[14,15].

To match the thermal expansion coefficient (TEC) of substrates, we prepared the polyimide blends with PMDA/TFDB and 6FDA/TFDB[16]. The TEC of these blends can be controlled. However, they cannot be applied to optical use due to local fluctuations of refractive index caused by microphase separation in the blends. We attempt copolymerization of 6FDA, PMDA, and TFDB to avoid the phase separation. This article gives the results of property control of new fluorinated copolyimides and their optical applications.

MATERIAL SYSTEM

The copolyimides were prepared by the reaction shown in Figure 1. The fluorinated copoly(amic acid)s were synthesized with TFDB and a mixture of PMDA and 6FDA in N,N-dimethylacetamide (DMAc). Then the poly(amic acid)s were spin-cast onto silicon wafers and converted into copolyimides by heating.

The properties of 6FDA/TFDB and PMDA/TFDB are shown in Table1 [17]. 6FDA/TFDB has a low refractive index, a low dielectric constant, and a low water absorption rate because of the four trifluoromethyl groups (fluorine content: 31.3 wt%). The TEC of PMDA/TFDB is lower than that of 6FDA/TFDB because of its rigid-rod structure.

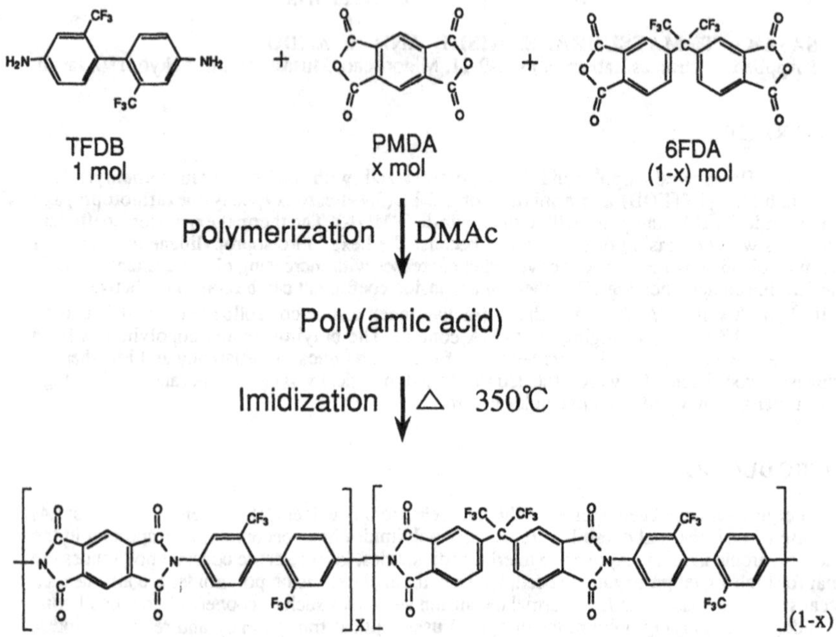

Fig. 1 Synthesis of fluorinated copolyimide

Table I Characteristics of Fluorinated Polyimides

	6FDA/TFDB	PMDA/TFDB
Fluorine Content (%)	31.3	23.0
Intrinsic Viscosity[a]	1.00	1.79
Polymer Decomposition Temp. (°C)[b]	569	610
Glass Transition Temp.[c]	335	>400
Dielectric Constant[d]		
Dry condition	2.8	3.2
Wet condition (50%RH)	3.0	3.6
Refractive Index[e]	1.556	1.647
Water Absorption Rate (%)[f]	0.2	0.7
Thermal Expansion Coefficient (°C^{-1})[g]		
1st run	3.8×10^{-5}	3.3×10^{-6}
2nd run	8.0×10^{-5}	-4.9×10^{-6}

[a] measured in poly(amic acid) solution
[b] 10% weight loss in N_2 atmosphere
[c] measured by DSC
[d] at 1 MHz
[e] λ=589.6 nm, 20°C
[f] after 3 days
[g] temperature range: 50-300°C

The 6FDA/TFDB film has greater transparency than the PMDA/TFDB film, because electron conjugation and intermolecular interaction are prohibited by the trifluoromethyl unit. The color intensity of the 6FDA/TFDA film is compared to those of PMDA/TFDB and the polyimide from PMDA and 2,2'-dimethyl-4,4,'-diaminobiphenyl (DMDB) films of the same thickness in Figure 2.

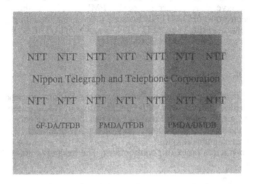

Fig. 2 Comparison of color intensity

In applying polyimides to optical components, the transparency in the near-infrared region , rather than in the visible region, is needed, because the wavelength of 1.3 and 1.55 μm are used in optical communication. The light absorption spectrum was measured for the 6FDA/TFDB solution in acetone-d6 (Figure 3).

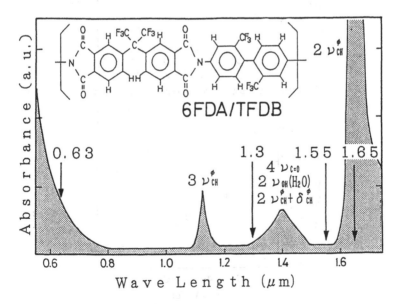

Fig. 3 Light absorption spectrum of 6FDA/TFDB

The absorption in the visible region is caused by electron transition, but the absorption in the near-infrared region is mainly due to the harmonics and their coupling of stretching vibrations at chemical bonds. Especially, C-H bonds and O-H bonds strongly affect the absorption. 6FDA/TFDB has low optical loss at the wavelengths of 1.3 and 1.55 μm, because of the low number of hydrogen atoms in the monomer unit, and the fact that are only the aromatic hydrogen atoms have sharp absorption peaks. PMDA/TFDB is thought to show the same light absorption spectrum as 6FDA/TFDB because they have almost the same number of hydrogen atoms in the monomer unit. Thus, 6FDA/TFDB and PMDA/TFDB are expected to be applied to opto-electronics as well as microelectronics.

PROPERTY CONTROL

Figure 4 shows the thermal expansion coefficient of the fluorinated copolyimides. The TEC of the copolyimides decreases with decreasing 6FDA content, because a hexafluoroisopropylidene group makes the main chain flexible. The copolyimide with 20% 6FDA content has almost the same CTE as copper, and the copolyimide with 10% 6FDA content has almost the same CTE as silicone dioxide.

Figure 5 shows the refractive index of copolyimides. The refractive index decreases with increasing 6FDA content.

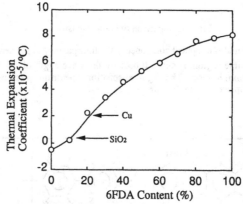

Fig. 4 Thermal expansion coefficient of copolyimides

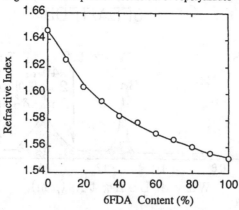

Fig. 5 Refractive index of copolyimides

Thus, in the copolyimides from 6FDA, PMDA and TFDB, the thermal expansion coefficient can be controlled between -0.5 x 10^{-5} and 8 x 10^{-5} $°C^{-1}$ and the refractive index can be controlled between 1.556 and 1.647 at 589.3 nm , by changing the 6FDA content.

Application to optical components

We have been investigating the application our polyimides and copolyimides to optical communication components. Optical interference filters are widely used in optical fiber communication systems.These filters, used for eliminating cross-talk in wavelength-division-multiplexing and so on, must be thin, from 20 to 50 μm thick, because they are embedded in a slot formed across the optical fiber (Figure 6). A conventional thin filter is fabricated by lapping the thick filter with multilayers of TiO_2 and SiO_2 on a glass substrate. The glass filter is expensive and difficult to handle. To solve these problems, we have applied our polyimides and copolyimides with high optical transparency in the near-infrared region and low thermal expansion. The multilayered interference filter consists of alternating TiO_2/SiO_2 layers and the PMDA/TFDB film or a copolyimide film with a high PMDA content. The filter was la ong-wavelength-pass filter. Figure 7 shows the transmission spectrum of the filter.

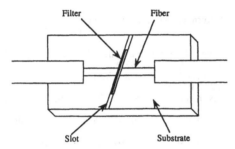

Fig.6 Schematic diagram of fiber pigtailed filter

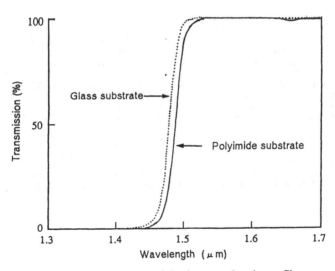

Fig. 7 Transmission spectra of the long-wavelength-pass filters

Transmission loss in the pass band was less than 0.1 dB and cross-talk attenuation of over 50 dB for 1.55 ±0.3 μm is obtained[18]. These values are almost the same as for a multilayer filter deposited on a glass substrate. Thus, multilayered interference filters deposited on polyimide film exhibited excellent optical properties and are cheaper than conventional filters.

We have been investigating polyimide optical waveguides using our polyimides and copolyimides with high optical transparency in the near-infrared region and refractive index control. These polyimides offer a promising prospect for a high performance optical waveguide. The results will be presented at another meeting.

CONCLUSION

New fluorinated copolyimides can be synthesized with TFDB and a mixture of 6FDA and PMDA. The TEC and the refractive index of the copolyimides can be controlled by changing the 6FDA content. The TEC is between -0.5×10^{-5} and 8×10^{-5} / °C, and the refractive index is between 1.556 and 1.647. The copolyimides can be used in optical filters because of their high optical transparency and low TEC.

REFERENCES

1. R. Rubner, H. Ahne, H. Kurn, and G. Kolodziej, Photographic Sci. Eng., 23, 303,1979
2. K. Mukai, A. Saiki, K. Yamanaka, S.Harada, and S.Shoji, IEEE J. Solid-State Circuits, SC-13(4), 462(1978)
3. W.H.Morita and S.R.Graves, Natl. SAMPE Symp. Exhib., Proc., 26, 402(1981)
4. B. Dexter, NASA Conf. Publ., CP-2079(1979)
5. D.M.Stoakley, A.K.St. Clair, and R.M.Baucom, 3rd International SAMPE Electronics Conference June 20-22, 224(1989)
6. S. Numata, K. Fujisaki, N. Kinjo, D. Makino, D. Proc. of the 2nd Intl. Conf. on Polyimides, 164(1987)
7. A.K.St. Clair, T.L.St.Clair, and K.I.Shevket, ACS Polymeric Materials Sci. and Eng., 51, 62(1984)
8. A.K.St.Clair, T.L.St.Clair, and W.S.Slemp, Proc. of the 2nd Intl. Conf. on Polyimides, 14 (1987)
9. T. Ichino, S. Sasaki, T. Matsuura, and S. Nishi, J. Polym. Sci., Polym. Chem. Ed., 28, 323(1990)
10. T. Matsuura, M. Ishizawa, Y. Hasuda, and S. Nishi, SPSJ 38th Symposium on Macromolecules,Japan, 38(3), 434(1989)
11. T. Matsuura, T. Ichino, M. shizawa, N. Yamada, Y. Hasuda, and S. Nishi, SPSJ 38th Symposium on Macromolecules, Japan, 38(3), 435(1989)
12. N. Yamada,Y. Hasuda, and S. Nishi, SPSJ 38th Symposium on Macromolecules, Japan, 38(3), 436(1989)
13. T. Matsuura, S. Nishi, M. Ishizawa,N. Yamada, and Y. Hasuda, Pacific Polymer Preprints, First Pacific Polymer Conference, 1, 87(1989)
14. F.W.Harris and S.L-C.Hsu, and C.C.Tso, Abstracts of papers, Part 1, The 1989 International Chemical Congress of Pacific Basin Societies, Honolulu, Hawaii, Dec. 17-22 1989, 07, 123
15. F.W.Harris, S.L-C.Hsu, and C.C.Tso, ACS Polymer Preprints, 31(1), 342(1990)
16. N. Yamada, T. Matsuura, and S. Nishi, SPSJ 38th Symposium on Macromolecules, Japan, 38(11), 4104(1989)
17. T. Matsuura, Y. Hasuda, S. Nishi, and N. Yamada, Macromolecules (in press)
18. T. Oguchi, J. Noda, H. Hanabusa, and S. Nishi, Electronics Letter (in press)

GROWTH OF POLYIMIDE FILMS BY CHEMICAL VAPOR DEPOSITION AND THEIR CHARACTERIZATION

Steven P. Kowalczyk*, Christos D. Dimitrakopoulos**, and Steven E. Molis*

*IBM Research, T. J. Watson Research Center, Yorktown Heights, N. Y., 10598
**Henry Krumb School of Mines, School of Engineering and Applied Sciences, Columbia University, New York, N. Y., 10027

ABSTRACT

A dry ultra high vacuum technique for the preparation of polyimide films is investigated as an alternative to standard wet processing techniques involving the use of solvents. This technique is based on the co-deposition of monomers (pyromellitic dianhydride and oxydianaline in this study) to form polyimide (poly (4,4′-oxydiphenylenepyromellitimide)). Various parameters to optimize film growth are investigated and the properties of the films are compared to films prepared by spin casting from solution. Films from 10Å to 15 μ were grown. Uniform films over 3 inch wafers were successfully grown. Many properties such as adhesion, dielectric constant and stress were similar to spun films. However, these films did exhibit more crystallinity but less orientation than the spun films. Finally, being a vapor phase process the films were conformal rather than planar.

INTRODUCTION

Polyimide has become an important dielectric material for microelectronic applications [1,2]. Among its useful properties are a low dielectric constant and good thermal stability [3]. The standard method of preparing films of polyimide is that of spinning polyamic acid from solution subsequently followed by thermal treatments to induce cycloimidization of the polyamic acid [4]. This is a simple process which has inherently good planarization properties, however, it is a wet process requiring solvents which conceivably might be disadvantageous for some applications. In this work we investigate a dry growth process, chemical vapor deposition (CVD). This process consists of co-deposition of the constituent monomers from a molecular beam of each monomer. This process was first studied by Salem et al [5]. Shortly after this, it was further developed for ultra high vacuum (UHV) studies of very thin films for in situ interfacial chemistry studies [6-11] and studies on the role of solvents in interfacial reactions [12]. In this paper we report on some recent studies on thick films of the polyimide, poly (4,4′-oxydiphenylenepyromellitimide) (PMDA-ODA) to compare the properties of these films to spun films. The main goals of this work are to clarify the key growth parameters for the growth of uniform, high quality films and to further characterize the properties of these films. The key growth parameters for producing good surface morphology and good uniformity were the relative monomer flux ratios and the rotation of the substrate. A number of key properties such as dielectric constant, adhesion, step coverage, refractive index, and stress were investigated and were found to be comparable to the properties of spun films.

CVD GROWTH OF POLYIMIDE FILMS

CVD Growth Reactor System

A stand alone cold wall CVD reactor dedicated to polyimide growth was constructed. A schematic of the system is given in Fig. 1 and is briefly described below. The growth chamber was a 6 way stainless steel UHV cross with 8 inch conflat flanges. The chamber is pumped with a 60 l/s ion pump attached on one side flange and isolatable with a gate valve. The chamber is roughed with a turbo-molecular pump backed by a mechanical pump. The system after bake out has a base pressure $<1\times10^{-9}$ torr measured with a nude ion gauge. The bottom flange is the source flange. Four sources (two each PMDA and ODA) are mounted on this flange in a MACOR™ block.

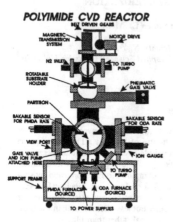

POLYIMIDE CVD REACTOR

Figure 1. A schematic of the CVD reactor system.

The sources are quartz Knudsen cells with a capacity of 2.8 ccm. The charges were high purity zone refined pyromellitic dianhydride (PMDA) (1,2,4,5-benzenetetracarboxylic anhydride) and 4,4'-oxydianiline (ODA) (4-aminophenyl ether). The sources, one PMDA and one ODA, are heated resistively to produce a sufficient vapor pressure of each monomer and the flux of each monomer is independently monitored by separate quartz crystal monitors which are mounted on another flange. After each time that the sources are refilled with charge they are recalibrated (all four sources). The top flange is the substrate flange and it is isolatable with a gate valve. Subtrates upto 3 inch diameter can be mounted using a ring clamp to hold the substrate in place. The source to substrate distance is 35cm. The substrate holder is mounted on a rotatable feedthrough which is driven with a variable speed motor. The substrate is typically rotated at 30-60 rpm for the present experiments. The present substrate holder is non-heatable, thus all growths reported here were done at room temperature and the curing was performed subsequently *ex situ* of the growth chamber.

CVD growth process

Before initiation of growth, the source fluxes of each of the monomers were stabilized to their calibrated stoichiometric ratio with the substrate shuttered. After the fluxes were stabilized, the sample shutter was opened. Typical pressure during growth for the fluxes employed was $\sim 5 \times 10^{-6}$ torr. Flux rates ranged from 2 Å/sec to 70 Å/sec with 5 Å/sec typical value employed. A variety of substrates were used with HF cleaned Si wafer the most commonly used. After growth the films were cured in either one of two ways. For most of these studies the films were cured in a standard curing oven under nitrogen with a stepped programmed cure to a final temperature of 400°C. A number of films were also cured under ultra-high vacuum (5×10^{-10} torr.).

The room temperature deposited films are polyamic acid as shown by previous work [5,6]. In our studies this was confirmed by X-ray Photoelectron Spectroscopy (XPS) and Fourier Transform Infra-red (FTIR) measurements. Overall XPS spectra were used to obtain the elemental composition of the films, while high resolution core-level spectra were used to verify the chemical state of the films. To convert the films to polyimide, the films were heated to cycloimidize the films. XPS and FTIR were used to monitor this condensation reaction and ascertain the completeness of the imidization.

CHARACTERIZATION OF CVD POLYIMIDE FILMS

We have characterized films grown by the above method with a number of techniques. In this section, we report on the characterization of properties of thick films, *i.e.* films in the thickness range 1μ to 20μ. Two of the first concerns were to demonstrate film uniformity and good morphology. Good uniformity for growth on 3 inch Si wafers was obtained as indicated by profilometry with Detak. The thickness variation across a three wafer of ~1% were observed. This uniformity was obtained by use of substrate rotation. Good morphology as judged by featureless scanning electron micrographs was also obtained. Here the crucial growth parameter was the relative flux ratio, which had to be controlled to the stoichiometric ratio. This is extremely important for thick films. In earlier studies of thin films, <1000 Å, good film growth was obtained without close control of the flux ratio [9,13]. An example of bad growth resulting from non-stoichiometric flux is shown in Fig 2. Initially the as deposited polyamic film was featureless, but upon curing structures appear on the surface as seen in Fig.2. XPS studies showed that the polyamic acid film was PMDA enriched. Fig. 3 compares XPS C1s spectra from two CVD films grown with different fluxes with that of a spun film. The interpretation of the features of the C1s spectra has been discussed previously [6,7,9,10]. Good fully cured films are essentially indistinguishable from spun films. FTIR microspectroscopy of the crystalline structure in the cured film indicates the presence of unreacted PMDA. However, FTIR spectra collected away from the crystalline structure do not exhibit vibrational absorption characteristics of the unreacted dianhydride. Thus what happens upon curing is that the enriched component, in this case PMDA phase separates and crystallizes.

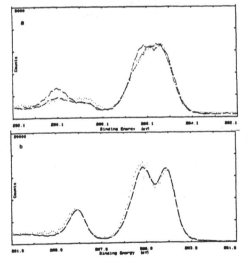

Figure 2. SEM micrograph of a film grown with non-stiochiometric flux ratio exhibiting crystallite formation.

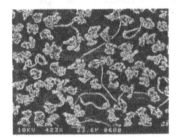

Figure 3. XPS C 1s spectra from: a) a spun polyamic acid film (dots) and two CVD films (lines) and b) a spun polyimide film (line) and a CVD polyimide film (dots).

The dielectric constant ε, of stoichiometric films were measured by fabricating metal-insulator-semiconductor structure with Al dots and measuring current-voltage characteristics. The value obtained for ε were in the range 2.93-2.97. The films had refractive index in the range 1.705-1.736. Initial adhesion studies with the 90° peel test were carried out on films prepared on Si and Cr, as well as a polyimide film spun onto a CVD polyimide film which had been partially cured. The peel strength obtained to Si was 86 g/mm and to Cr was 68 g/mm. Both values comparable to spun film values. The value of the peel strength for the spun polyimide to CVD polyimide was 68 g/mm. Stress measurements using the bending wafer technique give a value of 4.8 kpsi which compares well with data from spun films. Because the CVD process is a vapor process, films prepared by this technique are expected to be conformal, which is confirmed by the SEM micrograph (Fig 4.).

FTIR analysis of the cured CVD films indicates that this process results in polyimides which have less orientational order than the typical solution cast films, while having a greater extent of intermolecular packing order. Solution cast films have in-plane molecular orientation which is induced during the solvent evaporation process. This has a tendency to increase the intensities of absorptions which have their vibrational transition moment parallel to the chain axis, while decreasing the intensities of absorptions which have their vibrational transition moment perpendicular to the chain axis [14]. Transmission FTIR spectra of solution cast polyimide and CVD grown polyimide in the fully cured state are shown in Fig. 5. The 1,4 phenylene vibration of the ODA segment at 1500 cm^{-1} is polarized parallel to the chain axis while the carbonyl asymmetric stretching vibration of the PMDA segment occurring at 1725 cm^{-1} is a perpendicular vibration. The intensity of the 1500 cm^{-1} absorption relative to the 1725 cm^{-1} absorption is greater for the solution cast film than for the CVD film. This indicates that the chain axis of the solution cast polyimide is more highly oriented in the plane of the film than for the CVD polyimide. Intermolecular ordering in the CVD polyimide occurs to greater extent than in solution cast films. This may be detected by the greater intensity of the characteristic crystalline vibrations at 1117 and 1396 cm^{-1} and by the shift of the asymmetric imide carbonyl stretch to lower frequency [15].

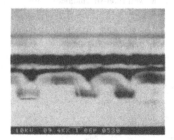

Figure 4. SEM micrograph of a cross section of a CVD film coverage over a structure on Si.

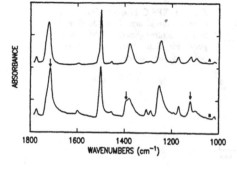

Figure 5. FTIR transmission spectra of: a) solution cast polyimide film and b) CVD grown polyimide film.

DISCUSSION AND CONCLUSIONS

The CVD technique was successfully used to grow films from about a monolayer thickness to greater than 15μ. For growth of thick films with good morphology, careful control of the relative flux ratios were required. Uniformity over 3 inch wafer was achieved with the aid of a rotating substrate holder. Most of the properties of the CVD polyimide films investigated here were quite comparable to those of spun films. However there was evidence that the CVD films were more crystalline , though less orientated than the spun films. Also because of the vapor phase nature of the process, the films are conformal rather than planar. Investigation of off stoichiometry growth showed that these films to undergo phase separation and crystallization of the excess component during the cure process, which leads to poor surface morphology.

ACKNOWLEDGEMENTS

We would like to acknowledge Paul Lauro for assistance in assembling the CVD reactor system. We also gratefully acknowledge Dr. Steven Cohen for the dielectric constant measurements and Dr. Martin Goldberg for assistance with the stress measurements.

REFERENCES

1. A. M. Wilson in *Polyimides*, edited by K. L. Mittal (Plenum, New York, 1984), Vol. 2, p.715.

2. R. J. Jensen, J. P. Cummings, and H. Vora, IEEE Trans. Comp. Hybrids Manufact. Technol. *CHMT 7*, 384 (1984).

3. M. I. Bessonov, M. M. Koton, V. V. Kudryavtsev, and L. Laius, *Polyimides*, (Consultants Bureau, N. Y, 1987).

4. C.E. Sroog, J. Poly. Sci. Macromol. Rev., *11*, 161 (1976).

5. J. R. Salem, F.O.Sequeda, J.Duran, W.Y.Lee, and R.M.Yang, J.Vac.Sci.Technol. *A4*, 369 (1986).

6. M. Grunze and R. N. Lamb,J. Vac. Sci. Technol. *A5*, 1695 (1987).

7. R. N. Lamb, J. Baxter, M. Grunze, C. W. Kong, and W. N. Unertl, Langmuir *4*, 249 (1988).

8. R. G. Mack, E. Grossman, and W. N. Unertl, J. Vac. Sci. Technol. *A8*, 3827 (1990).

9. S. P. Kowalczyk and J. L. Jordan-Sweet, Chem. Mater. *1*, 592(1989).

10. S. P. Kowalczyk in *Metallization of Polymers*, edited by E. Sacher, J.-J. Pireaux, and S. P. Kowalczyk (American Chemical Society, Washington, 1990), p.10.

11. S. P. Kowalczyk, S. Stafström, J. L. Brédas, W. R. Salaneck, and J. L. Jordan-Sweet, Phys. Rev. *B41*, 1645 (1990).

12. S. P. Kowalczyk, Y.-H. Kim, G. F. Walker, and J. Kim, Appl. Phys. Lett. *52*, 375 (1988).

13. S. P. Kowalczyk, unpublished results.

14. S. Molis, in *Polyimides: Materials, Chemistry, and Characterization*, edited by C. Feger, M. M. Khojasteh, and J. E. McGrath (Elsevier, Amsterdam, 1989), p.659.

15. H. Ishida, S. T. Wellinghoff, E. Bair, and J. L. Koenig, Macromol. *13*, 826 (1980).

POLY(ENAMINONITRILES):
NOVEL MATERIALS FOR USE IN MICROELECTRONICS

J. A. Moore* and Ji-Heung Kim
Department of Chemistry, Polymer Science and Engineering Program
Rensselaer Polytechnic Institute, Troy, NY 12180-3590

ABSTRACT

A series of new poly(enaminonitriles) were synthesized by the condensation polymerization of three different bis(chlorodicyano-vinyl) aromatic monomers with various aromatic diamines. These polymers have moderate to high molecular weight with intrinsic viscosities ranging from 0.35 to 0.77 dl/g and form flexible films. These polymers exhibit good solubility in many organic solvents and excellent thermal stability, retaining their weight up to 400 °C in air. Poly(enaminonitriles) undergo 'curing' reactions at high temperature (~300 °C) to provide insoluble, infusible material without emission of volatile by-products. Dielectric constants of these polymers ranged from 3.5 to 5.2 depending on the structure. Their thermal and dielectric properties are correlated to the structures.

EXPERIMENTAL

Monomer Synthesis

1,4- and 1,3-Bis(chlorodicyanovinyl)benzene and 2,6-bis(chlorodicyanovinyl)naphthalene monomers were synthesized using the modified procedure reported previously [1]. All the diamines used for this study were purified by standard procedures.

Polymer Synthesis

Equimolar quantities of the appropriate monomers and diamines were mixed, under nitrogen, at room temperature in dry N-methyl pyrrolidone (NMP). To this reaction mixture, two equivalents of [2.2.2]diazabicyclooctane (DABCO) as an acid acceptor were added. The resulting viscous yellow mixture was heated under nitrogen for 24 h at 70 °C and poured into vigorously stirred water. The precipitated polymer was dissolved in N,N-dimethylformamide (DMF) and reprecipitated into methanol/water. The polymers were dried at 120 °C in vacuo for 72 h.

Mat. Res. Soc. Symp. Proc. Vol. 227. ©1991 Materials Research Society

Characterization

All the polymers were characterized by IR, ^{1}H and ^{13}C nuclear magnetic resonsnce (NMR) spectroscopy. Viscosities of the polymer solutions were measured in an Ubbelohde viscometer. Thermal analysis (differential scanning calorimetry, DSC, thermogravimetric analysis, TGA, and thermomechanical analysis, TMA) of the polymers was carried out on a Perkin-Elmer 7 Series Thermal Analysis System. The glass transition temperature was taken as the mid point of the heat capacity change in DSC. The T_g values from the second runs of film samples are reported. A penetration probe with 200 mN of force was used to measure the softening point of film samples by TMA. The onset slope of the penetration curve was taken as T_g. The capacitance of structures composed of thin films upon which metal electrodes had been sputtered was measured with a precision bridge. The dielectric constants were obtained by dividing the capacitance by the electrode area.

RESULTS AND DISCUSSION

Polymerization

A series of poly(enaminonitriles) with various structures were prepared and characterized. The polymerization schemes are shown below. These polymers appeared to possess moderately high molecular weight judging from the intrinsic viscosity and good film-forming property.

These polymers have good solubilty in many organic solvents including THF, glymes, pyridine, and typical polar, aprotic solvents. The glass transition temperatures determined by both DSC and TMA are shown in Table I. The T_g of polymers ranged from 200 °C to 300 °C depending on the structure. As expected, flexible links such as ether or isopropylidine groups in the backbone lower the T_g. Also meta-linked structures further lower the glass transition by providing more flexibility to the chain. It was interesting to observe the decrease in the glass transition temperature achieved by simultaneously polymerizing two different monomers with ODA as shown in polymers cp75, cp50 and cp25. The copolymers showed lower T_g compared to both homopolymers. The polymers with irregular sequences of two different monomers seem to impart more flexibility to the chain, resulting in lower T_g values.

I	1,4-M	a		p-PDA
II	1,3-M	b		4,4'-ODA
III	NM	c		3,4'-ODA
		d		p-BAPB
		e		m-BAPB
		f		STBDA
		g		BIS-A
		h		BIS-M

Table I. INFLUENCE OF STRUCTURE ON PEAN PROPERTIES

#	Polymer		Tg (°C) DSC	TMA	$[\eta]$ $^{25\,°C}$ DMF
I b	1,4-M	/ 4,4'-ODA	274	271	0.72
I c		3,4'-ODA	197	202	0.37
I d		p-BAPB	255	238	0.51
I f		STBDA	250	234	0.41
I g		BIS-A			
I h		BIS-M	487	480	0.44
II a	1,3-M	/ p-PDA	278	287	0.55
II b		4,4'-ODA	253	252	0.47 *
II c		3,4'-ODA	201	205	0.38
II d		p-BAPB	228	230	0.51
II e		m-BAPB	200	201	0.35
II g		BIS-A	219	208	0.35
II h		BIS-M	197	198	0.58
IIIb	NM	/ 4,4'-ODA	280	281	0.59
cp75	75/25	"	258	255	0.77
	(NM/1,4-M)				
cp50	50/50	"	240	236	0.65
cp25	25/75	"	262	268	0.74

*Laser light scattering analysis reveals that this value corresponds to a $\overline{M}_W$ = 37,000 Daltons

Thermal Properties

Thermal stability of PEANs were measured by TGA and the results are summarized in Table II. All the polymers exhibited good to excellent thermal stability in both air and nitrogen. Most of the polymers decomposed completely in air above 600 °C but retain 70-80 % of their initial mass at 1000 °C in nitrogen. Even though direct comparisons of thermal stability between different structures are difficult because of the differences in the molecular weight and the content of low molecular weight oligomers, the general structural effects on the thermal stability in this series of polymers are clearly reflected. Flexible links such as ether and isopropylidine groups have a detrimental effect on thermal stability. More aromatic or rigid groups enhance the thermal stability.

Table II. THERMAL STABILITY OF PEANS

#	Polymer		10 wt% loss in N$_2$ in air (°C)		50 wt% loss in air (°C)
I b	1,4-M	/ 4,4'-ODA	566	544	661
I c		3,4'-ODA	609	509	643
I d		p-BAPB	509	456	512
I f		STBDA	618	525	677
I g		BIS-A	525	443	540
I h		BIS-M	487	480	653
II a	1,3-M	/ p-PDA	578	525	670
II b		4,4'-ODA	575	536	694
II c		3,4'-ODA	598	549	676
II d		p-BAPB	506	530	645
II e		m-BAPB	554	492	578
II g		BIS-A	527	440	531
II h		BIS-M	490	482	590
IIIb	NM	/ 4,4'-ODA	574	496	558
cp50	NM(50)&	"	565	533	642
	1,4-M(50)/				

Previous work showed that poly(enaminonitriles) could be cyclized without evolution of volatile by-products to a cured polymer containing some amount of aminoquinoline structures [2]. DSC analysis showed broad exothermic transitions with maxima at around 350 °C for all the polymers. When the samples were cooled and rescanned, no exotherms were observed. TGA did not show any weight loss in this temperature range indicating the occurrence of a thermally induced cyclization or curing reaction without generating volatile small molecules.

Dielectric Properties

The dielectric properties of a polymer are determined by the charge distribution in the macromolecules. The charge distribution in a polymer depends on a number of factors including polarity of the bonds, molecular configuration, and morphology. The enaminonitrile structure endows the polymer with good solubility in many organic solvents as well as imparting hydrolytic stability in contrast to the poly(amic acid) precursors to polyimides which are sensitive to hydrolysis. However, because of this highly polarized group on the backbone, the polymer showed relatively high dielectric constants as was reported earlier [3]. The dielectric constants of several of new polymers measured at two different temperatures are shown in Table III.

TABLE III. Dielectric Constant of PEANs at 100 KHz

#	Polymer	at 25 °C	at 200 °C
I d	1,4-M/p-BAPB	4.21	3.94
I f	1,4-M/STBDA	4.91	-
I h	1,4-M/BIS-M	3.67	3.56
II a	1,3-M/p-PDA	5.27	5.03
II d	1,3-M/p-BAPB	4.08	4.10
IIIb	NM/4,4'-ODA	4.69	4.95

The constants ranged from 3.5 to 5.2 depending on the structure. Polymer Ih from 1,4-M and BIS-M diamine showed the lowest value. This low value probably comes from the dipole dilution effect [3] which is obtained by introducing two, non-polar isopropylidine groups in the repeating unit. Compared to this example, polymers Id and IId, which contain two ether linkages, showed slightly higher values. Polymer IIa exhibited a rather high dielectric constant even though the structure is expected to be more rigid. The highly polar enaminonitrile groups are more concentrated in the same volume in this polymer which might account for the apparent high dielectric constant. Usually, higher values of the dielectric constant are expected at high temperatures because the polymer chain becomes more flexible and the polar groups can be oriented more easily. However, somewhat lower values were obtained from the measurements for some samples which might be caused by residual solvent left in the cast films. When the sample is heated to 200 °C, the residual polar solvent is removed which results in a net decrease in the dielectric constant.

In summary, a lower dielectric constant could be obtained by introducing less polar linking groups in the backbone. The isopropylidine group was effective in lowering the dielectric constant of polymer. The introduction of hexafluoroisopropylidine units in the enaminonitrile backbone seems to be a likely choice as the next design modification which should not only lower the dielectric constant but should also enhance thermal stability of these materials.

CONCLUSION

A series of new poly(enaminonitriles) was prepared and characterized. These polymers showed excellent thermal stability. The glass transition temperatures ranged from 200 °C to 300 °C by variations in the structures. The dielectric constants of several polymers were found to be in the range of 3.5 to 5.0 and were shown to be amenable to control by choice of the appropriate structure.

ACKNOWLEDGMENT

Measurements of the dielectric data were performed by Mr. James Mason of ATOCHEM America, King of Prussia, PA. Laser light scattering measurements were performed by Dr. T. Mourey of Eastman Kodak Co., Rochester, NY. Their generous help is appreciated. This work was supported by funds from IBM and the Office of Naval Research.

REFERENCES

1. J. A. Moore and D. R. Robello, Macromolecules, 19, 2667 (1986); 22, 1084 (1989).

2. J. A. Moore and P. G. Mehta, Proc. Polym. Sci. Mat. Eng., 60, 74 (1989).

3. J. A. Moore and Sang Youl Kim, Polym. Prep., 32 (1), 403 (1991).

PREPARATION OF POLYIMIDE-SILICA HYBRID FILMS

Masa-aki Kakimoto,* Atsushi Morikawa, Yoshitake Iyoku, and Yoshio Imai
Department of Organic and Polymeric Materials, Tokyo Institute of Technology,
Meguro Tokyo 152, Japan

ABSTRACT

Polyimide-silica hybrid films were successfully prepared by the sol-gel reaction starting from a mixture of tetraethoxysilane (TEOS), a solution of polyamic acid in N, N-dimethylacetamide and water of pH 7 and pH 3. The hybrid films were obtained by the hydrolysis-polycondensation of TEOS in the polyamic acid solution, followed by heating at 270°C . Fairly flexible films were obtained for silica contents up to 70 wt%. The films containing less than 8 wt% of silica were yellow and transparent, whereas the films with higher silica contents were yellow and opaque. The density of the silica in the hybrid films was estimated to be 1.65 and 1.69 g/cm^3 (pH 7 and pH 3). The ^{29}Si nuclear magnetic resonance spectrum indicated that the silica in the films consisted of non-hydroxy, monohydroxy, and dihydroxy siloxane structures. Silica particles with submicron diameter were observed in the hybrid films containing less than 8 wt% silica, whereas larger particle size around 5 μ m in the case of higher silica content. The decomposition temperature of the hybrid films increased with increasing silica content. The glass transition temperature of the hybrid films showed the minimum at 8 wt% of silica content. Tensile properties, such as elongation at break, tensile strength, and tensile modulus also exhibited the same tendency. The linear thermal expansion coefficient of the silica in the hybrid films was estimated to be 1.3 X 10^{-5} and 0.3 X 10^{-5} (pH 7 and pH 3), which suggested that the silica had a porous structure.

INTRODUCTION

The sol-gel process is a method for preparation of inorganic metal oxides under mild conditions starting from organic metal alkoxides[1]. The reaction consists of hydrolysis of the metal alkoxides, followed by polycondensation of the hydrolyzed intermediates. Scheme 1 shows the general process of preparation of glass by the sol-gel method. Tetraethoxysilane (TEOS) is hydrolyzed in alcoholic solution containing acidic water and DCCA (Drying Control Chemical Additive, typically dimethylformamide (DMF)). The viscosity of the solution increases, and the solution turns to the wet gel. Drying the wet gel at the elevated temperature afforded the dried porous gel. Finally, the glass is obtained by heating the dried gel at ca 1,000 °C which is about 1,000°C lower as compared to the ordinary preparation of quartz glass. The most important point of the sol-gel method for organic and polymer scientists is that the starting

Mat. Res. Soc. Symp. Proc. Vol. 227. ©1991 Materials Research Society

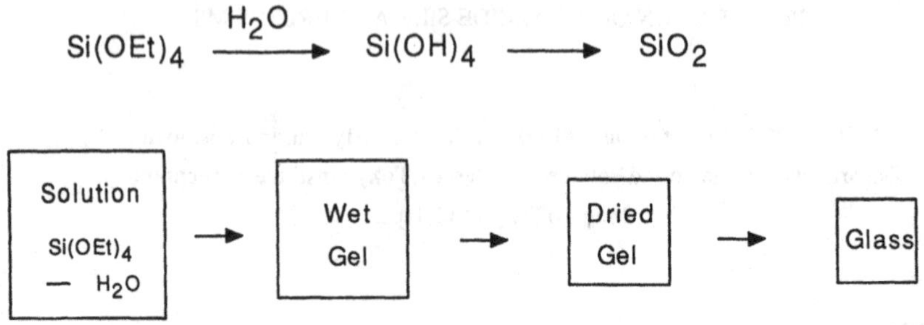

$$Si(OEt)_4 \xrightarrow{\;H_2O\;} Si(OH)_4 \longrightarrow SiO_2$$

| Solution Si(OEt)$_4$ — H$_2$O | → | Wet Gel | → | Dried Gel | → | Glass |

Scheme 1

metal alkoxides are organic compounds. Some organic chemicals can coexist in the starting solutions of the metal alkoxides to prepare organic-inorganic hybrid materials. On the other hand, the dried gels of such materials can not be converted to the glass, because organic chemicals must burn away during producing the final dense glass heating at such high temperature.

Recently, silicon oxide (silica) and polymer hybrid materials have been synthesized[2] from TEOS and polymers such as poly(oxytetramethylene)[3-6], poly(oxyethylene)[7], sodium poly(4-styrene sulfonate)[8], perfluorosulfonic acid ionomer[9, 10], poly(ether ketone)[11], poly(dimethylsiloxane)[5, 12, 13], polysiloxane elastmers[14-19], and polyoxazoline[20, 21]. Some of these hybrid-materials showed transparent and tough properties.

Polyimides are known as reliable high temperature polymers, especially in the aerospace and electronics industries. As shown in eq. 1, polyimide 4, which is insoluble in organic solvents, can be prepared from the soluble precursor polymer, polyamic acid 3, by heating at 300°C [22]. If the hybrid materials of thermally stable polyimide and inorganic silica are successfully prepared, they must be accepted as new high performance materials.

In this paper, we report the successful preparation of the silica-polyimides hybrid materials by the sol-gel process of TEOS in polyamic acid solution and subsequent heating of the resulting films.

EXPERIMENTAL

Typical procedure for preparation of polyimide-silica hybrid films

First, polyamic acid 3 was prepared from pyromellitic anhydride (PMDA) 1 and bis(4-aminophenyl) ether (ODA) 2 in N,N-dimethylacetamide (DMAc) as described in the literature[22]. The inherent viscosity of 3 was over 1.20 dL/g which was sufficiently high molecular weight for preparation of tough polyimide films. In a flask, 1.00 mL of tetraethoxysilane (TEOS) and 1.00 mL of water were added to 10.00 g of 10 wt% solution of polyamic acid 3 in DMAc. The

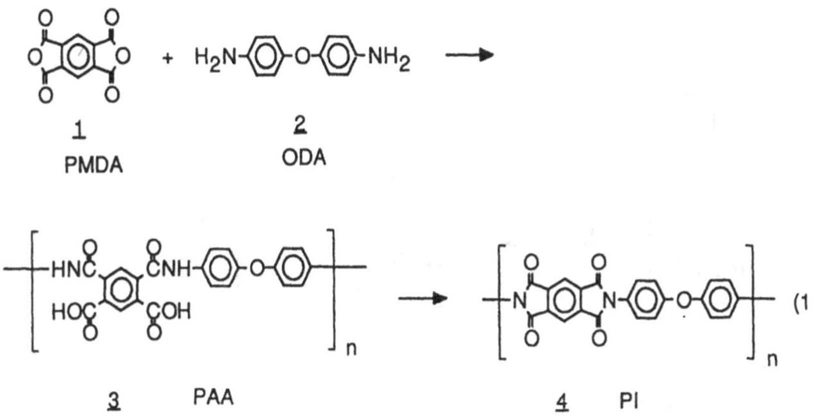

1   2

PMDA   ODA

(1

3   PAA   4   PI

heterogeneous solution was stirred for 6 h until the solution became homogeneous. The film was prepared by casting the solution on a glass plate. After the film had been dried at $60\,^{\circ}C$ for 12 h, the film was further heated at $270\,^{\circ}C$ for 3 h under nitrogen.

When the sol-gel reaction proceeded under acidic condition, aqueous hydrochloric acid of pH 3 was used instead of pure water.

RESULTS AND DISCUSSION

Polyimide-silica hybrid films were prepared as follows (Scheme 2): First, TEOS was reacted with water in the polyamic acid solution, second, the resulting homogeneous mixture was cast onto a glass plate to prepare the polyamic acid-silica hybrid film, and finally the polyimide-silica hybrid films were obtained by heating the precursor film at $270\,^{\circ}C$. As it has been known that the hydrolysis of TEOS is accelerated under acidic condition, preparation of the hybrid films in the presence of aqueous acid was also examined.

Although the sol-gel reaction required water to hydrolyze TEOS, polyamic acid 3 was also susceptible to hydrolysis by water. When a solution of polyamic acid 3 in DMAc was stirred in the presence of the same amount of water as in the case for the sol-gel reaction, even after 24 h, polyamic acid 3 retained enough high molecular weight as reflected by only a small decrease in the inherent viscosity, as shown in Fig. 1.

Table 1 summarizes the results of the preparation of polyimide-silica hybrid films. Silica (SiO_2) content in the Table denotes the value calculated by the assumption that the sol-gel reaction proceeded completely. All the solutions of the precursor, polymeric mixture before casting on a glass plate were homogeneous. The hybrid films having silica content up to 70 wt% were obtained as free standing films. The films of entries 1 and 2 were transparent and yellow even after the films were heated at $270\,^{\circ}C$, whereas the films containing more than 10 wt% silica

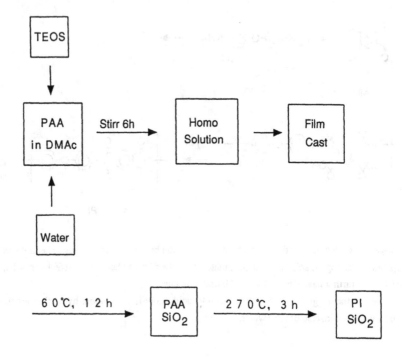

Scheme 2

Table I. Preparation of polyimide-silica hybrid films

Entry	PAA[a] (g)	H2O (mL)	TEOS (mL)	Silica[b] (wt%)	Remarks[c]
1	10.0	1.0	0	0	T
2	10.0	1.0	0.10	3	T
3	10.0	1.0	0.30	8	T
4	10.0	1.0	0.50	13	O
5	10.0	1.0	1.00	22	O
6	10.0	1.0	1.50	30	O
7	10.0	1.0	3.00	46	O
8	10.0	1.0	4.50	56	O
9	10.0	1.0	6.00	63	O
10	10.0	1.0	8.00	70	O

a) 10 wt% DMAc solution. b) Calculated silica content in the hybrid films.
c) T: Transparent, O: Opaque.

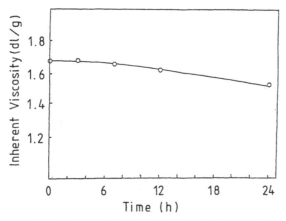

Fig. 1. Influence of water on the inherent viscosity of polyamic acid in DMAc.

were opaque and yellow.

The chemical structure of the matrix polymer was confirmed by means of IR spectroscopy. Fig. 2 shows the IR spectra measured after the film of entry 6 (30 % SiO$_2$) was successively heated at a) 100 °C for 30 min, b) 200 °C for 30 min, and c) 300 °C for 30 min. The conversion of polyamic acid 3 to polyimide 4 was confirmed a decrease in the amide carbonyl absorption at 1650 cm^{-1} and an increase in the imide carbonyl absorptions at 1720 and 1780 cm^{-1}. In addition, the appearance of two absorptions at 1100 and 830 cm^{-1} indicated the formation of silica molecules in the matrix polymers.

As observed in the ^{29}Si-NMR spectrum measured with polydimethylsilane (-34 ppm) as the external standard (Fig. 3), the silica in the present hybrid film after heated at 300 °C had three kinds of silicon species, i. e., non-hydroxy, monohydroxy, and dihydroxy species, suggesting that conversion of the present silica to usual silica glass was incomplete.

The density of the hybrid films increased with increasing silica content as shown in Fig. 4. The density of silica was estimated to be 1.65 and 1.69 g/cm^3 by extrapolation of the curve in the cases of pH 7 and pH 3 water, respectively. The fact that these values

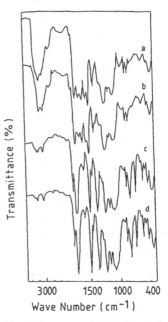

Fig. 2. IR spectra of polyimide-silica hybrid films. (a): Before heating. (b), (c), and (d): after heating at 100 °C, 200 °C, and 300 °C for 1h, respectively.

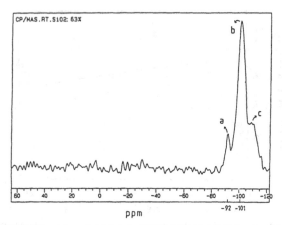

Fig. 3. ²⁹Si NMR spectrum of polyimide-silica hybrid films (silica content was 63 %), a) dihydroxy, b)monohydroxy, and c) non-hydroxy silicones.

were smaller than that of the usual silica glass of 2.22 g/cm³ suggested that the silica in the present hybrid films had fairly porous structure.

The fracture surface of the hybrid film was observed by the scanning electron microscope (SEM). The dispersed silica particles could be seen as the white beads having diameters of 3-7 μ m in the SEM photographs in both the cases of pH 7 and pH 3. The particle size increased with increasing silica content. No particles were observed in SEM photographs of the hybrid films containing 8 wt% of silica

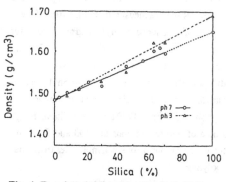

Fig. 4. Density of polyimide-silica hybrid films.

which were transparent in bulk. Photographs of the sliced samples using the transmission electron microscope (TEM) showed the small particles possessing diameters of 0.1 μ m and 0.025 μ m in the cases of pH 7 and pH 3, respectively.

Fig. 5 shows typical examples of the thermogravimetry curves of the hybrid films prepared in air, at pH 7. The results are summarized in Table 2. The decomposition temperature increased slightly with increasing silica content, and the weight residue at 800°C was almost proportional to the silica content. The dynamic mechanical behavior of the hybrid films indicated that those containing higher silica content exhibited higher values of storage modulus (Fig. 6). The glass transition temperatures (Tg) determined by the peak temperature

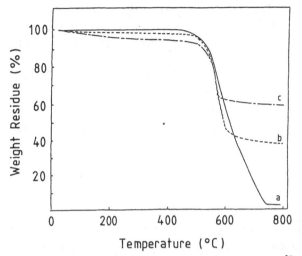

Fig. 5. TGA curves of polyimide-silica hybrid films at a heating rate of 10°C /min in air. (a), (b), (c): Polyimide films containing 0, 30, and 70 wt% of silica, respectively.

of the loss modulus curves are also summarized in Table 2. The relationship between the Tg and silica content is shown in Fig. 7. The hybrid films containing lower silica content exhibited lower Tg than that of the original polyimide film, showing the minimum in entry 3, despite that the phase separated structure could be observed by the SEM photograph in the films of entries 4, 5, and 6. The lowering of Tg suggested compatibility of the silica having low Tg and the polyimide. These phenomena may be explained if the silicon species had low molecular weight. Assuming that TEOS was polymerized incompletely in the polymer matrix under the low TEOS concentration, transparency of the films as

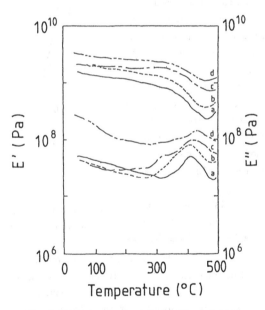

Fig. 6. DMA curves of polyimide-silica hybrid films. (a), (b), (c), (d): Polyimide films containing 0, 8, 30, and 63 wt% of silica.

Table II. Thermal Properties of Polyimide-Silica Hybrid Films

Entry	Silica[a] (wt%)	Td[b] (°C)	Ash[c] (%)	M[d] (X10^9Pa)	Tg[e] (°C)
1	0	470	3	1.9	419
2	3	475	8	1.5	410
3	8	480	12	1.6	396
4	13	477	17	2.2	403
5	22	478	26	2.0	405
6	30	492	35	1.2	412
7	46	510	49	2.2	425
8	56	513	49	4.4	412
9	63	515	58	3.0	430
10	70	505	57	6.0	419

a) Calculated silica content in hybrid films. b) Decomposition temperature determined by TG in air at a heat ing rate of 10 °C/min. c) Residual ash at 800 °C. d) Storage modulus at 100 °C. e) Determined by DMA at a heating rate of 2 °C/min at 10 Hz.

well as the Tg behavior in entries 2 and 3 could be explained by the compatibility, whereas lower content of low molecular weight silica might lower the Tg in the phase-separated films in entries 4, 5, and 6. The storage modulus obtained from Fig. 6 is plotted against the silica content as shown in Fig. 8. A steep increase in the modulus for films containing greater than 40 wt% silica becomes evident.

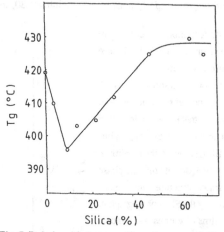

Fig. 7. Relationship between Tg and content of silica in polyimide-silica hybrid films.

Fig. 9 shows the relationship between the linear thermal expansion coefficient (TEC) and silica content. TEC of silica in the hybrid films was estimated to be 1.3×10^{-5} and 0.3×10^{-5} by extrapolation of the straight line for pH 7 and pH 3, respectively. Since the TEC of usual silica glass is known to be 5.5×10^{-7}, the TEC of the present silica was two orders of magnitude larger. This observation may also indicate that the silica in the hybrid films is porous in nature. In comparison of the pH condition, more dense particles are obtained under acidic condition.

Typical stress-strain curves of the hybrid films are shown in Fig. 10. Relationship between silica content and the tensile factors are plotted in Fig. 11. The films prepared at pH 3, showed higher elongation and lower tensile strength and modulus. The curve in tensile modulus indicated the same phenomena of the storage modulus in Fig. 8. In general, the modulus of composite materials increases in proportion to the content of the filler. The slow increase of the modulus at low silica content may be explained by existence of oligomeric silicon oxides which are compatible with polyimide matrix.

In conclusion, polyimide-silica hybrid films, which were fairly flexible even in the case of 70 wt% silica content, were successfully prepared by the sol-gel reaction starting from TEOS in polyamic acid

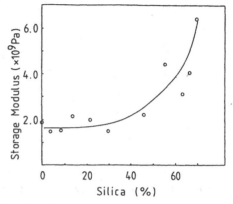

Fig. 8. Relationship between storage modulus and content of silica in polyimide-silica hybrid films.

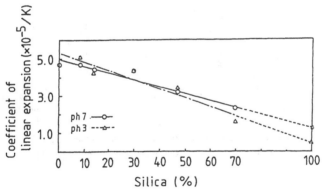

Fig. 9. Relationship between thermal expansion coefficiency and content of silica in polyimide-silica hybrid films.

solution and successive heating of the films to form polyimide. Thus, the sol-gel process is very useful to prepare new high temperature polymer-inorganic hybrid materials. Two different kind of silica particles seemed to exist in the hybrid films judging from the electron microscope observation. One of them was the submicron size particle observed in lower silica content films, whereas much larger particles persisted in the case of higher silica content hybrid films.

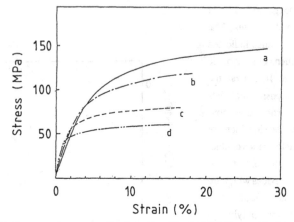

Fig. 10. Stress-strain curves of polyimide-silica hybrid films. (a), (b), (c), (d): Polyimide films containing 0, 22, 30, and 46 wt% of silica, respectively.

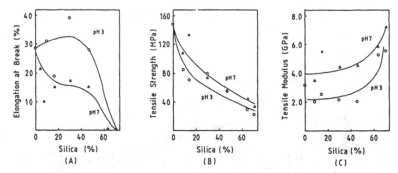

Fig. 11. Relationship between tensile properties and silica content in polyimide-silica hybrid films. A) Elongation at break, B) Tensile Strength, C) Tensile Modulus

REFERENCES

1. L. L. Hench, J. K. West, Chem. Rev., 90, 33 (1990).

2. G. L. Wilkes, H-H Huang, R. H. Glaser, in Silicon-Based Polymer Science. A Comprehen sive Resource, edited by J. M. Zeigler, F. W. G. Fearon, Advances in Chemistry Se ries, 224, (American Chemical Society, 1990), p. 207.

3. D. Ravaine, A. Seminel, Y. Charbouillot, M. Vincens, J. Non-Cryst. Solids, 82, 210 (1986).

4. H-H. Huang, G. L. Wilkes, Polym. Bull., 18, 455 (1987).

5. R. H. Glaser, G. L. Wilkes, Polym. Bull., 19, 51 (1988).

6. H-H. Huang, G. L. Wilkes, J. C. Carlson, Polymer, 30, 2001 (1989).

7. M. Fujita, K. Honda, Polym. Commun., 30, 200 (1989).

8. K. Nakanishi, N. Soga, J. Non-Cryst. Solids, 108, 157 (1989).

9. K. A. Mauritz, R. M. Warren, Macromolecules, 22, 1730 (1989).

10. I. D. Stefanithis, K. A. Mauritz, Macromolecules, 23, 2397 (1990).

11. J. L. W. Noell, G. L. Wilkes, D. K. Mohanty, J. E. MacGrath, J. Appl. Polym. Sci., 40, 1177 (1990).

12. H-H. Huang, B. Oriler, L. Wilkes, Polym. Bull., 14, 557 (1985).

13. H-H. Huang, B. Orier, G. L. Wilkes, Macromolecules, 20, 1322 (1987).

14. Y. -P. Ning, M. -Y. Tang, C. -Y. Jiang, J. E. Mark, W. C. Roth, J. Appl. Polym. Sci., 29, 3209 (1984).

15. J. E. Mark, C. -Y. Jiang, M. -Y. Tang, Macromolecules, 17, 2613 (1984).

16. C. -Y. Jiang, J. E. Mark, Makromol. Chem., 185, 2609 (1984).

17. Y. -P. Ning, J. E. Mark, J. Appl. Polym. Sci., 30, 3519 (1985).

18. J. E. Mark, Y. -P. Ning, C. -Y. Jiang, M. -Y. Tang, W. C. Roth, Polymer, 26, 2069 (1985).

19. G. S. Sur, J. E. Mark, Makromol. Chem., 187, 2861 (1986).

20. Y. Chujo, E. Ihara, H. Ihara, T. Saegusa, Macromolecules, 22, 2040 (1989).

21. Y. Chujo, E. Ihara, S. Kure, K. Suzuki, T. Saegusa, Polymer Preprints, 31 (1), 59 (1990).

22. C. E. Sroog, in Macromolecular Synthesis, Coll. Vol. 1, edited by J. A. Moore, (John Wiley & Sons, New York, 1977), p. 295.

IMIDE-ARYL ETHER KETONE BLOCK COPOLYMERS

J. L. HEDRICK,* W. VOLKSEN* AND D. K. MOHANTY**

*IBM Research Division, Almaden Research Center, 650 Harry Road, San Jose, California 95120-6099
**Department of Chemistry and Center, for Applications in Polymer Science, Central Michigan University, Mount Pleasant, Michigan 48859

ABSTRACT

Imide-aryl ether ketone block copolymers were prepared and their morphology and thermal and mechanical properties investigated. The key feature of this copolymerization is the preparation of soluble aryl ether ketimine oliogmers which may be subsequently hydrolized to the aryl ether ether ketone form. A bis(amino) aryl ether ketimine oligomer was prepared via a nucleophilic aromatic substitution reaction with a molecular weight of 6,000 g/mol. The oligomer was co-reacted with 4,4'-oxydianiline (ODA) and pyromellitic dianhydride (PMDA) diethyl ester diacyl chloride in N-methyl-2-pyrrolidone (NMP) in the presence of N-methylmorpholine. The copolymer compositions, determined by H-NMR, of the resulting amic ester based copolymers ranged from 8 to 20 wt% aryl ether ketimine content. Prior to imide formation, the ketimine moiety of the aryl ether ketimine block was hydrolyzed (p-toluene sulfonic acid) to the ketone form producing the aryl ether ether ketone block. Solutions of the copolymers were cast and cured to effect imidization, producing clear films. The copolymers displayed good thermal stability with decomposition temperatures in excess of 450 °C. Multiphase morphologies were observed irrespective of the co-block type or composition.

INTRODUCTION

Rigid aromatic polyimides (i.e., PMDA/ODA polyimide) have gained prominence as packaging materials in the manufacturing of microelectronic components and devices. These materials are generally processed from a soluble poly(amic-acid) precursor and subsequently cured to effect imidization, producing films which exhibit high thermal and dimensional stability, low thermal expansion coefficients and good mechanical properties. It has been shown that the properties of such rigid and semi-rigid polyimides result from their ordered or liquid crystalline type morphology.[1] However, a manifestation of this ordered morphology and the absence of a T_g is poor melt processability (planarization) and auto-adhesion (i.e., the adhesion the polyimide to itself). This latter feature is a major deterrent in the successful fabrication of multilayer circuitry.

Recently, several approaches in the modification of the auto-adhesion characteristics of rigid and semi-rigid polyimides have been described. Imide aryl ether phenylquinoxaline and imide-aryl ether benzoxazole statistical or random copolymers have been prepared and their morphology and adhesion characteristics investigated.[2,3] In each case, the incorporation of the co-diamine containing either a preformed phenylquinoxaline or benzoxazole moiety resulted in significantly improved auto-adhesion. However, wide angle x-ray diffraction measurements showed that the "liquid crystalline" like ordering observed in the PMDA/ODA polyimide persists in the copolymers where the phenylquinoxaline or benzoxazole compositions were less than 50 wt.%. At higher compositions the ordering vanishes due to hindrance of interchain packing, and coincident with this loss in ordering is the clear development of a T_g and the loss of the high temperature dimensional stability. Thus, these copolymers showed improved adhesion with the retention of the ordered morphology over a very narrow composition range.

Another approach in modifying the adhesion characteristics of polyimide without sacrificing the ordered morphology and properties associated with this morphology involves the preparation of multiphase block polymers derived from polyimide and amorphous engineering thermoplastics.[4-6] Several thermoplastic co-blocks were investigated including aryl ether phenylquinoxalines and aryl ether benzoxazoles, and in each case, sequentially cast and cured block polymers showed significantly improved auto-adhesion. However, the adhesion of fully cured dry films was minimal except in the random copolymers which showed a T_g.

We are currently investigating the use of thermopolastics (i.e., poly(ether-imide)) to laminate the rigid polyimide films. To facilitate the adhesion of the rigid polyimide to the thermoplastic, a block polymer of polyimide and poly(ether ether ketone) (commercially available as PEEK®), 1, was prepared. Since PEEK and poly(ether-imide) are miscible, a thermodynamic driving force for adhesion should be realized. In this paper we will describe the synthesis of the block copolymers, the evaluation of the thermal and mechanical measurements.

$$ \left[\underset{}{\bigcirc} - O - \bigcirc - \overset{\overset{O}{\|}}{C} - \bigcirc - O \right]_n $$

1

EXPERIMENTAL

N-[bis(4-Fluorophenyl) Methylene]Benzenamine. To a 1000 mL round bottomed flask, aniline (0.1 mol, 9.1 g) and 4,4′difluorobenzophenone (0.1 mol, 20.0 g) were added along with 300 mL of toluene and 150 g of molecular sieves. The reaction mixture was gently refluxed for a period of 24 h. The sieves were removed by filtration and washed with methylene chloride. The solvents were removed from the mixture by rotary evaporation at reduced pressure to afford the crude product yield 65 %. The resulting crude solid was crystallized from hexane; mp 107-109 °C; IR(kBR) 1619 cm^{-1} (C = N); MS m/e (% of base peak) 293 (100 %, m$^+$), 198 (53 %), 77 (59 %). Anal. Calc′d for $C_{19}H_{13}F_2ON$: C, 77.80; H, 4.46; F, 12.95; N, 4.77. Found: C, 77.69; H, 4.55; F, 12.90; N, 4.73.

The poly(aryl ether ether ketimine) oligomers were prepared according to a published procedure.[7,8] The amic ester-aryl ether ketone copolymers were prepared by the co-reaction of the aryl ether ketone oligomers with ODA and PMDA diethyl ester diacyl chloride in NMP in the presence of N-methylmorpholine according to literature procedures.[4-6, 9-14]

RESULTS AND DISCUSSION

The synthesis of poly(aryl ethers) is based on the nucleophilic displacement of an aryl halide with a phenoxide in polar aprotic solvents. The aryl halide is activated by an electron-withdrawing group like carbonyl or sulfone.[15,16] In addition, these activating groups can accept a negative charge lowering the activation energy for the displacement through a Meisenheimer Complex (Scheme 1). Aryl ether ether ketone oligomers are not readily prepared since solubility is realized only in strong acids or diphenylsulfone at elevated temperatures.[15,16] This insolubility makes the synthesis and characterization of function oligomers and subsequent transformations (i.e., copolymerization) difficult. We have used an another synthetic route to the preparation of aryl ether ether ketone oligomers, 1, which involves the derivitization of one of the monomers to afford oliogomer solubility and after subsequent transformations (i.e., copolymerization), a deprotection step yields the aryl ether ether ketone block. It has been demonstrated that the ketone moiety in 4,4′-difluorobenzephenone can be derivatized with aniline to produce a ketimine.[7,8] This disrupts the trigonal bond arrangement which provides the planar zig-zag chain packing, precluding crystallization. Interestingly, it has been demonstrated that the ketimine moiety is sufficiently electron with drawing to activate aryl fluorides towards nucleophilic aromatic displacement. In addition, the ketimine can accept the negative charge developed in the transformation through a Meisenheimer complex which lowers the activation energy for the displacement (Scheme 1). This allows polymerization in common aprotic dipolar solvents (i.e., NMP). The ketimine group may be quantitatively hydrolized in the polymer form to produce the parent structure, 1.

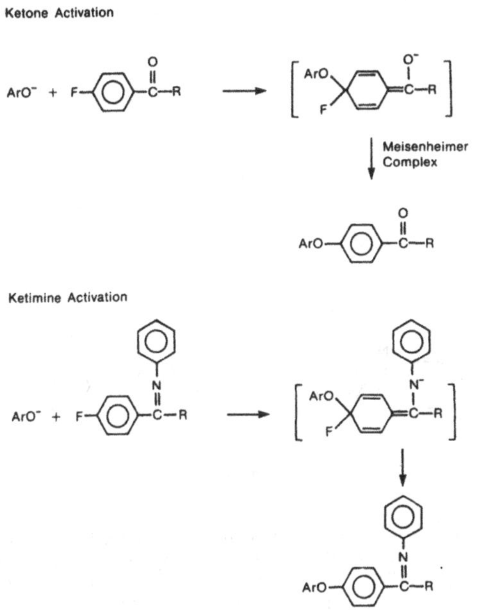

Scheme 1

The synthesis of the bis(amino) aryl ether ketimine oligomer was carried out in an analogous fashion to other poly(aryl ether) synthesis (Scheme 2). The ketimine functional 4,4′-bisfluoride was reacted with hydroquinone and 3-aminophenol in a NMP/toluene solvent mixture in the presence of K_2CO_3 via a nucleophilic aromatic substitution reaction. It has been shown that the ketimine is hydrolytically stable in basic environments.[7,8,17] The subsequent oligomer, 2, was isolated in a methanol/water (50/50) mixture and dried in a vacuum oven. 3-Aminophenol was used as a monosubstituted monomer to control both the molecular weight and end group functionality as in the previous case, and the molecular weight, as determined by H-NMR, was in good agreement with that predicted by the charge for 2 (Table I)[18].

We have used an alternative synthetic procedure for the preparation of imide containing copolymers based on a poly(amic alkyl-ester) intermediate to the polyimide.[9-14] In contrast to a conventional poly(amic-acid) synthesis, a poly(amic-alkyl ester) precursor offers more synthetic flexibility due to improved solubility and greater structural variety in both the polyimide backbone and the co-block type, composition and molecular weight. This hydrolytically stable precursor may be isolated, characterized and purified. Furthermore, since imidization occurs at substantially higher temperatures, molecular mobility is enhanced which will minimize the control of the resultant morphology by kinetic factors.

The copolymer synthesis involved the incremental addition of PMDA diethyl ester diacyl chloride to a NMP solution of 4,4′-oxydianaline (ODA) and 2 in the presence of N-methylmorpholine (Scheme 3). The solids content was maintained between 12 and 13

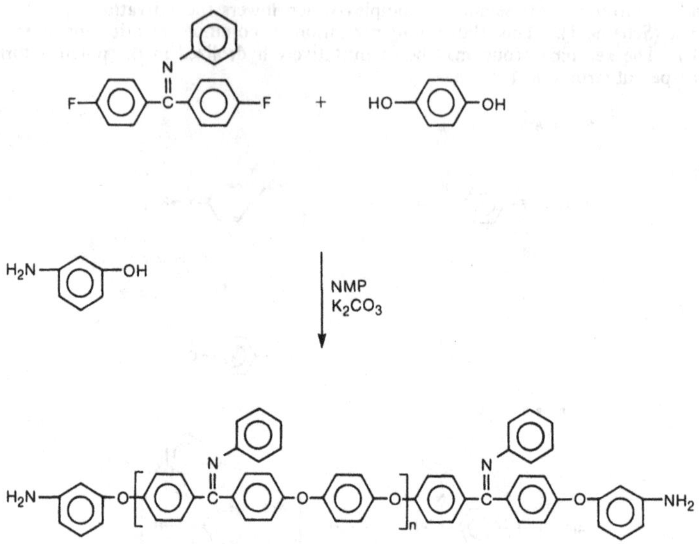

Scheme 2

wt%. High molecular weight was readily achieved as judged by the dramatic increase in viscosity, and the resulting amic-ethyl ester based copolymers were isolated in an excess methanol/water mixture (50/50). The polymers were then subjected to both a water and methanol rinse and then dried under vacuum (30°C). If required the copolymers were also subjected to a chloroform rinse to remove possible poly(aryl ether ketimine) contamination.

The imide-aryl ether ketimine copolymers were prepared with compositions below 20 wt% and using the meta-isomer of PMDA diethyl ester diacyl chloride so as to maintain the solubility after the subsequent hydrolysis to the amic ester-aryl ether ketone copolymer.[9] The composition of aryl ether ketimine in the block copolymer (3a and b) was comparable to that of the charge (Table II), and the molecular weights, as judged by the viscosity measurements, were high.

TABLE I

Characteristics of the Aryl Ether Ketimine Oligomer

Sample No.	$< M_n >$, g/mol		T_g
	Theory	Actual	°C
2	7,000	6,500	145
PEEK®			145,340*

*Melting Point

TABLE II

Characteristics of Amic Ester-aryl Ether Ketimine Copolymers

Sample No.	Block Length, g/mol	Co-Block Composition, wt% Theory	Actual	$[\eta]_{NMP}^{25°C}$ dL/g
3a	6,500	10	9	0.52
3b	6,500	20	16	0.65

Prior to imide formation, the imide-aryl ketimine copolymers were converted to the imide-aryl ether ether ketone analog by hydrolysis of the ketimine moiety with para toluene sulfonic acid hydrate (PTS) according to a literature procedures.[7,8] Copolymers 3a and 3b were dissolved in NMP and heated to 50 °C, and subjected to excess PTS for 8 h. The reaction mixtures were isolated in excess water and then rinsed with methanol and dried in a vacuum oven to afford the imide aryl ether ether ketone copolymers 4a and b, respectively.

Solutions of the copolymers (3a and b and 4a and b) were cast and cured (350°C) to effect imidization, yielding copolymers series (5a and b and 6a and b), respectively. The solutions at ambient temperature were somewhat cloudy. Upon heating, usually in cast film form, the copolymer solutions became clear at ~80°C. Upon imidization, clear tough films were obtained indicative of minimal homopolymer contamination. The thermal analysis for the copolymers are shown in Table III together with polyimide homopolymers to facilitate comparison. No detectable T_g was observed for the polyimide homopolymer or for any of the copolymer series, providing no insight as to the morphology of the block copolymers. Table III also contains the thermal stability, as determined by the polymer decomposition temperature, PDT, for the block copolymers and a polyimide control. The PDT's were comparable to polyimide. The incorporation of the aryl ether ketone blocks into polyimide did not adversely affect the thermal expansion coefficient (TEC), and these values ranged from 32 to 55 ppm (Table III).

The imide-aryl ether ketimine (5a + b) and imide aryl ether ether ketone (6a + b) copolymers showed multiphase morphologies irrespective of the co-block type or composition. The dynamic mechanical spectra for 5a and 6a and 5b and 6b are shown in Figures 1 and 2, respectively. The aryl ether ketimine containing copolymers 5a and b show transitions at ~150 °C and 365 °C. The lower transition, associated with the ketimine containing block, is nearly identical to that of the oligomer used in the synthesis, indicative of minimal phase mixing in the aryl ether ketimine phase. Likewise the transition at 365 °C is identical to the transition observed in the parent polyimide homopolymer, consistent with high phase purity. Markedly different behavior was observed after conversion of the aryl ether ketimine block into the aryl ether ether ketone form. Although the T_g of 1 is lower than that of 2a, in the block copolymer the aryl ether ether ketone T_g is substantially higher than 2 and broad. The polyimide transition in copolymers 6a and b is somewhat lower than the polyimide homopolymer. This behavior is consistent with poor phase purity or phase mixing, which may have resulted from the high temperature processing conditions due to the limited solubility or from an increase in the viscosity of the system due to the rigid nature of 1. Crystallization of the aryl ether ether ketone component in the block copolymer is precluded due to the size scale of the domains (<500 Å). Furthermore, in each case, a reflection at 15.5 Å was observed by WAXS measurements, corresponding to the order in the polyimide block, however, the reflection was weak and diffuse consistant with the poor phase purity.

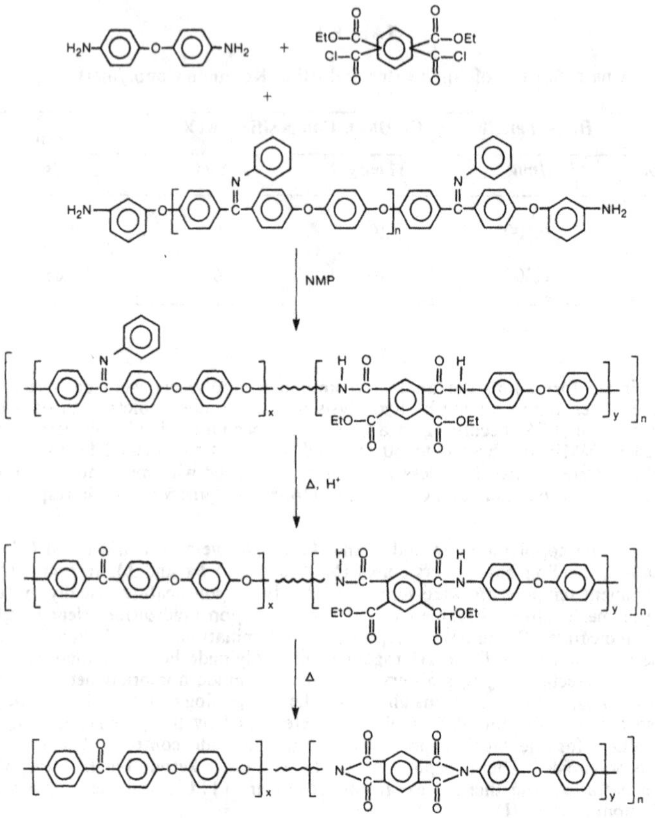

Scheme 3

TABLE III

Thermal Characteristics of Imide Containing Copolymers

Sample No.	Co-Block Length g/mol	Co-Block Composition, wt.%	Polymer Decomposition Temp., °C	TEC, ppm
5a	6,500	13	450	32
5b	6,500	19	480	42
6a	6,500	13	450	—
6b	6,500	14	500	40
PMDA/ODA Polyimide			480	35

*Co-block compositions change with imidization.

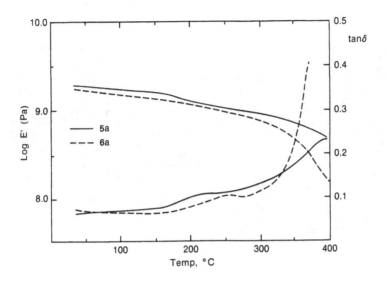

Figure 1. Dynamic mechanical behavior of copolymers **5a** and **6b**.

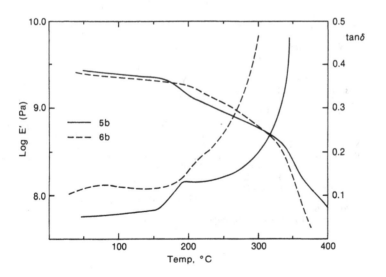

Figure 2. Dynamic mechanical behavior of copolymers **5b** and **6b**.

88

SUMMARY

Imide-aryl ether ketone block copolymers were successfully prepared and their thermal and mechanical properties and morphology investigated. Bis(amino) aryl ether ketimine oligomers were prepared with a molecular weight of 6,000 g/mol. The oligomer was co-reacted with ODA and PMDA diethyl ester diacyl chloride in NMP in the presence of N-methylmorpholine. The poly(amic ethyl ester) intermediate to the polyimide was found to be more versatile than the poly(amic acid) analog for the preparation of the copolymers since it can be isolated characterized and washed to remove homopolymer contamination prior to imidization. In addition, the poly(amic ester) is soluble in a variety of NMP based solvent mixtures at high solids compositions, and since imidization occurs at a substantially higher temperature than the poly(amic acid) analog, vitrification occurs after solvent loss. This latter feature is important is developing phase separated microstructures with high phase purity. Prior to imidization, the aryl ether ketimine containing block copolymers were hydrolyzed to the aryl ether ether ketone co-block. Solutions of the copolymers were cast and cured (350 °C) to effect imidization, affording clear films with tough mechanical properties and moduli comparable to the parent polyimide. Multiphase morphologies were observed for both the imide-aryl ether ketone and imide-aryl ether ether ketone block copolymers. The thermal stability of the copolymers was good with decomposition temperatures in the 450 °C range.

REFERENCES

1. T. P. Russell, *J. Polym. Sci., Polym. Phys. Ed.* **22**, 1105 (1986).
2. J. L. Hedrick, T. P. Russell, J. W. Labadie, J. G. Hilborn and T. D. Palmer *Polymer* **31**, 2384 (1990).
3. J. L. Hedrick, T. P. Russell, J. W. Labadie, and T. D. Palmer, *Polymer*, 000 (1991).
4. J. L. Hedrick, J. W. Labadie and W. Volksen, *SAMPE Proceedings* **4**, E14 (1990).
5. J. L. Hedrick, J. Hilborn, T. D. Palmer, J. W. Labadie and W. Volksen, *J. Polym. Sci.: Part A: Polymer Chem.* **28**, 2255 (1990).
6. J. W. Labadie and J. L. Hedrick, *SAMPE Proceedings* **4**, 49 (1990).
7. D. K. Mohanty, R. C. Lowry, G. D. Lyle and J. E. McGrath, *Int. SAMPE Symp.* **32**, 408 (1987).
8. D. K. Mohanty, J. S. Senger, C. D. Smith and J. E. McGrath, *Int. SAMPE Symp.* **33**, 970 (1988).
9. W. Volksen, *Macromolecules* in preparation (1991).
10. S. N. Kharkov, Ye P. Krasnov, S. N. Lavrona, S. A. Baranova, V. P. Akesovova, and A. S. Chengolya, *Vysokomol Soyed* **13**, 833 (1971).
11. V. V. Korshak, S. V. Vinogradova, Ya S. Vygodskii, and Z. V. Gerashenko, *Vysokmol Soyed* **13**, 1190 (1971).
12. V. V. Kudryavtsev, M. M. Koton, T. K. Meleshko and V. P. Sklizkova, *Vysokomol Soyed* **A17(8)**, 1764 (1975).
13. Ye D. Molodtsova, G. I. Timofeyeva, S. S. A. Pavolva, Y. S. Vygodskii, S. V. Vingradova and V. V. Korshak, *Vysokomol Soyed* **19**, 346 (177).
14. S. Nishizaki and T Monwaki, *J. Cllem. Soc. Jpn. Ind. Chem. Soc.* **71**, 559 (1968).
15. R. N. Johnson, A. G. Farnham, R. A. Clendinning, W. F. Hale, and C. N. Merriam, *J. Polym Sci. A-1*, **5**, 2375 (1967).
16. T. E. Atwood, D. A. Barr, G. G. Faasey, V. Ju. Leslie, A. B. Newton and J. B. Rose, *Polymer* **354** (1977).
17. "The Chemistry of Carbon-Nitrogen Double Bond." S. Patai, Ed. (Interscience, 1970), p. 64.
18. M. Jurek and J. E. McGrath, *Polymer* **30**, 1552 (1989).

PREPARATION AND CHARACTERIZATION OF
THERMOPLASTIC/THERMOSETTING POLYIMIDE BLENDS

Y.Yamamoto, T.Toyoshima and S.Etoh
Materials & Electronic Devices Laboratory, Mitsubishi Electric
Corp., Amagasaki, Hyogo, Japan

ABSTRACT

Three kinds of thermoplastic/thermosetting polyimide blends have been prepared using three kinds of commercially available soluble polyimides(AL1051, PI2080, Ultem) and N,N'-(methylenedi-p-phenylene) bismaleimide(BMI). The DSC thermograms of the polyimide blends showed exotherms at higher temperatures compared to BMI alone. The cure characteristics of the blends changed significantly according to the thermoplastic polyimide component. The fracture surfaces of PI2080 and Ultem blends, examined under a scanning electron microscope, showed distinct nodular morphological features. However, in the case of AL1051, which was derived from a aliphatic acid, this kind of phase separation was not observed. The mechanical properties of the blends were studied by using a thermomechanical analyzer. The blends yielded films with good mechanical properties even at high temperatures. The BMI component did not have any significant effect on the glass transition temperatures. However it showed improved mechanical properties at high temperatures. The blends were brittle compared with the thermoplastic polyimides. The tensile and electrical properties of the blends and their coefficients of thermal expansion were also studied.

INTRODUCTION

Condensation type aromatic polyimides show excellent thermal, chemical and dielectric performance. In addition, they remain strong and tough over a wide temperature range. However this type of polyimide is difficult to process. Several thermoplastic polyimides have been developed to overcome the processing problems[1]. The thermoplastic polyimides are soluble and show good toughness and thermooxidative stability. However,their mechanical properties at high temperatures are poor. On the other hand, thermosetting polyimides based on bismaleimides are brittle, but give high glass transition temperatures[2]. It is anticipated that thermoplastic/thermosetting polyimide blends will avoid these drawbacks. It has been previously reported that thermoplastic polyimide PI2080/BMI blends show good mechanical properties at high temperatures as a composite matrix resin[3].

In this study, three kinds of thermoplastic/thermosetting polyimide blends have been prepared, using three kinds of commercially available soluble polyimides(AL1051, PI2080, Ultem) and N,N'-(methylenedi-p-phenylene)bismaleimide(BMI). The cure characteristics, mechanical properties and morphology of the blends were investigated. The blends yielded films with good mechanical properties even at high temperatures. The tensile and electrical properties of the blends and their coefficients of thermal expansion were also studied.

Mat. Res. Soc. Symp. Proc. Vol. 227. ©1991 Materials Research Society

EXPERIMENTAL

Materials and sample preparation

Polyimide blends were prepared by dissolving soluble polyimides(AL1051:Japan Synthetic Rubber Co.Ltd., PI2080:The Upjhon Company, Ultem1000:General Electrics) with N,N'-(methylenedi-p-phenylene) bismaleimide(BMI) in N-methyl-pyrrolidinone. The thermoplastic polyimides and BMI were used as received. Their chemical structures are shown in Fig.1. Films were cast from the blend solution and dried at 150°C for 15 minutes. The dried films were cured at 200°C for 2 hours and then postcured under various conditions.

Instrumentation and methods

Thermal polymerization of the blends were investigated with a Perkin-Elmer DSC-2C at a heating rate of 10 °C/minute under nitrogen atmosphere. Samples fractured in liquid nitrogen were coated with a thin layer of Au and examined in a scanning electron microscope. Glass transition temperatures of cured blends were measured by using a Perkin-Elmer TMS-1 thermomechanical analyzer with a penetrating probe at a heating rate of 20 °C/minute. Dynamic mechanical analysis of films was performed on a Iwamoto Viscoelastic Spectrometer VES-F3. The tensile tests were performed on an Instron Machine with a crosshead speed of 2 millimeters/minute. Electrical properties were measured with a Yokokawa Hewlett Packard 4274A multifrequency LCR meter at 1KHz. Measurement of thermal expansion coefficient was performed on a Shinku Riko TMA 1500 thermal mechanical analyzer at a heating rate of 5 °C/minute.

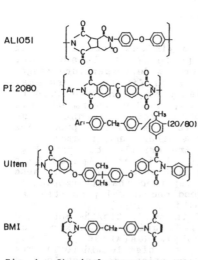

Fig. 1 Chemical structures of thermoplastic polyimides and BMI.

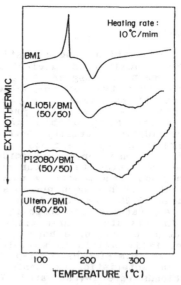

Fig. 2 DSC thermograms for the polymerization of BMI and Polyimide blends.

RESULT AND DISCUSSION

Cure characterization

The cure characteristics of the blends were studied by means of differential scanning calorimetry(DSC). Fig.2 shows DSC thermograms for the polymerization of the blends. The DSC thermograms of the polyimide blends showed exotherms at higher temperatures compared to BMI alone. The cure characteristics of the blends changed significantly according to the base thermoplastic polyimides. AL1051/BMI and PI2080/BMI blends showed two exotherm peaks. It was assumed that vitrification occurred during the heating up process because of the high glass transition temperatures of AL1051 and PI2080 as the thermoplastic polyimide components.

The mechanical responses of the blends were studied by using a thermomechanical analyzer. Fig.3 shows the effect of postcuring temperature on the TMA scans of the AL1051/BMI blend. It is necessary to postcure at high temperatures to obtain a high glass transition temperature. The glass transition temperature of the blends became almost the same as that of the thermoplastic polyimide when postcured at 300°C.

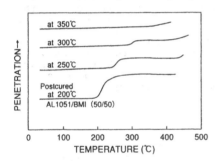

Fig. 3   Effect of postcuring temperature on the TMA scans of AL1051/BMI blend(50/50).

Morphology

The fracture surfaces of the blends, examined under a scanning electron microscope, are shown in Fig.4. The PI2080 and Ultem blends showed distinct nodular morphological features. The nodules of the Ultem blend are especially large. It was presumed that the thermoplastic polyimides precipitated

Fig. 4   Fracture surfaces of polyimide blends.

out in the form of particles during cure. The fracture surface morphology of the AL1051/BMI blend did not show phase separation. AL1051 is derived from a aliphatic acid, so the solubility parameter of AL1051 is similar to that of cured BMI. It is likely that the AL1051/BMI blend has a molecular solution network.

Thermomechanical properties

Postcure of the blends resulted in a hard film with a high glass transition temperature. Fig.5 shows the TMA scans of the thermoplastic polyimides and their blends postcured at 300 °C for 1 hour. The glass transition temperatures of the blends correspond closely with those of the thermoplastic polyimides. The BMI component did not have any significant effect on the glass transition temperatures when postcured completely. However the BMI component decreased the penetration of a probe into the blend films. It should be noted that the blend of BMI improve mechanical properties of thermoplastic polyimides at high temperatures, especially in the case of AL1051/BMI blend which did not show the phase separation.

The dynamic mechanical properties of the blends were studied over the temperature range of 50 °C to 400 °C. The dynamic storage modulus of the thermoplastic polyimides and their blends are shown in Fig. 6. The AL1051/BMI and PI2080/BMI blends show higher modulus than those of AL1051 and PI2080 at temperatures beyond their glass transition temperatures. In the case of AL1051/BMI, a glass transition phenomena was not observed. Ultem and its blend were not testable; they broke in the glass transition region.

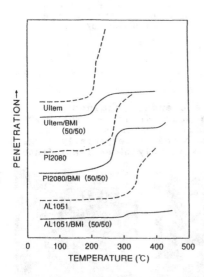

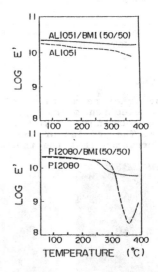

Fig. 5   TMA scans of thermo-plastic polyimides and their blends with BMI.

Fig. 6   Dynamic storage modulus of thermoplastic polyimides and their blends with BMI.

Tensile, thermal and electrical properties

The tensile properties of the thermoplastic polyimide and their blends are shown in Table 1. In general the BMI component raised the tensile modulus. The thermoplastic polyimide and BMI blends were brittle compared with the thermoplastic polyimides. The BMI component did not significantly affect the strength, however, it reduced elongation percentage. Only AL1051, which is originally a flexible polymer, showed good toughness when blended with BMI.

The coefficients of thermal expansion of the blends and their electrical properties are shown in Table 2. The BMI component did not affect the coefficients of thermal expansion except AL1051/BMI 50/50. The blends, as well as the thermoplastic polyimides, showed good electrical properties.

Table 1 Tensile properties of thermoplastic polyimides and their blends with BMI.

Polyimides	E (kg/mm²)	σ (kg/mm²)	elongation (%)
AL1051/BMI			
(100/0)	236.8	10.0	22.0
(65/35)	293.0	10.9	10.6
(50/50)	282.4	11.4	9.5
PI2080/BMI			
(100/0)	273.7	13.1	10.2
(65/35)	282.3	11.6	8.4
(50/50)	351.7	12.0	4.9
Ultem/BMI			
(100/0)	297.0	12.2	7.7
(65/35)	265.7	11.5	8.6
(50/50)	356.8	11.6	5.3

Table 2 Coefficients of thermal expansion and electrical properties of thermoplastic polyimides and their blends with BMI.

Polyimides	CTE (ppm/K)	ε (1 kHz)	tan δ (1 kHz) (%)
AL1051/BMI			
(100/0)	52.7	3.3	0.31
(65/35)	54.6	3.1	0.20
(50/50)	63.4	3.3	0.19
PI2080/BMI			
(100/0)	55.6	3.1	0.17
(65/35)	52.3	3.2	0.22
(50/50)	53.3	3.3	0.13
Ultem/BMI			
(100/0)	54.5	3.0	0.14
(65/35)	54.4	3.0	0.16
(50/50)	54.5	—	—

CONCLUSION

Three kinds of thermoplastic/thermosetting polyimide blends have been prepared using three kinds of commercially available soluble polyimides(AL1051, PI2080, Ultem) and N,N'-(methylenedi-p-phenylene)bismaleimide(BMI). The blends showed exotherms at higher temperatures compared to BMI alone. The cure characteristics of the blends changed significantly according to the thermoplastic polyimide components. The fracture surfaces of PI2080 and Ultem blends showed distinct nodular morphological features. However, in the case of AL1051, which was derived from a aliphatic acid, this kind of phase separation was not observed. The blends yielded films with good mechanical properties even at high temperatures. The BMI component did not have any significant effect on the glass transition temperatures. However it showed improved mechanical properties at high temperatures. The blends were brittle compared with the thermoplastic polyimides, but showed almost the same coefficients of thermal expansion as those of the thermoplastic polyimides. The blends also showed good electrical properties.

REFERENCES

1. Alberino, L.M, Farrisey, W.J.Jr. and Rose, J.C., U.S.Patent 3,708,458(1972).

2. Grundshober, F. and Sambeth, I, U.S.Patent 3,380,964(1964).

3. Yamamoto, Y., Satoh, S. and Etoh, S., 30th National SAMPE Sym., 903(1985).

HIGHLY TEMPERATURE RESISTANT SILICONE LADDER POLYMERS

Hiroshi Adachi, Etsushi Adachi, Shigeyoshi Yamamoto and Hirozho Kanegae

Manufacturing Development Laboratory, Mitsubishi Electric Corporation Amagasaki, Hyougo Japan.

ABSTRACT

High purity high molecular weight poly (phenylsilsesquioxane) (PPSQ) has been synthesized and specimen films prepared by a solvent casting method. The decomposition temperature of the material is 550 °C in air. A PPSQ specimen cured at 350 °C for 60 min has a weight loss of 1% when held at 460 °C for 60 min. A specimen initially dried at 200 °C for 30 min then held at 460 °C loses twice as much. The main constituent of the volatile material from PPSQ pyrolyzed at 500 °C for 60 min is benzene, which is 5.7% of the specimen weight. The tensile strength for the cured specimen is 8.0×10^9 dyne/cm^2 at 25 °C, and 5.5×10^9 dyne/cm^2 at 250 °C. The strength of the dried specimen is 4.0×10^9 dyne/cm^2 at 250 °C and about one half that at 25 °C. The thermal expansion coefficient for the cured specimen is $11\text{-}14 \times 10^5$ /°C over a temperature range from 5 °C to 350 °C. Residual stress is 2.8×10^8 dyne/cm^2 at 25 °C and one order lower at 350 °C.

INTRODUCTION

Highly temperature resistant polymers have been widely studied for application to ICs. Polyimide resins [1], silicone ladder polymers [2] and others have been reported. Among these polymers, polyimide resins have been widely used in the semiconductor industry for passivation, buffer coating and interlayer insulation [1].

Polyimide materials are thermosetting resins, so polyimide films are prepared by spin-coating a poly (amic acid) solution onto silicon wafers and curing, typically at 350 °C for 60 min. On this curing process the imidization reaction takes place and as a result highly temperature resistant materials are formed. During imidization, there is weight loss and film shrinkage; furthermore some by-products besides water are formed. Another problem is that polyimide films may have poor adhesion to inorganic films such as the Si wafer, SiO$_2$, or SiN.

With the increasing density of ICs, there is a fear that these disadvantages will become very important problems for the manufacture and reliability of ICs. Therefore, a high quality polymer is required for future integrated circuits. The silicone ladder polymers are highly temperature resistant, especially poly (phenylsilsesquioxane) (PPSQ) [3,4]. PPSQ materials are expected to be without the above disadvantages, and the reasons for this are as follows:

1): The polymers exhibit the behavior of thermoplastic resins
2): Its affinity for inorganic films is better than that of an organic resin, owing to Si-O bonds in the main chain.

However, the properties of PPSQ, especially the mechanical properties, are not well known. It is difficult to prepare the specimen films necessary for mechanical testing, as PPSQ film is very brittle.

Mat. Res. Soc. Symp. Proc. Vol. 227. ©1991 Materials Research Society

We have synthesized pure PPSQ with molecular weight of 10^5 and prepared films by solvent casting methods. Property measurements have been made on films dried at 200 °C and on films dried and then cured at 350 °C. The purpose of this study has been primarily:

1): to estimate the thermal properties
2): to determine the temperature dependance of mechanical behavior
3): to estimate the effects of cross-linking on properties.

EXPERIMENTAL

Materials

PPSQ resin was prepared in two steps. First, the starting material, phenyltrichlorosilane, was hydrolyzed in an aliphatic solvent at 0 °C. The products of the hydrolysis are a prepolymer with weight average molecular weight of 1000 and hydrochloric acid. After removal of the acid layer, the solution of hydrolyzate was sufficiently neutralized by washing several times. The prepolymer has Si-OH as its terminal group, easily characterized by IR [5].

Next, the high polymer was prepared from the prepolymer, using an alkaline condensing agent under reflux for 24 hours. Dehydration condensation took place between the terminal Si-OH groups of the prepolymer. This process gave a polymer with M_w over 10^5 and with Si-OH as terminal groups.

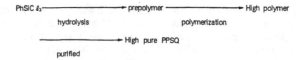

The crude product was purified by a dissolution - reprecipitation method. Figure 1 shows the GPC curve of the purified PPSQ. The polydispersity is less than 3.3. This is a narrow distribution compared to silicone ladder polymers prepared using other methods. After the purification, a highly pure polymer was obtained. Sodium and potassium ions were detected at well below 1 ppm and radioactive elements were detected at well below 1 ppb. PPSQ is not reactive with water, and is stable on storage at room temperature.

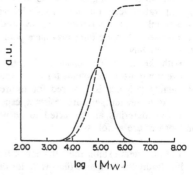

Figure 1 GPC curves for purified PPSQ. Mw= 165,000; Mw/Mn = 3.3

Figure 2 shows the H^1-NMR spectrum of high molecular weight PPSQ. The broad peak at 7 ppm can be assigned to

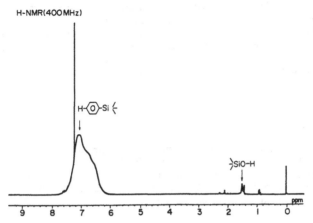

H-NMR(400MHz)

H—⟨O⟩—Si ⊱

⟩SiO—H

ppm

Figure 2 H¹-NMR spectrum for PPSQ with weight average Molecular weight of 10⁵.

protons of the phenyl group. The small peak at 1.5 ppm can be assigned to the proton of the Si-OH group. A C¹³ NMR spectrum shows only carbon atom attached to Si atom.

Samples to be used for evaluation of mechanical properties (dynamic mechanical analysis and quasi-static mechanical properties) were prepared from a solution of high molecular weight PPSQ by solvent casting. The solution was dried in a frame to evaporate the solvent very slowly at room temperature. It took 14 days for the solvent to evaporate completely. The film obtained was cut into strips with widths if 3-5 mm, thicknesses of 0.1-0.3 mm and lengths of 20-40 mm. Since the PPSQ film is very brittle, the strips had to be prepared with great care. Strips were annealed at 150°C for 60 min and then at 200°C for 30 min to obtain dried specimens. TO prepare cured specimens, the dried films were further annealed at 350°C for 60 min. The dried specimens could be dissolved in organic solvents, but the cured specimens could not. This indicates that crosslinks are formed from dehydration of terminal Si-OH groups during curing.

Thermal Behavior

Thermogravimetric analysis was performed with a Dupont Instrument TGA-951. Measurements were made on both dried and cured specimens over a temperature range from 25°C to 900°C with a heating rate of 10 or 20°C/min in air. The weight loss of specimens held at 430°C and 460°C for 60 min was measured. Analysis of evolved gas from the dried PPSQ specimen pyrolyzed at 500°C under N_2 was made by GC-MS (JEOL;JMS-DX303). The conditions for gas chromatography were: column temperature 60°C to 230°C, heating rate, 8°C/min.

Mechanical Properties

Dynamic viscoelastic properties were measured for both dried and cured films with a Rheometrics RSA-2 at 16 Hz from 25°C to 400°C. Specimens for these measurements were usually 0.2 mm thick, 5 mm wide and 40 mm long. Tensile stress-strain behavior and thermal expansion coefficient were measured in a tensile tester (Seikou densi industry TMASS-20) using specimens 0.2 mm thick, 3 mm wide and 20 mm long. Tensile tests on both dried and cured specimens were made at 25, 100, 180 and 250°C at an strain rate of 0.02 s^{-1}. Thermal expansion was measured over a range from 25°C to 300°C at 3°C/min. The specimens were then held at 350°C for 60 min., cooled to room temperature ant then re-heated to 350°C.

Residual Stress

Warping of silicon wafer was measured at room temperature using a stress tester (FSM-8900TC) as a function of film thickness of PPSQ. The film was spun on the wafer, then dried at 200°C. Residual stress for dried PPSQ films 6 μm thick was determined over a temperature range from 25°C to 350°C using a heating rate of 10°C/min. The specimens were held at 350°C for 60 min then cooled to room temperature.

RESULTS

Thermal Decomposition Temperature

Thermogravimetric analysis of the cured PPSQ film shows that the thermal decomposition temperature is 500°C, and the 5% weight loss temperature is 550°C. PPSQ film is more thermally resistant, by about 50°C, than the common polyimide. Figure 3 shows the weight loss of the PPSQ specimens, both dried and cured. When heated to 460°C for 60 min, the weight loss was 1.0% for the cured specimen and 3.0% for the dried specimen. Cross-linking thus improves the thermal stability.

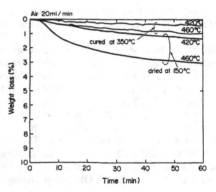

Figure 3 Weight loss for dried and cured specimens of PPSQ held at 420°C and 460°C for 60 min. in air

Pyrolysis

Figure 4 shows the CG result for volatile materials from PPSQ pyrolyzed at 500°C for 60 min. A large peak (a) is observed at long retention time and some small peaks (b,c,d) at shorter retention times. The CG-MS analysis of these materials shows that the large peak is for benzene and the small peaks for CO_2, C_3H_6, and water. No silicon compounds were observed.

The amount of volatile materials formed from the PPSQ specimen is 4.4% at 30 min and 5.7% at 60 min, as listed in Table 1. The amount of benzene increases with heating time; other gases are in very small amounts, and the amounts are independent of heating time. This implies that the elimination reaction of the Si-Ph bond is the main mechanism of thermal decomposition of PPSQ.

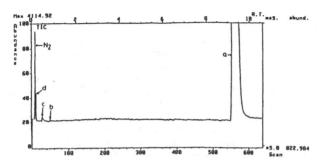

Figure 4 CG spectrum for volatiles from PPSQ specimen

Table I Amount of volatile materials from pyrolyzed PPSQ

Component	amount (%)	
	30 min	60 min
CO_2	0.22	0.21
C_3H_6	0.09	0.09
benzene	4.31	5.71

Dynamic Mechanical Properties

Figure 5 shows the temperature dependance of the tensile storage modulus E and loss tangent δ for both dried and cured PPSQ films. At room temperature the storage modulus for the dried specimen is 1.8×10^{10} dyne/cm²; this is larger than that for the cured specimen. e The storage modulus gradually decreases for both specimens on heating until at 250 °C the modulus becomes constant. This constant value is 3×10^9 dyne/cm² for the cured specimen and 2×10^9 dyne/cm² for the dried specimen. The loss tangent δ shows a broad peak from 25 °C to 250 °C in both specimens. These results show that the PPSQ specimens soften up to 250 °C.

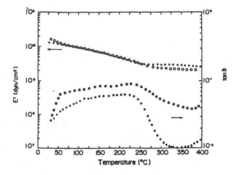

Figure 5 Storage modulus and loss tangent δ for dried and cured films of PPSQ at 16 Hz from 25 °C to 400 °C at a heating rate of 5 °C/min

Stress-Strain Behavior

Figure 6 shows the stress-strain curves for the PPSQ specimens, both dried and cured. Both specimens show brittle fracture, and for both specimens, strain at break increases from about 0.4% to 3% as the temperature of testing increases. The strain at break of the cured specimen is always larger than that of the dried specimen. The tensile strength is about 8.0×10^9 dyne/cm^2 below 200 °C for both specimens. At 250 °C the strength of the cured film is 5.5×10^9 dyne/cm^2 and that of the dried film is 4.0×10^9 dyne/cm^2. The dried films become softer while on the other hand the cured films remain stiffer. The temperature dependance of the modulus determined from the slope of the stress-strain curve corresponds to the temperature dependance of the storage modulus. That is, the Young's modulus of the cured PPSQ specimen is lower than that for the dried PPSQ specimens at low temperature. Above 250 °C the modulus of the cured specimen is the larger.

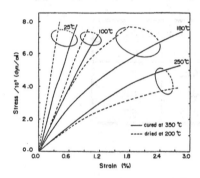

Figure 6 Stress-strain curves for dried and cured specimens of PPSQ at indicated temperatures and a strain rate of 0.02 s^{-1}.

These results imply that cross-linking caused by curing at 350 °C improves the strength and stiffness at temperatures above 250 °C.

Thermal Expansion

Figure 7 shows the thermal expansion for both the dried and the cured specimens over a temperature range from room temperature to 350 °C. The dried specimen exhibited two straight lines which crossed at about 220 °C. On the other hand, the cured specimens show a slowly downward curve. The thermal shrinkage on curing at 350 °C is 0.1%. Thermal expansion coefficient determined from the slope of curves in Figure 7 is 11-14 $\times$ 10^5 for both specimens below 220 °C. Above 220 °C that for the dried specimen decreases to 9×10^5 so cross-linking made the thermal expansion coefficient larger above 220 °C.

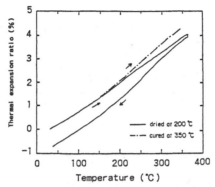

Figure 7 Thermal expansion for a dried film of PPSQ. The specimen was heated to 350°C at 3°C/min, held at 350°C for 60 min, cooled to room temperature and re-heated to 350 °C.

Residual Stress

The film thickness dependance of the warping of coated Si wafers indicates that the residual stress in the PPSQ films is 2.8×10^8 dyne/cm^2 at room temperature. Figure 8 shows the temperature dependance of residual stress for a 6 μm thick dried PPSQ film; some hysteresis is observed. The residual stress at 350°C is about ten times lower than that at 25°C. There is no change in the residual stress during cure. The difference in stress between dried films and cured films is 1×10^8 dyne/cm^2. This can be explained by the difference in thermal expansion coefficient between the two types of sample.

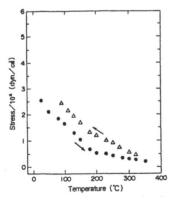

Figure 8 Temperature dependence of residual stress for 6 μm films of dried PPSQ. The film was heated to 350 °C for 60 min, then cooled to room temperature. The stress was measured *in situ.*

CONCLUSION

Determination of thermal properties and the temperature dependence of mechanical properties have been made using PPSQ specimens with two thermal histories; dried at 200 °C and cured at 350 °C. Weight loss of the specimens held at 460 °C for 60 min in air is 1% for the cured specimen and 3% for the dried specimen. Tensile strength of the films is 8×10^9 dyne/cm^2 at 25 °C. At 250 °C the strength of cured films is 5.5×10^9 dyne/cm^2, of dried films, 4×10^9 dyne/cm^2. These results imply that cross-linking improves the thermal resistance and the mechanical properties at temperatures above 200 °C.

REFERENCES

1.   D. Makino IEEE Electron Insulation Magazine 4, 65 (1988)

2.   J.F. Brown Jr., L.H. Vogt Jr., A. Katchman, J.W. Eustance, K.M. Kiseerand and K.E. Krantz, J. Amer. Chem. Soc., 82 6194 (1960)

3.   Z. Xinsheng, S. Lionghe, L. Shuging and L, Yizhen, Polymer Degradation and Stability, 20, 157 (1988)

4.   H. Adachi, E. Adachi, O. Hayashi and K. Okahashi, Rep. Prog. Polym. Phys. Jap., 28 261 (1985)

5.   J.F. Brown Jr., J. Polymer Sci., C-1 83, (1963).

REAL-TIME FT-IR STUDIES OF THE REACTION KINETICS FOR THE POLYMERIZATION OF DIVINYL SILOXANE BIS-BENZOCYCLOBUTENE MONOMERS

T. M. Stokich, Jr., W. M. Lee and R. A. Peters

Dow Chemical USA, Central Research, Midland, MI.

ABSTRACT

The thermal polymerization reaction of divinyl siloxane bis-benzocyclobutene (DVS bis-BCB) was monitored in-situ with FT-IR spectroscopy in order to follow specific chemical changes and determine the reaction order and rate constants at temperatures from 150° to 210°C. FT-IR spectra were obtained at regular intervals throughout the reaction with a Nicolet 170SX spectrophotometer.

Monomeric DVS bis-BCB contains mixed stereo and positional isomers of 1,3-bis(2-bicyclo[4.2.0]octa-1,3,5-trien-3-ylethenyl)-1,1,3,3-tetramethyl disiloxane (CAS 117732-87-3). It polymerizes via Diels-Alder cycloaddition reactions between vinyl groups and an intermediate $\underline{o}$-quinodimethane formed by first-order, thermally initiated ring openings of the benzocyclobutene rings. Gaseous byproducts are not produced; therefore, the cure is easier to manage than are cures for polyimides which evolve water in polycondensation reactions. The DVS bis-BCB has four reactive elements per monomer unit and, thus, polymerizes into a very highly cross linked and solvent resistant network.

With the FT-IR methodology, the reaction was easily monitored through the points of gel formation and vitrification. With the exception of DSC (i.e., calorimetry) which does not sense specific chemistry, other methods were not successful in following the reaction after a gel was formed. We have found that the polymerization was first-order until vitrification occurs; the gelation alone had no apparent effect on the reaction rate.

DVS bis-BCB is under development at Dow as high performance dielectric material for multilayer interconnect coating applications for the microelectronics industry. Methodology reported here is employed in developing effective cure management strategies.

INTRODUCTION

DVS bis-BCB, which was patented at Dow in 1985 [1,2], has the monomer structure depicted in I. Commercial interest in the material centers on its ease of processing via thermal polymerization, its low dielectric constant (2.7) and dielectric loss (0.0008), its hydrophobicity and its resistance to solvents when fully polymerized and crosslinked. These properties result from its hydrocarbon/siloxane structure and from having four sites per monomer unit for crosslinking during polymerization. For applications in microelectronics as a thin film dielectric coating material, its ability to planarize over circuitry in simple, single stage spin-coating operations is also attractive.

Mat. Res. Soc. Symp. Proc. Vol. 227. ©1991 Materials Research Society

Currently the DVS bis-BCB is sold as a partially polymerized (B-staged) product, dissolved in mesitylene at 35%, 55% and 62% w/w concentrations. B-staged DVS bis-BCB is ~40% polymerized. Depending on concentration, solutions of B-staged DVS bis-BCB have a viscosity of from 15 to 250 cP, providing flow properties suitable for spin-coating the pre-polymer to produce uniform, defect-free coatings on silicon or aluminum. By controlling the spin time time and speed, film

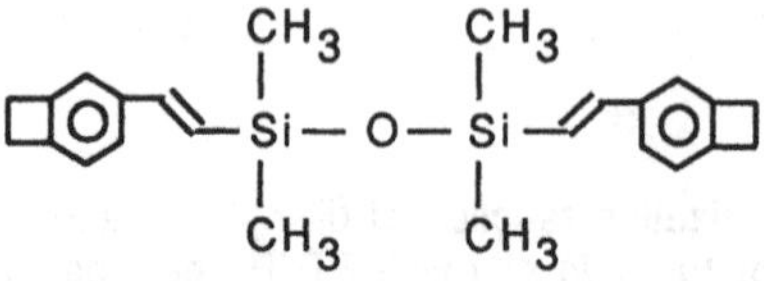

I

DVS bis-BCB

thicknesses from 300 Å to 25 microns have been fabricated in the Dow clean room, reproducibly. Without catalysts or initiators, the user completes the cure simply by heating the coated substrate under nitrogen at temperatures from 210° to 250°C for one hour. This report discusses the chemistry and the kinetics of that cure.

Chemistry involved in the polymerization is currently understood to proceed through two steps [3,4], the first of which involves a thermally driven ring opening reaction given by

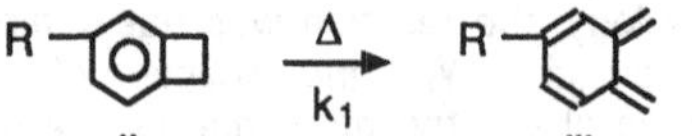

$$\qquad(1)$$

II   III

BCB   o-quinodimethane

where II is the BCB reactant, III is an o-quinodimethane intermediate, R is the remainder of the monomer or polymer and k_1 denotes the rate constant. The second step in the polymerization poses at least two alternatives. If olefinic material is available, the reaction proceeds as indicated by

$$\qquad(2)$$

III   IV   V

where V, a tetrahydronaphthalene group, is the polymer linking element and it probably is the result of a Diels Alder cycloaddition reaction between the more reactive diene in III and the dienophile IV. A and B simply denote other extended elements in the network and k_2 is the bimolecular rate constant. Plainly, there are four ways to select a diene from the structure in III, but only one of them will lead to the more stable aromatic product which is shown.

Other second-step reactions, those between two BCBs, are also well known to occur [3], but they are not completely characterized in terms of the products or mechanisms. In general, these reactions are denoted as

in which VI and VII denote the bonding currently known to exist in product mixtures that develop when no olefins are available to react with the o-quinodimethane.

Polymerization reactions of DVS bis-BCB monomers can, in principle, react either as indicated by (2) or (3) since there is an abundance of both BCB and olefinic species. For this study, the objectives are to monitor several bond groups during polymerization and to estimate the degree of competition between the different types of reactions. Kinetic information is presented, based on the rate of change of the infrared absorbance which is assumed to be linear in concentration. The findings expand upon, but are consistent with, results reported earlier which were based on calorimetric measurements [3].

MATERIALS

Materials used in these experiments included the neat monomer of DVS bis-BCB (a low viscosity liquid at room temperature) and a sample of B-staged pre-polymer. The B-staged material was dissolved in mesitylene (55% solids) and was a specimen from the commercial product designated as XU13005.01.

METHODS

Sample Preparation

Drop-sized samples of the monomer were placed directly onto polished KBr plates and then were covered with another FT-IR cell window. The resultant film thicknesses were approximately 2-5 microns. Two of the samples were degassed under a moderate vacuum (~50 mm Hg) for one hour prior to covering them and subjecting them to the in-situ analysis. A third sample of monomer was placed in the cell without degassing.

Samples of the B-staged pre-polymer in mesitylene were spin coated onto polished KBr plates, to a nominal pre-polymer thickness of 3 microns. After spin coating, they were heated to 150°C under nitrogen for half an hour to remove any remaining solvent. The coated plates were then cooled to room temperature and covered with another KBr window.

<u>Apparatus and Procedure</u>

Polymerization reactions were monitored *in-situ* with a Nicolet Model 170SX FT-IR spectrophotometer, operating at 2 cm^{-1} resolution. A Globar source and an MCT-B detector were used in the instrument setup. Experiments were done under automated control, meaning that the instrument was programmed to trigger the data acquisition at periodic intervals. Seventy spectra were collected in each run. Each spectrum represents the averaging of 50 to 400 interferograms to enhance the signal-to-noise ratio. The triggering time interval between *successive* spectra was programmed as follows: 3 minutes for a run at 210°C (3.5 hours overall), 14 minutes for a run at 190°C (16.3 hours overall), 1 hour for a run at 170°C (70 hours overall) and 3 hours for a run at 150°C (210 hours overall). Under automated control, the timing accuracy was better than ±1 second between sequential spectra.

These lengthy monitoring periods are not those that are used in normal cures of the DVS bis-BCB. In fact, Dow recommends curing the B-staged material at 250°C for one hour (under nitrogen). However, at 250°C, the reaction produces 95% of the attainable cross-link bonds within ~20 minutes, and the reaction is simply too fast to monitor with the current FT-IR setup. Therefore, to obtain kinetics data over a wide range of temperature, it was necessary to use lower temperatures and much longer reaction times.

<u>Temperature Control</u>

All the reactions were monitored isothermally. Temperature control was provided by a heater cell coupled with an LFE Systems controller (Model 2012) and calibrated J-type thermocouples. The cell system provided an 800 Watt heating capability. An external set of KBr windows (outside those defining the sample gap) was installed in the cell in order to limit temperature gradients caused by convection currents. Cell temperature calibrations, revealed that the short term cyclic drifts were ±1°C of the set point and long term drifts were much less than one degree.

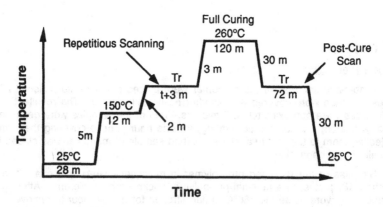

Figure 1. Temperature profile for the DVS bis-BCB cure schedules.

Each polymerization study was conducted in accord with the temperature program in Figure 1, and the timing for the data acquisition was synchronous with it. A first stage (28 minutes at 25°C) provided a delay, allowing for a nitrogen purge of the spectrometer. The second, at 150°C, elevated the temperature closer to the ultimate set point, without initiating significant reaction, in order to minimize temperature over-shoot in the subsequent step. The third step to a temperature T_r (run temperature) initiated the polymerization and was maintained throughout the FT-IR monitoring period. A forth step to 260°C completed the cure of any material remaining unreacted, and the fifth step (back to T_r) was used to take a reference spectrum of the fully cured material at the same temperature as that of the kinetics run. Spectra collected at this last point in the program are denoted as "post-cure".

Data Reduction

FT-IR data were reduced to absorbance (Figure 2), relative to a background collected for the blank cell (KBr windows in place) at each temperature. Separate calibrations were done, revealing that the absorbances of key bands were linear functions of the path length. For calibrations covering path lengths from 0.7 to 30 microns, the linear region extended up to an absorbance of 1.5 or higher. A confirmation of the expected linearity between absorbance and concentration could not be made directly since there were no alternative means for the concentration measurement. Nonetheless, such behavior is reasonable to expect.

Accordingly, absorbance measurements, $A_{\tilde{v}}$, are represented as [5]

$$A_{\tilde{v}} = \varepsilon_{\tilde{v}} bc_{\tilde{v}} + A_{0,\tilde{v}} . \tag{4}$$

The parameters here are: ε, the extinction coefficient; b, the path length; c, the concentration and A_0, the contribution of absorbances which are not due to the chromophore of interest. A subscript of $\tilde{v}$ denotes the wavenumber position in the spectrum; the use of it with concentration is meant *only* to indicate that concentrations of different bond groups can be monitored independently (can refer to vinyl, BCB, tetrahydronaphthalene or other groups). For these studies, the concentrations are time dependent and corresponding fractions of materials present at time t are defined as

$$f_{\tilde{v}} \equiv \frac{(c_{\tilde{v}})_t}{(c_{\tilde{v}})_{t=0}} = \frac{(A_{\tilde{v}})_t - (A_{\tilde{v}})_{\infty}}{(A_{\tilde{v}})_{t=0} - (A_{\tilde{v}})_{\infty}} \qquad \text{(for reactants)} \tag{5}$$

and

$$f_{\tilde{v}} \equiv \frac{(c_{\tilde{v}})_t}{(c_{\tilde{v}})_{\infty}} = 1 - \frac{(A_{\tilde{v}})_t - (A_{\tilde{v}})_{\infty}}{(A_{\tilde{v}})_{t=0} - (A_{\tilde{v}})_{\infty}} \qquad \text{(for products)} \tag{6}$$

where the subscript, ∞, refers to measurements made on the post-cure scans.

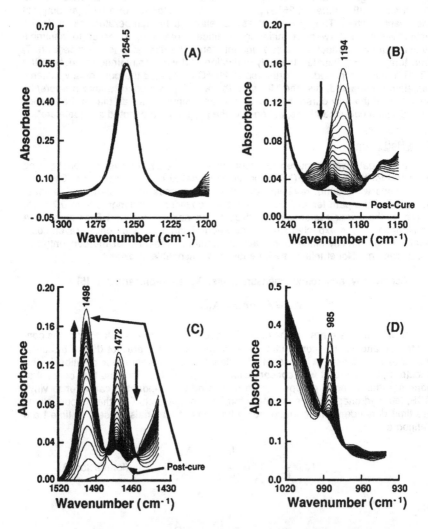

Figure 2. Selected regions of FT-IR spectra for the cure of DVS bis-BCB monomer at 190°C. Insets each contain 35 spectra which were collected *in-situ* (one every 28 minutes for 16.3 hours). Arrows denote the progression of the spectral bands with time.

RESULTS AND DISCUSSION

For selected regions of the FT-IR spectrum, Figure 2 shows several important changes that occur during the thermal polymerization. All figure insets refer to the reaction conducted at 190°C; each one shows a superposition of 35 spectra (every other one from the set collected in the run). At *all* temperatures, the *same* spectral changes were observed, except for the reaction rates.

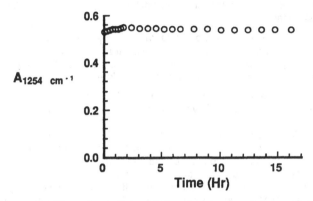

Figure 3. Time dependence of the absorbance of the methyl rocking mode at 1254.5 cm^{-1} during a thermal cure of DVS bis-BCB monomer at 190°C.

Figure 2 (A) shows a strong absorption (1254.5 cm^{-1}) for the rocking mode of the methyl groups attached to the silicon atoms. The corresponding time dependence is provided in Figure 3. Nearly unaffected by the polymerization, the peak provides an internal reference showing that there is little temporal variation in the number of chromophores in the beam path (evidently, no material escapes from the gap). There is, however, a small absorbance increase initially, followed by a decrease of similar magnitude. The cause of this is unknown. To eliminate spurious effects of this type in the remaining analysis, absorbances for other bands were *normalized* relative to the silicon-methyl reference band by means the relation

$$(A\tilde{v},t)_{\text{Normalized}} = \left\{ \frac{(A_{1254.5})_{t=0}}{(A_{1254.5})_t} \right\} (A\tilde{v},t)_{\text{Measured}} \, . \qquad (7)$$

A vibration of the BCB group, shown in Figure 2 (B), is also found in the spectrum as an isolated peak. The specific mode of motion has not been ascertained with certainty. Its position in the spectrum is consistent with aromatic C-H rocking (which is sensitive to substitution) [6]; however, other group modes are also possible. The absorption is found in several model compounds containing BCB and it is not found in vinyl-containing model compounds which do not contain BCB.

During the reaction, the peak progressively decreases and, at post-cure, it is nearly eliminated.

Similarly, Figure 2 (C) shows two absorptions (1498 and 1472 cm⁻¹) which, respectively, are for the tetrahydronaphthalene group being formed and the BCB group being consumed. The vibrational modes are ring bending modes for the aromatic ring; they too are sensitive to the ring substitution. At 1472 cm⁻¹, the peak progressively decreases (virtually disappears in the post-cure scan), while the peak at 1498 cm⁻¹, which is absent in the monomer (initial scan), progressively develops into a strong absorbance.

Lastly, the out-of-plane bending mode of the vinyl moiety appears as a moderate-to-strong absorbance in the monomer spectrum (see Figure 2-D, at 985 cm⁻¹). Progressively, the absorbance is depleted during the reaction, and in the post-cured material, only ~5% or less of the original peak remains.

Numerous other informative bands exist in the DVS bis-BCB spectrum which will not be shown in this report. Several of them undergo changes during the polymerization. The changes found in the C-H and Si-C stretching bands are very prominent but could be consistent with any of the chemistry depicted in Equations (1) or (2) and, thus, are not definitive. The absorbance for a strong Si-O stretch found at 1050 cm⁻¹ is slightly reduced during the polymerization (~10%, after normalization). How this finding relates to the reaction chemistry is , as yet, unclear. It is likely that the Si-O bond group in DVS bis-BCB is not actually the species causing those absorbance changes, since unresolved, overlapping absorptions are highly probable. For instance, similar polymerization studies conducted at Dow on Diketone bis-BCB monomers have shown that without the Si-O connector group, a moderately strong absorption exists at 1067 cm⁻¹ which vanishes during a thermal cure [7]. The RAMAN spectrum of DVS bis-BCB does, in fact, contain a well resolved band at 1068 cm⁻¹. It appears that the FT-IR simply can not isolate this band and that there are no reasons for concern over the fate of the Si-O linkage during polymerization.

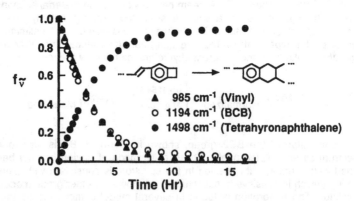

Figure 4. A comparison of the of the fractions of vinyl, BCB and tetrahydro-naphthalene groups present at time t during the polymerization of DVS bis-BCB monomer at 190°C.

Beyond the qualitative findings, FT-IR spectra provide very useful quantitative information on the reaction kinetics. Figure 4, for instance, shows the time dependences for the fractions of the vinyl, BCB and tetrahydronaphthalene bond groups during the polymerization of DVS bis-BCB at 190°C. Within the error of measurement, the BCB and vinyl moieties disappear concurrently, suggesting rather strongly that they react with one another as expected for the Diels-Alder chemistry in Equation (2); there apparently is very little material proceeding along other pathways such as those of Equation (3). If the latter were an important consideration, the vinyl and BCB moieties would disappear at different rates, ultimately to different degrees of completion. Likewise, as expected, the tetrahydronaphthalene linkages evolve with a complimentary time dependence. Nowhere else in the spectrum does there appear any evidence for other types of bonds being formed, at least in appreciable concentration; therefore, the predominant network structure being formed is very likely to be that shown below in VIII.

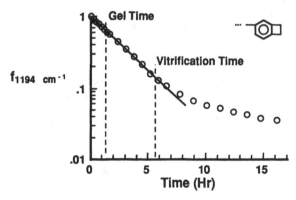

VIII

The reaction order is also readily obtained from these measurements. Figure 5, for instance, shows a semi-log plot of the fraction remaining at time t of the BCB groups (1194 cm^{-1}). Until those groups are depleted by ~90%, the rate is clearly first-order, so a unimolecular step such as that shown in Equation (1) is rate limiting.

Figure 5. A rate plot demonstrating first-order kinetics for the polymerization of the BCB monomer. T=190°C. Vitrification apparently changes the rate.

Thereafter, the rate slows appreciably. Note that also depicted on this plot are the gel and vitrification times for the reaction. Both these times were determined experimentally by CJ Carriere [8] in other experiments in which the shear modulus, G', was monitored as a function of time during polymerization (at the same temperature, under nitrogen). The gel time is the time when the equilibrium modulus, G_e, becomes finite, which occurs when an extended network exists. For DVS bis-BCB the gelation normally occurs when f is near 0.67 (it can appear later, at lower f, if impurities contaminate the monomer). In all reactions studied thus-far, the gel-point has had no effect on the reaction rate, suggesting that local mobilities are virtually unaffected. The vitrification time (likewise determined from measurements of G') is the time when the material becomes a glass at the temperature of the reaction. Clearly, it is this time which correlates extremely well with the change in reaction rate. A loss of free volume associated with the glass transition is apparently adequate to reduce the mobility locally.

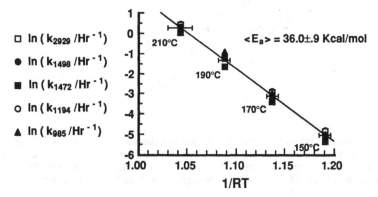

$\square$ $\ln (k_{2929} /Hr^{-1})$

$\bullet$ $\ln (k_{1498} /Hr^{-1})$

$\blacksquare$ $\ln (k_{1472} /Hr^{-1})$

$\circ$ $\ln (k_{1194} /Hr^{-1})$

$\blacktriangle$ $\ln (k_{985} /Hr^{-1})$

$<E_a> = 36.0\pm.9$ Kcal/mol

Figure 6. Arrhenius plot for the rate constants determined from the thermal polymerization of DVS bis-BCB. Symbols denote the following: $\blacktriangle$, the vinyl group; $\circ$ and $\blacksquare$, the BCB group; $\bullet$, the tetrahydronaphthalene group; and $\square$, an absorption band corresponding to the methylene asymmetric -CH stretch.

First-order rate constants were obtained from the initial portions of the polymerization reactions. Expressions for the time dependences are given by

$$f_{\tilde{\nu}} = \exp(-k_{\tilde{\nu}}\, t) \qquad \text{(for reactants, when } f_{\tilde{\nu}} \geq 0.85) \qquad (8)$$

and

$$1 - f_{\tilde{\nu}} = \exp(-k_{\tilde{\nu}}\, t) \qquad \text{(for products, when } f_{\tilde{\nu}} \leq 0.85). \qquad (9)$$

where $k_{\tilde{\nu}}$ has replaced the earlier notation of k_1 and it simply denotes the rate constant obtained from the *band* of interest in the FT-IR spectrum. From plots of $\ln(f_{\tilde{\nu}})$ or $\ln(1 - f_{\tilde{\nu}})$ (as appropriate) versus time, the rate constants were obtained by means of linear regression. Five absorbance bands were examined in this analysis for temperatures from 150° to 210°C. Absorbance data for the 210° rate constants was obtained using the B-staged pre-polymer and, for the other temperatures, it was obtained using the neat monomer. The results of those calculations are shown in Figure 6, where all rate constants are in units of hours^{-1}; R (on the x-axis) denotes the gas constant and T is temperature in °K. For reference, the data points are also annotated with the Centigrade temperature. The temperature dependence of the reaction rate is clearly of Arrhenius form (an activated process) and the activation energy is 36.0±0.9 Kcal/mole. To within the measurement error, each bond group provided the same activation energy, suggesting then that each one evolved via the same rate-limited chemical process as, of course, was expected.

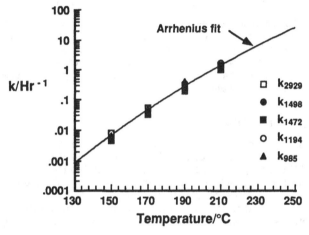

Figure 7. Temperature dependence of the first-order rate constant for the polymerization of DVS bis-BCB monomer or B-staged pre-polymer.

The Arrhenius equation was used to interpolate or extrapolate these rates to other temperatures:

$$k_{\tilde{v}, T} = k_{\tilde{v}, T_0} \exp\left\{ -\frac{E_a}{R} \left(\frac{1}{T} - \frac{1}{T_0} \right) \right\} \qquad (10)$$

Terms in the expression are: $k_{\tilde{v}, T}$, the rate constant at temperature T; $k_{\tilde{v}, T_0}$, the rate constant at temperature T_0; E_a, the activation energy and R, the gas constant. With T_0=190°C (463°K) and $k_{\tilde{v}, T_0}$=0.27 hr^{-1}, one obtains the curve shown in Figure 7 (symbols are for the measured values). Estimates are therefore available for the reaction rate at temperatures more practical for curing the DVS bis-BCB materials. The activation energy obtained here via FT-IR is identical to that obtained earlier via calorimetric techniques [3].

CONCLUSIONS

In-situ FT-IR measurements have provided the means to monitor specific chemistry in the thermal polymerization of DVS bis-BCB and other monomers. We have found that the reaction follows first-order kinetics, which is not a new finding for this system. However, we also found this to be true for each of the relevant bond groups, which is a new contribution. Secondly, it was determined that the BCB and vinyl moieties are consumed concurrently, making it very likely that the tetrahydro-naphthalene linkage is overwhelmingly the principal product throughout the reaction. Substantiation was found in monitoring the development of the product.

The reaction rate temperature dependence clearly follows an Arrhenius form and knowledge of this behavior makes it convenient to compute rates of reaction for any other temperature of interest, which is useful in planning curing procedures.

Finally, through this investigation, we found that thermally cured DVS bis-BCB (cured under nitrogen) results in a polymer product having the same FT-IR spectrum regardless of the cure temperature for temperatures from 150° to as high as 310°C. While kinetics could not be monitored at the higher temperatures, it is clear that the processing window for curing the material is not highly restrictive with respect to the temperature program.

ACKNOWLEDGEMENTS

The authors thank the following individuals for assistance in this project: CJ Carriere for his direct contributions by providing the gel and vitrification times; RA Kirchhoff and KJ Bruza for providing guidance in understanding the chemistry ; CE Mohler for investigating the spectra of model compounds to aid in the interpretation; DC Burdeaux for spin coating substrates in his Dow clean room facility; and Dow, Michigan Division for providing the materials. We also gratefully acknowledge CL Putzig and RA Nyquist for guidance in the spectral interpretation and development of methodology.

REFERENCES

1. RA Kirchhoff, U.S. 4,540,763 (Sept 10, 1985).

2. AK Schrock, U.S. 4,812,588 (Mar 14, 1989).

3. KJ Bruza, CJ Carriere, RA Kirchhoff, NG Rondan and RL Sammler, J. Macromolecular Sci., in press,1990.

4. RA Kirchhoff, CE Baker, JA Gilpin, SF Hahn and AK Schrock, 18th International SAMPE Symposium, 1986, pp. 478-489. (See also: IL Klundt, Chem. Rev., 1970, **70** (4); 471; KP Thrummel, Acc. Chem. Res., 1980, **13**, 70-76 and W Oppolzer, Synthesis (Reviews), 1978, 793).

5. JW Robinson (editor), Practical Handbook of Spectroscopy, CRC Press, Boston, 1991, 534-535.

6. CJ Pouchert, The Aldrich Library of Infrared Spectra, Edition. III, Aldrich Chemical Co., 1981, 559.

7. KJ Bruza, PJ Bonk, RF Harris, RA Kirchhoff, TM Stokich, Jr., RL McGee, RA DeVries, 36th International SAMPE Symposium, Apr 1991, **36** (Book I), p. 457.

8. CJ Carriere, unpublished results.

Characterization, Structure and Properties

INTRAMOLECULAR CHARGE TRANSFER IN AROMATIC POLYIMIDES

JOSEPH M. SALLEY,* TAKAO MIWA,** AND CURTIS W. FRANK*
*Department of Chemical Engineering, Stanford University, Stanford, CA 94305-5025
**Hitachi Research Laboratory, Hitachi Ltd., 4026 Kuji-cho Hitachi-shi Ibaraki-ken, 319-12, Japan

ABSTRACT

The use of polyimides in integrated circuits to planarize complex topography and the need to decrease the dimensions of electronic devices have motivated us to gain a better understanding of the details of the polymer microenvironment. UV/Vis absorption experiments suggest that intramolecular charge transfer (ICT) exists and therefore might utilized as intrinsic, nonperturbing probes. The transfer of charge is believed to occur upon exposure of ultraviolet light as a result of donation from the phenyl ring (from the diamine fragment) to the adjacent imide ring. Furthermore, the transfer of charge should be a function of the torsional angle between the planes of the donor and acceptor segments.

In this study, we first investigate the nature of the ICT by conducting UV/Vis absorption measurements on a series of model compounds. A comparative analysis of 10^{-5} M solutions of the models in dioxane provides evidence that intramolecular charge transfer occurs when the compounds are exposed to 264 nm (ultraviolet) light. Similarly, a comparison of an analogous series of polyimides demonstrates that the same phenomenon occurs in the more complex polymeric systems when spin-cast films are exposed to 326 nm (ultraviolet) light.

INTRODUCTION

Over recent years, polyimides have been an important topic of research because of their excellent thermal and mechanical properties. Of particular interest is the characterization of these polymers by photophysical methods to better understand their fundamental nature and to correlate molecular structure with bulk properties. One such area of concern involves the study of both intermolecular and intramolecular charge transfer.

The first to suggest that intermolecular charge transfer (CT) complexes exist in the aromatic imide systems were Dine-Hart and Wright, who based their conclusions on the donor/acceptor nature of the alternating segments, and the color, solubility, and fusibility of their models.[1] More recently, others have rationalized that these complexes effect bulk properties such as T_g and photoconductivity.[2,3] Still others have proposed that the formation of intermolecular CT complexes depends on the intermolecular chain distance and that the complexes can be utilized as probes of chain ordering that occurs during the curing process.[4,5]

Similarly, intramolecular charge transfer (ICT) is believed to exist in polyimides and is thought to be a function of the polymer chain conformation;[4] recent semi-empirical calculations support this theory.[6] It is believed that upon exposure to UV radiation, the molecules are excited

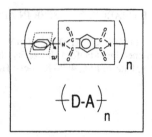

Figure 1: Donor/Acceptor nature of polyimide

to a higher energy state such that electron density is transferred from the (amine) phenyl ring to the adjacent imide ring. Furthermore, the transfer of charge should be a function of the torsional angle ω, defined as the angle between the planes of the donor and acceptor segments (Figure1). The transfer of charge should therefore be maximized in the limit that ω approaches 0° and minimized in the limit that ω becomes 90°.

Assuming that the transfer of charge occurs and that it depends on the torsional angle between the donor (amine phenyl ring) and acceptor (imide ring), then it may be possible to utilize the ICT as a probe of the polymer chain conformation. Our UV/Vis absorption study was therefore designed to determine if an ICT can be observed and to elucidate the details of the mechanism of such a transition. In order to do this, we first examined a series of model compounds. In a dilute solution environment, we believe that these trimer systems should be void of any solute-solute intermolecular interactions and should therefore be good models for predicting the trends of the intramolecular electronic properties of the polymers. The models systems are subsequently compared to a cognate polymer series.

EXPERIMENTAL

The model compounds N,N'-diphenyl-3,3',4,4'-benzophenonetetracarboximide (I), N,N'-dibenzyl-3,3',4,4'-benzophenonetetracarboximide (II), N,N'-dicyclohexyl-3,3',4,4'-benzo-phenonetetracarboximide (III), and N,N'-bis(2-methylphenyl)-3,3',4,4'-benzophenonetetra-carboximide (IV) were prepared following the method of Dine-Hart[1] and Ishida.[7] The benzo-phenonetetracarboxylic dianhydride (BTDA) was provided by Hitachi, and the amines and anhydrous 1-methyl-2-pyrrolidinone (NMP) were purchased from Aldrich. One equivalent of the dianhydride was added to two equivalents of the amine (dissolved in NMP). The temperature was maintained between 15 and 20 °C upon addition of the dianhydride, then the monomers were allowed to react at room temperature under N_2 for two hours to yield a clear red or yellow amic acid solution (10-15 wt%). Typically, subsequent heating of the amic acid at 150 to 170 °C for two hours yielded first a dark brown solution and then the imide precipitate. The solution was allowed to cool to room temperature; the model compound was then filtered, washed with NMP and then with distilled water, and dried. The above procedure was slightly modified for the more soluble compounds, (eg., Model IV). These compounds gave no precipitate upon heating or cooling, but were easily recovered by the addition of water to the cooled imide solution. The structure and purity of the compounds were confirmed by NMR (Varian XL-400) and IR (Perkin Elmer 1710). Solutions of the model compounds were made with spectrophotometric grade 1,4 dioxane purchased from Aldrich.

Polymers A-E (Figure 7) were prepared by condensation polymerization of the dianhydride (Hitachi) and the diamines (Aldrich). Prior to synthesis, the diamines were purified via vacuum sublimation, and the BTDA was baked at 200 °C to negate the effects of any hydrolysis that occurred during storage. After purification of the monomers, the diamine was dissolved in NMP. An equimolar amount of BTDA was then slowly added to the solution while the temperature was maintained between 15 and 20 °C. The monomers were allowed to react at room temperature (under N_2) for two hours. The viscosity of the polyamic acids was adjusted to 14 poise, and the polymers were spin-coated on quartz wafers at 2000 rpm for 120 seconds. The polymers were then baked at 80 °C in air for one hour to remove the solvent, then imidized and cured under N_2 at 200 °C for one hour, then 350 °C for one hour. The UV/Vis absorption spectra of the models in solution and of the polyimides on wafers were obtained with an HP 8452A diode array spectrophotometer.

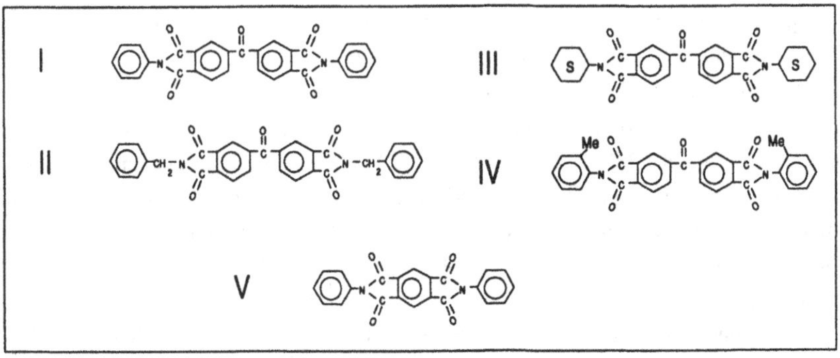

Figure 2: Model compounds

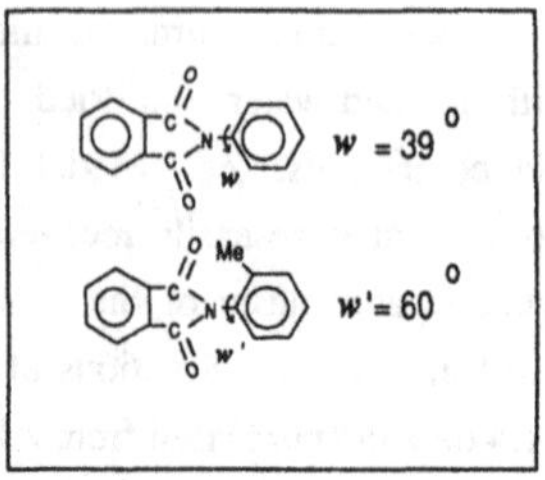

Figure 3: MM2 Results

RESULTS AND DISCUSSION

Model Compound Study

Assuming that the intramolecular charge transfer (ICT) occurs as a result of donation from the (amine) phenyl ring to the adjacent imide ring in a π-π^* transition, the primary structures of Models I-IV (Figure 2) were varied such that the transfer of charge would be allowed (I), totally inhibited (II and III), or partially inhibited (IV). Model I should allow for the charge transfer since the donor and acceptor are immediately adjacent to each other. Models II and III, however, were designed such that the transfer of charge should be completely inhibited. A methylene "spacer" group was inserted in Model II, thereby separating the donor and acceptor segments, and a saturated ring was used in place of a donor ring in compound III. Finally, Model IV has contiguous donor and acceptor fragments like Model I, but a methyl group was placed in the ortho position of the phenyl ring. This methyl group should, by steric hindrance with the carbonyl group, cause the torsional angle ω' in the minimum energy conformation of Model IV to be greater than the angle ω in Model I. In fact, a molecular modeling simulation of phenylphthalimide and o-methylphenylphthalimide (Figure 3) was conducted using the MM2 parameters of CHEM-3D+. The simulation predicts that, upon addition of the methyl group, the torsional angle increases from 38° to 60° in the respective minimum energy conformations. Thus the transfer of charge should be partially inhibited in Model IV with respect to the ICT of Model I because of a smaller overlap of π orbitals from the phenyl and imide rings.

Solutions of the Models I-IV were prepared in dioxane at concentrations of 10^{-5} M. At such dilute concentrations, we expect that solute-solute interactions should be prohibited, and all observed behavior should arise from intramolecular phenomena. The absorption spectra of the compounds were obtained and are shown in Figures 4-6.

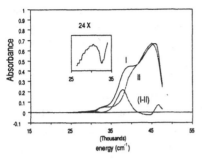

Figure 4: UV/Vis absorption spectra of Models
I, II, and difference (I-II) in dioxane

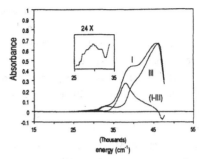

Figure 5: UV/Vis absorption spectra of Models
I, III, and difference (I-III) in dioxane

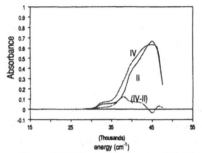

Figure 6: UV/Vis absorption spectra of Models
IV, II, and difference (IV-II) in dioxane

Figure 4 shows the absorption spectra of Models I, II, and the difference spectra of the models. One can understand the difference spectra from the Additivity Principle which states that, to a good approximation, the absorption due to chromophores separated by one or more saturated carbons (eg., Model II) is additive.[8-10] In contrast, we might expect interactions between two adjacent chromophores as in the case of Model I where charge transfer is believed to occur. In fact, Mulliken theory predicts that, when the donor and acceptor are in the proper configuration for charge transfer, a new absorption band will appear that is characteristic of neither the isolated donor nor the isolated acceptor.[11] We therefore believe that the Gaussian-like peak in the difference spectrum at 37,900 cm[-1] (264 nm, 4.7 eV) corresponds to an ICT band found in Model I and inhibited in Model II. A close inspection of the long wavelength "tail" region reveals a similar peak of much lower intensity at 30,300 cm[-1] (330 nm, 3.8 eV) that may correspond to a similar transition (inset of Figure 4). (Nonzero values in the difference spectrum above 41,000 cm[-1] probably arise from small shifts in the spectra; steep slopes in this region tend to magnify differences due to small deviations in the peak positions.)

Similarly, we compare Model I to Model III. We expect that Model III should also be void of the charge transfer since the donor ring has been replaced by a saturated ring. However, because of the absence of a phenyl ring in Model III, we do not expect the difference technique to yield results as "clean" as those obtained in the first case. In fact, the difference spectrum of Models I and III (Figure 5) reveals the same transitions at 37,900 cm^{-1} and 29,600 cm^{-1}, with a "perturbation" at the high energy edge of the large peak; this "perturbation" is probably a consequence of the lack of the phenyl chromophore in Model III. Likewise, a transition at 37,900 cm^{-1} appears in the difference spectrum of Models IV and II (Figure 6). Again we see a small "perturbation" at the higher energy edge of the ICT band that may be due to effects of the methyl group. However, the important feature is that in Model IV, where the torsional angle is larger than that of Model I, the ICT band appears to have diminished. Also, the small peak at low energy is not apparent as in the other cases.

We can also compare our experimental data with the results of LaFemina[8] who used the semi-empirical program CNDO/S3 to compute the electronic structure of the Model V (N,N' diphenyl-pyromellitimide). His calculation predicts an absorption band of charge transfer character at 4.7 eV which corresponds nicely with the higher energy (4.7 eV) transition in Models I and IV . Since the pyromellitimide moiety has a greater electron affinity than the benzophenonetetracarboximide, we might expect the predicted transition in V to be red-shifted with respect to the observed transitions in Models I and IV. It is possible, however, that the difference in acceptor fragments is offset by solvent effects present in our experimental work.

In light of these results, we believe that the aromatic imide systems can be described as follows. Upon exposure to UV light, the molecule undergoes a $\pi \to \pi^*$ transition whereby electron density from the phenyl ring is transferred to the imide ring. Furthermore, as the torsional angle between the planes of the donor and acceptor rings is increased, the overlap between the π orbitals of the respective segments decreases, and the probability of the transition concomitantly decreases. The charge transfer can be inhibited either by separating the donor and acceptor fragments (with an alkyl spacer, for example) or by eliminating the donor ring (eg., replacing the phenyl ring with a saturated ring).

Polymer Studies

In order to extend the concepts of our model compound studies to the polymer systems, a series of five polyimides was designed so that the ICT would be selectively allowed, partially inhibited or completely inhibited. Figure 7 shows the UV/Vis absorption spectrum of polymer A (BTDA/PDA) and the difference spectra of polymer A with the other BTDA-based polymers B through E. Despite the more complex behavior observed in the polymers, well-structured peaks are observed in the difference spectra remarkably similar to the difference peaks in the cognate model compound systems. We note, however, that the position of the large ICT band is

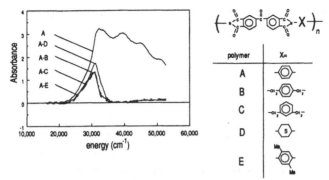

Figure 7: Absorption spectrum of polymer A and
difference spectra of A with B, C, D, and E

red-shifted from 37,900 cm^{-1} in the model compounds to 30,700 cm^{-1} (326 nm) in the polymer systems. This difference can perhaps be accounted for by the finite length of the model system. Since Huckel theory predicts that the energy gap between the molecular orbitals is less for a large conjugated system than for a small conjugated system, we expect the excitation energy to be red-shifted in going from the models to the polymers. Furthermore, the fact that the transition at 30,700 cm^{-1} appears to be the largest, low energy transition suggests that it is a HOMO --> LUMO type transition.

In general, the intensity of a peak in the difference spectra corresponds to the degree of ICT inhibition in polymers B-E. It is therefore interesting to note that the relative peak heights in the difference spectra indicate that the inhibition capabilities of the polymers B, C, D, and E are approximately the same. We note that polymer D does deviate from the others; again, this is probably due to the lack of a phenyl ring rather than an increased ability to inhibit the charge transfer. It is of particular interest to note that the methyl-substituted polymer (E) seems to inhibit the ICT as much as the "spacer" polymers (B and C) whereas the methyl-substituted Model IV only partially inhibited the ICT. We believe that this large inhibition indicates that the methyl groups strongly effect the conformation of the polymer chain, causing the phenyl and imide rings to deviate considerably from coplanarity; the further restriction on rotation invoked by the polymer matrix then precludes the chain from having intersegmental torsional angles sufficiently large to allow any ICT to occur. In addition, we note that there is virtually no difference between the meta and para isomers of the "spacer" polymer (B&C, Figure 7).

Thus from the polymer systems we have observed transitions in the difference spectrum that are in good qualitative agreement with the model compound studies. The nature and mechanism of the ICT in the polymers is therefore analogous to that of the ICT found in the model compounds. The polymer systems, however, do have several distinctive features. The ICT bands are red-shifted with respect to the model studies, as one would predict from Huckel Theory; the dimethyl-substituted polymer seems to have a greater ability to inhibit the ICT than does the corresponding model compound; and finally, the peaks in the difference spectra are skewed toward lower energy.

III. CONCLUSIONS

A comparative analysis of a series of model compounds in dilute solution has not only proven that intramolecular charge transfer occurs in imide systems, but it has effectively "isolated" the ICT band and provided much information about the mechanism of the transfer. By then comparing the models to the corresponding set of polyimides, we found that at least one such transition occurs in the BTDA-based aromatic polymers. We conclude that, upon absorption of ultraviolet light, π-π^* transitions of charge transfer character occur as a result of donation of electron density from the (amine) phenyl ring to the imide ring. Furthermore this charge transfer can be significantly reduced by introducing substituted groups in the ortho position of the (amine) phenyl ring and inhibited by separating the donor and acceptor segments with an alkyl "spacer" or by eliminating the (amine) phenyl ring. Future work will demonstrate similar behavior in a different series of models and polymers.

ACKNOWLEDGEMENTS

We would like to thank Hitachi Limited for financial support for the project, Dr. Robert Roginski for useful discussions, and Professor Alice P. Gast for the use of the spectrophotometer.

REFERENCES

1. R.S. Dine-Hart and W.W.Wright, Die Makromolekulare Chemie, 143, 189-206 (1971).
2. M. Fryd, in Polyimides: Synthesis, Characterization, and Applications, edited by K.L. Mittal, Plenum Press (New York,1984), pp.377-383.
3. S.C. Freilich, Macromolecules, 20, 973-978 (1987).
4. E. Wachsman and C.W. Frank, Polymer, 29, 1191 (1988).
5. M. Hasegawa, M. Kochi, I Mita, and R Yokota, Eur. Polym. J., 25, (4), 349-354 (1989).
6. J. LaFemina, G. Arjavalingam, and G.Hougham, J.Chem. Phys., 90(9), 5154-5160 (1989).
7. H Ishida, S.T. Wellinghoff, E Baer, and J.L. Koenig, Macromolecules, 13, 826-834, (1980).
8. R. Shriner, R.C. Fuson, D.Y. Curtin, and T.C. Morrill, The Systematic Identification of Organic Compounds, John Wiley & Sons (New York, 1980), pp. 421-425.
9. E. Braude, J. Chem. Soc., 1902 (1949).
10. M. O'Shaughnessy and W. Rodebush, J. Amer. Chem. Soc., 62, 2906 (1940).
11. R.S. Mulliken, Molecular Complexes (John Wiley & Sons, New York, 1969).

CT FLUORESCENCE AND MOLECULAR AGGREGATION OF A POLYIMIDE AS FUNCTION OF IMIDIZATION CONDITION.

M.HASEGAWA, H.ARAI, *K.HORIE, **R.YOKOTA and ***I.MITA,
Toho University, *The University of Tokyo, **Institute of Space and
Astronautical Science, ***Dow Corning Japan, Ltd.

ABSTRACT

The emission mechanisms of solid PI(BPDA/PDA) derived from biphenyltetracarboxylic dianhydride (BPDA) and p-phenylenediamine (PDA) were examined with the absorption and fluorescence spectra of model compounds (denoted by M). M(BPDA/CHA) (CHA: cyclohexyl amine) fluoresces at ca. 430 nm in hexafluoro-2-propanol(HFP) solution, while M(BPDA/AN) (AN: aniline) does not. PI(BPDA/PDA) film does not show the monomer fluorescence of biphenyldiimide unit, but shows only intermolecular CT fluorescence peaking at 530-540 nm. This suggests that for PI(BPDA/PDA) film and PI(BPDA/AN) in solution the local excited state of biphenyldiimide units is deactivated owing to intramolecular charge-transfer(CT).

The intermolecular CT fluorescence reflecting sensitively molecular packing of PI chains was used to monitor isothermal imidization process of poly(amic acid)(PAA) of BPDA/PDA. The fluorescence of PAA(BPDA/PDA) peaking at 490 nm decreases rapidly and disappears at 30-40% conversion, then the fluorescence of PI(BPDA/PDA) peaking at 540 nm increases gradually during isothermal imidization. The fluorescence intensity at 540 nm increases rapidly as imidization proceeds when imidized at higher temperature. A kinetic study on isothermal imidization shows that the vitrification is strongly related to the reorientation of polymer chains and the final PI structures.

INTRODUCTION

Aromatic polyimides emit intermolecular CT fluorescence which reflects sensitively molecular packing of PI chains.[1] We have already shown that the fluorescence is also sensitive to the miscibility of PI/PI blends.[2,3] In order to confirm the availability of this method, it is necessary to elucidate the excited state and emission mechanisms of aromatic PIs in detail. The problem on electronic state of aromatic PIs is also important in the fields of photoconductors, u.v. light- and radiation-resistance, photoresists, and ablative photodecomposition by excimer laser. In this work, the excited state and CT emission mechanism of PI are discussed with electronic spectra of model compounds. It is very important to examine the relation between the morphologies of final PIs and imidization reaction mechanisms. The CT fluorescence is also used for monitoring isothermal imidization.[4]

EXPERIMENTAL

PAAs were prepared by condensation of diamine and dianhydride in DMAc. Thin PI films were prepared by thermal imidization of spin-coated PAAs on a quartz plate at 250 ℃ for 2 h. The model compounds were prepared by thermal imidization of the corresponding amic acids, followed by recrystallization and vacuum drying.

The fluorescence and u.v.-vis absorption spectra were measured at room temperature in air using a fluorescence spectrophotometer (Hitachi, model 850) and a u.v.-vis spectrophotometer (Jasco, model UVIDEC-660). The phosphorescence of model compound was measured in a degassed rigid solution at 77 K in a cryostat (oxford). The excitation and fluorescence spectra were corrected for wavelength-dependent intensity of lamp and for wavelength-dependent sensitivity of detectors respectively, using Rhodamine B standard solution for 200–600 nm and ethylene glycol solution of methylene blue for 600–720 nm. Absorbances of the samples used are about 0.1. The degree of imidization was estimated by the absorbance ratio of 1774 cm^{-1} band (C=O stretching in imide rings) to that for the PI annealed at 300 ℃ for 20 min (i=100%) using 1517 cm^{-1} band (C=C stretching in benzene rings in PDA) as an internal standard, using an IR spectrophotometer (Jasco, model IR-700).

RESULTS AND DISCUSSION

Electronic Spectra of PI(BPDA/PDA) and Model Compounds

PI(BPDA/PDA) which has rigid main chain and is emissive is focused. Fig 1 (a)–(d) shows the absorption spectra of PI(BPDA/PDA) thin film and model compounds in HFP. M(BPDA/AN) is insoluble to common organic solvents except for dichloroacetic acid and HFP. In Fig 1(a), PI(BPDA/PDA) has broad and structureless band from 400 to 250 nm and a shoulder at 350 nm. In Fig 1(b), both M(SA/AN) (SA: succinic anhydride) and M(SA/PDA) have no absorption longer than 280 nm. Therefore, PDA moieties are not excited by 350 nm light used for the characterization of the molecular packing of PIs. From the comparison of M(BPDA/CHA) and M(PA/CHA) (PA: phthalic anhydride) in Fig 1(c), conjugation effect is observed in biphenyl bond because the absorption intensity of M(BPDA/CHA) is two times stronger than that of M(PA/CHA) and M(BPDA/CHA) has a long wavelength tail. The comparisons of M(BPDA/AN) with M(BPDA/CHA) and of M(PA/AN) with M(PA/CHA) indicate that the conjugation between benzimide unit and aromatic ring adjacent to nitrogen atom is small. Accordingly these results led to the conclusion that 350 nm light is mainly absorbed biphenyl diimide moieties. Judging from the work by Ishida et al.[5] and the magnitude of ε, the bands at 300-320 nm and the absorption tail at longer wavelength for M(BPDA/AN) and M(BPDA/CHA) are assigned to π,π^* transition due to biphenyldiimide and n,π^* transitions due to imide carbonyl group, respectively.

Fluorescence spectra of the model compounds were measured in HFP solution. The model compounds are non-fluorescent except for M(BPDA/CHA) and M(PA/CHA). The fluorescence peaking at 430 nm for M(BPDA/CHA) is observed in HFP solution but is observed neither for PI(BPDA/PDA) nor for M(BPDA/AN). Non-fluorescent nature of M(BPDA/AN) is considered to be due to an effective deactivation via intramolecular charge-transfer (D^+A^- state) at excited state between biphenyldiimide unit and aromatic ring adjacent to nitrogen atom, although the conjugation at ground state between them is small. The phosphorescence measurement is also carried out for M(BPDA/CHA). M(BPDA/CHA) in degassed toluene at 77 K shows phosphorescence around 550 nm (τ_p=2-3 s) and structured fluorescence at 380 nm (the fluorescence red-shifts up to 430 nm in HFP owing to the solvent effects) as shown in Fig 2. On the basis of a criterion for aromatic carbonyl compounds, such long lifetime suggests that the lowest triplet excited state (T_1) is due to π,π^*.

Fig 3 shows energy diagram for photophysical process in solid PI(BPDA/PDA). Biphenyldiimide units in PI(BPDA/PDA) chains are first excited by 350 nm light. First, Let us consider the case of an isolated PI(BPDA/PDA) chain. From the local excited state of biphenyldiimide unit (S_1), intersystem crossing to T_1, fluorescence emission, and intramolecular charge-transfer (D^+A^-) followed by deactivation compete. But the phosphorescence of PI(BPDA/PDA) film is not observed at room temperature in air. The fact that no monomer fluorescence of biphenyldiimide units is observed for PI(BPDA/PDA) film suggests that there is a rapid deactivation process such as intramolecular CT. The observation of intermolecular CT fluorescence of PI(BPDA/PDA) film suggests that there is a lowest excited level for the intermolecular CT complex.

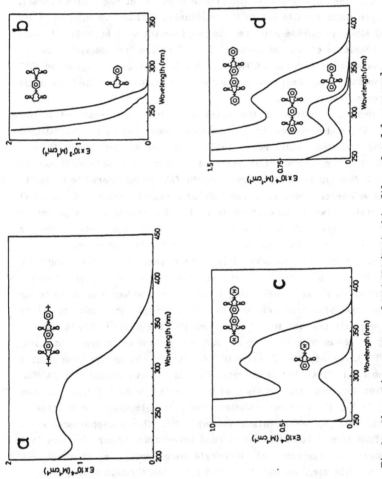

Fig 1. Absorption spectra of PI (BPDA/PDA) thin film and model compounds in HFP solution.

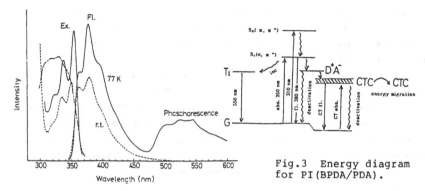

Fig.2 The fluorescence, phosphorescence
and excitation spectra of M(BPDA/CHA)
in degassed rigid solution at 77K and
room temperature.

Fig.3 Energy diagram
for PI(BPDA/PDA).

Isothermal Imidization of Poly(amic acid) Monitored by Intermolecular CT Fluorescence

Isothermal imidization of the solid PAA was examined using the intermolecular CT fluorescence of the PI. Fig 4 shows the change in fluorescence spectra during imidization at 170 °C. As imidization proceeds, the fluorescence of the PAA peaking at 490 nm reduces rapidly and becomes very weak around 30-40% conversion, and then the PI fluorescence peaking at 530-540 nm increases gradually. Fig 5 shows the change in the fluorescence intensity during imidization at various imidization temperature (T_i). The fluorescence intensity begins to increase gradually from around 50% conversion, showing that the intermolecular CT complex formation occurs. It is clearly shown that when imidized at 150 °C the intensity increases only slightly as the reaction proceeds, while at higher T_i such as 270 °C the increase in the intensity become remarkable. The fact that the molecular packing of the final PI is strongly affected by T_i suggests that the molecular mobility of the polymer chains which changes during imidization is closely related to the molecular packing of the final PI. It is also noted that from about 60% conversion the intensity-conversion curves begin to branch and the intensity at a certain conversion depends strongly on T_i. If the CT fluorescence does not provide information for intermolecular aggregation but intramolecular information, only one intensity-conversion curve must be depicted regardless of T_i (do not branch). As the reaction proceeds, the T_g of the polymer increases gradually, and exceeds the T_i at certain conversion (i_g) where the polymer is in glassy state. The more T_i increases, the more i_g will increase, too. it is difficult to determine experimentally the T_g of the partial imidized PAAs by thermal analysis such as TMA, DSC or dynamic mechanical analysis because imidization occurs due to heating. According to Laius et al.,[6] the values of i_g were estimated from a kinetic study. The arrows marked in Figure 5 denote i_g at 150 and 200 ℃, respectively. In the case of T_i=150℃, the vitrification occurs at only initial stage while for 200℃ it takes place at about 60% conversion. It seems that the molecular packing of the final PI is already determined before the vitrification (below i_g), because above i_g molecular motion enough to rearrange the polymer chains is inadequate. In other words, the degree of increase in the fluorescence intensity (molecular packing) after the vitrification depends strongly on the amount of i_g.

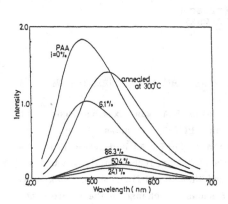

Fig.4 The change in the fluorescence spectra with the progress of imidization at 170*C.

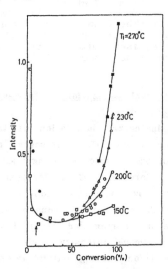

Fig.5 Fluorescence intensity as a function of conversion.

REFERENCES

1. M.Hasegawa, M.Kochi, I.Mita and R.Yokota, Eur.Polym.J., 25, 349 (1989).
2. M.Hasegawa, M.Kochi, I.Mita and R.Yokota, Polymer, in press.
3. M.Hasegawa, PhD thesis, The University of Tokyo, 1991.
4. M.Hasegawa, H.Arai, I.Mita and R.Yokota, Polym.J., 22, 875 (1990).
5. H.Ishida, S.T.Wellinghoff, E.Baer and J.L.Koenig, Macromolecules, 13, 826 (1980).
6. L.A.Laius, M.I.Bessonov and E.S.Florinskii, Polym.Sci.U.S.S.R., 13, 2257 (1971).

A FLUORESCENCE STUDY OF POLYIMIDE CURE KINETICS

D.A. HOFFMANN[1], H. ANSARI[2] AND C.W. FRANK[2]*
[1]Department of Materials Science and Engineering and Department of
[2]Chemical Engineering, Stanford University, Stanford, CA 94305

ABSTRACT

Isothermal annealing studies reveal long term increases in charge transfer fluorescence in spin-cast BTDA-ODA/MPD films. Due to the amorphous nature of these materials, the increase in charge transfer complex (CTC) population is attributed to thermally activated hindered rotation leading to local segmental correlations. This ordering process is driven by the non-equilibrium structure of the initially imidized film. Increased in-plane orientation at higher spin speeds produces a stress activation effect, lowering the energy barrier for segmental rotation.

INTRODUCTION

Due to their excellent thermal stability and dielectric properties, polyimides are well suited for use as interlayer dielectrics and passivation coatings. In these applications, polyimide films are formed by spin-casting solutions containing the soluble poly (amic acid) precursor, followed by thermal imidization. This is a complex process involving chain orientation, solvent evaporation, chemical reaction and structural relaxation. Consequently, the relationships between final film structure and processing conditions are of considerable practical significance.

Recently, Wachsman et al. [1] demonstrated the sensitivity of the electronic structure of spin-cast polyimide films to processing conditions and thermal history using fluorescence spectroscopy. Polyimide fluorescence arises from charge transfer complexes (CTC's) that form between donor (diamine) and acceptor (diimide) segment pairs. Both intermolecular and intramolecular CTC's can occur; the former with the formation of sandwich structures between aromatic rings on neighboring chains and the latter due to rotation of adjacent segments toward coplanar orientations. In either case, the presence of CTC's is an indicator of microstructure and an increase in fluorescence intensity is interpreted as a rise in the CTC population with molecular ordering. The fluorescence intensity is sensitive to anneal [1,2] and imidization [2] temperatures, and bi-axial stress [3]. The purpose of the present work is to extend these previous studies into the long term physical aging regime and to investigate the effects of spin speed on subsequent structural relaxation.

EXPERIMENTAL

Polyimide films were derived from thermal imidization of Du Pont PI-2555, a 19% solution of poly (amic acid) (PAA) precursor in 1-methyl-2-pyrrolidinone (NMP). The PAA precursors are formed from polycondensation of 3,3',4,4'-benzophenone tetracarboxylic dianhydride (BTDA), oxydianiline (ODA) and meta-phenylene diamine (MPD), present as a minor component [4]. Samples were prepared by spinning the PAA solution on 3" quartz wafers for 2 minutes at 1000, 2500 or 8000 rpm. at room temperature. The samples received an initial cure of 150° C for 30 minutes (B-stage) prior to the high temperature cure. All thermal treatments were done under nitrogen in a convection oven. Following each cure study, sample thickness was measured using a Dektak profilometer. The film thicknesses for 1000, 2500 and 8000 r.p.m. films are approximately 5.5, 2.6 and 1.3 μm, respectively.

Fluorescence emission spectra were recorded using a Spex Fluorolog 212 spectrophotometer with excitation wavelengths of 365 nm and 485 nm, corresponding to the characteristic polyimide excitation bands [1]. Emission spectra obtained with 365 nm and 485 nm excitation will be referred to as CT1 and CT2, respectively. All measurements were made at room temperature using 2 mm slit widths. Fluorescence intensity was normalized with

Mat. Res. Soc. Symp. Proc. Vol. 227. ©1991 Materials Research Society

respect to lamp intensity fluctuations using a rhodamine-B standard solution, and multiplied by correction factors to account for variations in PMT sensitivity, film thickness and absorption. UV/Vis absorbance was measured using a Cary 3 double beam spectrophotometer with a bare quartz wafer in the reference path.

RESULTS

The CT1 and CT2 fluorescence intensity from individual samples, prepared at 1000, 2500 or 8000 r.p.m., was measured at intervals during the course of an extended anneal. The 1000 and 8000 r.p.m. results are plotted as a function of anneal time at 300 and 350° C in Figures 1 and 2, respectively. In these long term experiments, imidization is complete after the initial cure interval, and the increasing fluorescence intensity is attributed to an increase in the concentration of CTC's with physical aging. From these figures it is evident that fluorescence intensity increases in a similar manner with anneal in all samples and that the time scale decreases with increasing temperature. The fluorescence intensity initially increases linearly and levels off at long times, showing that the CTC population is eventually stabilized. The CT1 intensities are higher, but increase at about the same rate as the CT2 intensities. The difference in intensities is attributed to the higher extinction coefficient of the CT1 absorption chromophore. The large intensity fluctuation at long times at 350° C may be due to a competition between thermal degradation and ordering processes. At 300° C, intensity increases more rapidly and reaches a higher plateau with increasing spin speed; and at 350° C, the intensity plots for different spin speeds converge. This demonstrates a different temperature dependence for the rates of fluorescence intensity increase at different spin speeds.

The initial increase in fluorescence intensity was studied in greater detail at 300, 325 and 350° C. At each temperature, 3-4 cure intervals were selected within the early annealing period: 2-50 hours, 2-14 hours and 1-8 hours at 300, 325 and 350° C, respectively. For each time, temperature and spin speed, a separate group of three samples was studied. The rate of change of fluorescence intensity, $\Delta I_{fc} / \Delta t$, was obtained from linear data fits. Assuming zeroeth order kinetics, activation energies, E_a were obtained from,

$$\Delta I_{fc} / \Delta t = k_f = k^0{}_f \, e^{-E_a / RT} .$$

From plots of $\ln (\Delta I_{fc} / \Delta t)$ vs. $1/T$, the activation energies shown in Table 1 were obtained. The differences in CT1 and CT2 activation energies are within experimental error ($\sim \pm 20\%$) showing that the molecular rearrangements corresponding to the observed fluorescence intensity in either band follow similar kinetics. The decrease in E_a with increasing spin speed is significant and demonstrates an effect of the initial film structure on subsequent ordering processes.

Table 1. Fluorescence Activation Energies (kcal/mole ± 20%) in Early Cure

Spin Speed (rpm)	Ea CT1	Ea CT2
1000	39	37
2500	30	27
8000	25	22

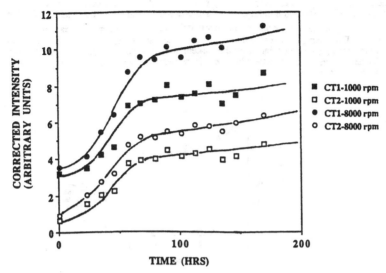

Figure 1. CT1 and CT2 fluorescence intensity during anneal at 300° C from samples spin-cast at 1000, 2500 and 8000 rpm.

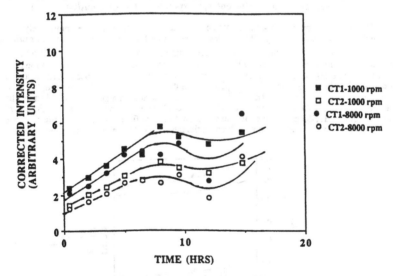

Figure 2. CT1 and CT2 fluorescence intensity during anneal at 350° C from samples spin-cast at 1000, 2500 and 8000 rpm.

To further investigate the effect of spin speed, four samples of different thickness were prepared from precursor solutions of varying viscosity using 2500 r.p.m. spin speed. In addition three samples were prepared from precursor solutions of the same viscosity using spin speeds of 1000, 2500 and 8000 r.p.m. All samples were cured for 13 hours at 325° C. The thickness dependence of the absorbance per unit thickness, ε', of these samples is shown in Figure 3. Decreasing thickness, either by lowering the viscosity of the PAA precursor solution or by increasing spin speed, increases the concentration of absorbing species. This is consistent with a density gradient within the film due to increased packing near the polymer/substrate or polymer/air interfaces.

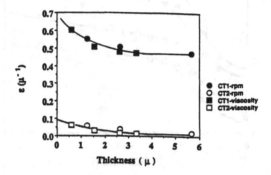

Figure 3. Absorbance per unit thickness as a function of thickness in samples prepared using different spin speeds and PAA solution viscosities.

The effects of spin speed were isolated by comparing samples that were immediately imidized following spin casting with samples that were dried under vacuum for 14 hours prior to B-stage cure. As shown in Figure 4, the effect of spin speed on ε' is reduced in the vacuum dried samples. Figures 5 and 6 show the increase in fluorescence intensity with cure time at 325° C for these samples. In these figures it is apparent that the effect of spin speed on fluorescence intensity is lost in the vacuum dried samples. When the samples are dried in vacuum, the solvent evaporates more slowly and the flexible PAA chains can reorient to a greater extent than in films that are immediately thermally imidized. The effect of chain relaxation is to decrease the absorber concentration and reduce the effect of spin speed on the initial film structure. The fluorescence results show that the initial structure influences the rate of CTC formation during subsequent annealing.

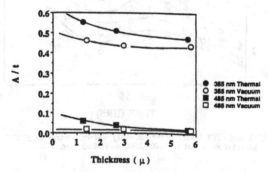

Figure 4. Effect of relaxation prior to imidization on average absorbance, A/t.

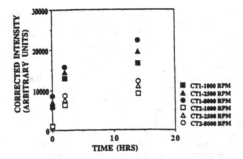

Figure 5. Fluorescence intensity during anneal in films imidized immediately following spin casting.

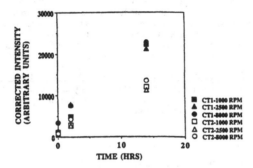

Figure 6. Fluorescence intensity during anneal in vacuum dried films.

DISCUSSION

The long term annealing results show that the CTC population, and therefore the polyimide morphology, is stabilized only after extended heat treatment at the temperatures studied. Previously, Wachsman argued that a "quasi-equilibrium" is reached after a relatively short heat treatment at any temperature, and that additional cure at a higher temperature was required to induce further changes in structure [1]. Numata proposed that the glass transition temperature, T_g, was approximately equal to the highest imidization temperature [5]. Our results indicate that structural rearrangements persist over long times (15-200 hours) at a single temperature, and that the maximum degree of order actually decreases with increasing annealing temperature. Owing to the flexibility imparted by the BTDA ketone linkage, BTDA-ODA/MPD has a low T_g (~310 °C) compared with other polyimides and molecular ordering in the 300-350° C temperature range is not surprising. The eventual stabilization of film structure is consistent with volume relaxation in glassy polymers [6]. As segments pack, their mobility is reduced, impeding further ordering. This study demonstrates that fluorescence measurements can be used to follow long term physical aging in polyimides.

Given that long term ordering processes occur, the nature of the polyimide morphology is of interest. BTDA-ODA/MPD is less likely to form ordered phases than more rigid polyimides [7,8]. WAXD results reported in the literature show an absence of long range order both before and after annealing of other BTDA based polyimides [9,10]. We have performed similar studies on BTDA-ODA/MPD and observe an amorphous structure at all stages of anneal and all spin speeds. Both Wachsman [1] and Hasegawa [2] have suggested that fluorescence intensity is

136

related to the state of the amorphous regions, consistent with our results. The local ordering involves a rotation of individual segments into orientations approaching coplanarity with neighboring segments [1]. The magnitude of the activation energies in Table I is consistent with hindered rotation leading to CTC formation. We therefore attribute the increase in fluorescence intensity to local segmental correlations resulting in CTC formation in the amorphous phase.

Physical aging is affected by the initial structure of the spun film following imidization. During spin casting, the chains are forced into strained configurations due to hydrodynamic forces and loss of mobility with solvent evaporation [11,12]. The resulting structure contains significant in-plane orientation of chain backbones [13]. Subsequent imidization increases chain rigidity, raising the T_g and further reducing chain mobility. At this point, the structure is thermodynamically equivalent to a polyimide film that has been equilibrated at some very high temperature and quenched. The driving forces for structural rearrangement arise from the decrease in free energy with packing and the relaxation of strained chain conformations.

The effect of spin speed on fluorescence intensity demonstrates the dependence of molecular ordering on the initial state of the film. Although the influence of spin speed on the structure of the imidized film has not been reported, we expect more in-plane chain orientation in films prepared using higher spin speeds. The decreasing activation energy with increasing spin speed then reflects a less hindered segment rotation in more oriented films. The results in Figure 3 indicate an increase in density near the film boundaries consistent with improved packing in thinner films, regardless of spin speed. However, the results in Figure 4 show that the chains are strained during spinning. If the films are allowed to relax prior to imidization, the chains recoil, producing a less dense structure. As shown in Figures 5 and 6, this initial relaxation eliminates the effect of spin speed on subsequent ordering processes. We propose that local segment rotation leading to CTC formation is favored in more oriented films. This is similar to the effect of deformation on crystallization in polymers [14] Orientation decreases the entropy and increases the enthalpy, raising the free energy of the initially imidized film. The increase in configurational free energy prior to annealing results in the reduced energy barrier for segment rotation in films prepared using higher spin speeds.

REFERENCES

1   E.D. Wachsman and C.W. Frank, Polymer 29, 1191 (1988).
2   H. Hasegawa, K. Masakatsu, I. Mita and R. Yokota, Eur. Polym. J. 25, 349 (1989).
3   P.S. Martin, E.D. Wachsman and C.W. Frank in Polyimides, edited by C. Feger, M.M. Khojasteh and J.E. McGrath (Elsevier, New York, 1989) pp. 371-378.
4   E.D. Wachsman, P.S. Martin and C.W. Frank, A.C.S. Symp. Ser. 407, 26 (1989).
5   S. Numata, K. Fujisaki and N. Kinjo in Polyimides, edited by K. L. Mittal, (Plenum Press, New York, 1984) Vol. 1, pp. 259-271.
6   L.C.E. Struick, Physical Aging in Amorphous Polymers and Other Materials, (Elsevier Scientific Publishing Co., Amsterdam ,1978)
7   P.J. Flory, Macromolecules, 11, 1141 (1978).
8   N. Takahashi, D.Y. Yoon and W. Parrish, Macromolecules, 17, 2583 (1984).
9   M. Kochi, S. Isoda, R. Yokota, I. Mita and Kambe in Polyimides, edited by K. L. Mittal (Plenum Press, New York, 1984) Vol. 2, pp. 671-681.
10 S. Numata, S. Oohara, K. Fujisaki, J. Imaizumi and N. Kinjo, J. Appl. Polym. Sci. 31, 101, (1986).
11 Y. Cohen and S. Reich, Polym. Sci. 19, 599 (1981).
12 Kosbar, L. L., Kuan, S. W. J., Frank, C. W. and Pease, R. F. W. ACS Symp. Ser. 381, 95 (1989).
13 T.P. Russell, H. Gugger and J.D. Swalen, J. Polym. Sci. 21, 1745 (1983).
14 V. N. Kuleznev and V.A. Shershnev, The Chemistry and Physics of Polymers, (Mir Publishers, Moscow, 1990), p. 195.

ACKNOWLEDGEMENT

The authors thank Ford Motor Company for financial support and Dr. J.E. Anderson for helpful discussions. We also thank Mike Springman and Du Pont for generously providing PI-2555 for this research.

RIGID-ROD AND SEGMENTED RIGID-ROD POLYIMIDES: GEL/SOL
AND LIQUID CRYSTALLINE TRANSITIONS, FIBERS AND FILMS

STEPHEN Z.D. CHENG, FRED E. ARNOLD, JR., MARK EASHOO, SONG-KOO
LEE, STEVE L.C. HSU, CHUL JOO LEE AND FRANK W. HARRIS
Institute and Department of Polymer Science, University of
Akron, Akron, OH 44325-3909

ABSTRACT

Organo-soluble rigid-rod and segmented rigid-rod polyimides
and their copolyimides exhibit isotropic solutions in hot $\underline{m}$-cre-
sol, but form gels upon cooling. A lyotropic liquid crystal
phase is observed below the gel/sol transition. Mechanical gel
formation is caused by liquid-liquid phase separation, while the
liquid crystal phase may be formed through a nucleation process
after gelation. High performance fibers can be spun from the
hot isotropic solutions using a dry-jet wet spinning method.
After the fibers are drawn at high temperatures, they display
tensile strength higher than 3.2 GPa and an initial modulus
higher than 130 GPa. In particular, the fibers retain relatively high mechanical properties at elevated temperatures. Solution casted films exhibit very low thermal expansion coefficients and dielectric constants. Their structure, morphology and
property relationships will also be discussed.

INTRODUCTION

Aromatic polyimides exhibit excellent electrical and
mechanical properties, along with high thermal and thermo-
oxidative stability. The polymers also display chemical and
solvent resistance, good adhesive properties and light and
dimensional stability. This unique combination of properties
leads to wide applications in films, fibers, coatings, adhesives, and matrix materials in polymer composites. The polyimides are usually processed in their poly(amic acid) precursors
due to their insolubility in conventional solvents. A typical
example is poly(4,4'-oxydiphenylenepyromellitimide)(PMDA-
ODA)[1,2].
A family of organo-soluble, rigid-rod and segmented, rigid-
rod aromatic polyimides have been recently synthesized in our
laboratory[3]. These polyimides were prepared in refluxing $\underline{m}$-
cresol in a one-step process where the intermediate poly(amic
acids) were not isolated. Upon cooling, the polyimide solutions
set to gel-like structures, followed by a formation of a
lyotropic liquid crystal phase[4]. These polyimides exhibit
excellent fiber and film-forming tendencies. High-performance
fibers have been prepared[5], while films show excellent dielectric properties as well as dimensional stability[6]. In this
report, our studies on gel/sol, liquid crystalline transition,
films and fibers of these polyimides are presented. The chemical structures of these polyimides are as follows:

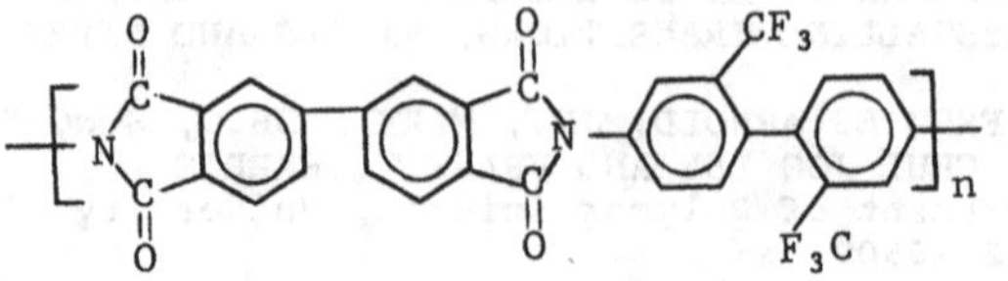

BPDA-PFMB

and

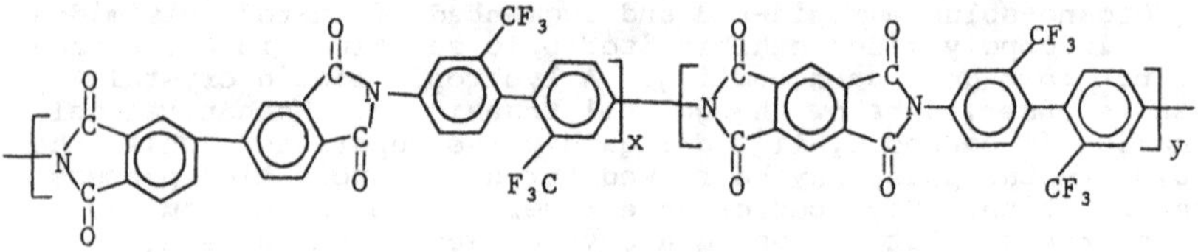

(BPDA-PFMB)$_x$-(PMDA-PFMB)$_y$

where ratios of x/y are 50/50, 70/30 and 85/15.

GEL/SOL AND LIQUID CRYSTALLINE TRANSITIONS

The polyimides can be dissolved in hot m-cresol to form
homogeneous solutions. When temperature is decreased, the
solvation power is drastically reduced, and mechanical gels are
formed. This can be judged by the observations where steady
flow of the systems stop although the concentrations of the
polyimides only range from 1% to 15%. On the other hand, from
solution differential scanning calorimetry (SDSC) experiments a
first-order transition can be found as shown in Figure 1. At
the same time, polarized light microscopy (PLM) observations
show lyotropic liquid crystal textures based on its optical
properties (Figure 2). These textures disappear at the same
temperatures as observed in the transitions of SDSC during
heating, indicating that the liquid crystal phase corresponds to
the first-order transition observed.

We ask ourselves whether this liquid crystal phase is
responsible for the formation of the mechanical gel. It requires further study of the structure formation kinetics of the
liquid crystal phase and the kinetics of mechanical gelation.
Figure 3 shows development of the heat of transition as a
function of time at different temperatures. It is clear that
with increasing the temperature this development slows dowm. If
one studies the formation kinetics of mechanical gel formation,
as shown in Figure 4, at a constant supercooling (ΔT-T_{gs}-T_a
where T_{gs} is the gel/sol temperature and T_a, annealing temperature), faster kinetics can be observed. This indicates that the
mechanical gelation forms first, and the liquid crystal structure develops later.

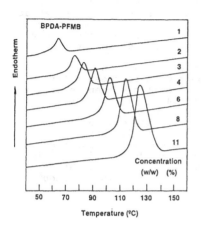

Figure 1. Set of SDSC heating curves for different concentrations of BPDA-PFMB.

Figure 2. PLM patterns for BPDA-PFMB (8%) at 40°C.

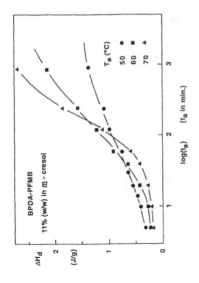

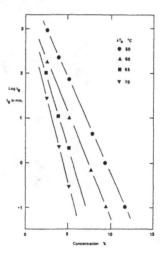

Figure 3. Relationship between the heat of transition and time at different temperatures for BPDA-PFMB (11%).

Figure 4. Relationship between the mechanical gelation time and concentration of the BPDA-PFMB at different supercooling.

POLYIMIDE FIBERS

The polymers (10%-15% (w/w) concentration in hot m-cresol) remain completely in solution until they are cooled below about 150°C. Fibers were dry-jet spun, from an isotropic solution and then coagulated in a water/methanol bath. As-spun fibers can be drawn at different temperatures (>300°C) and different draw ratios (up to 10 times) in air.

Figure 5 shows a wide angle X-ray diffraction (WAXD) pattern for highly drawn BPDA-PFMB fibers with fourteen diffraction spots observed. The crystal unit cell has been determined to be a monoclinic structure with a=1.540 nm, b=0.990 nm, c=2.025 nm and γ=56°[5].

As-spun BPDA-PFMB fibers have a very low degree of crystallinity (about 10%), while highly drawn fibers exhibit a crystallinity of about 50%. Figure 6 shows a relationship between the crystal orientation factor and draw ratio. It is evident that with increasing the draw ratio the orientation factor increases first, and reaches a plateau value of about 87%.

Figure 5.   WAXD fiber pattern for BPDA-PFMB fibers.

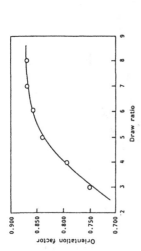

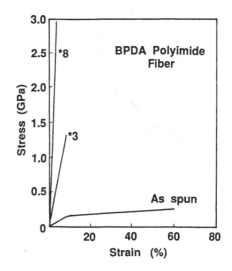

Figure 6. Relationship between crystal orientation factor and draw ratio.

Figure 7. Stress-strain curves for the fibers at different draw ratio.

For as-spun fibers, the tensile properties are low. By increasing the draw ratio, a remarkable increase in strength and modulus can be observed. At the draw ratio of ten, the tensile strength reaches 3.2 GPa and the initial modulus calculated from the ratio between tensile stress and strain is 130 GPa. The tensile properties at different draw ratios are shown in Figure 7. This behavior is different than that displayed by fibers spun from lyotropic liquid crystal states, such as Kevlar fibers and PBZT fibers. In those cases, as-spun fibers show relatively high strength and modulus. The successive annealing process mainly affects the fiber modulus. Since the polyimide fibers are spun from an isotropic solution, the chain molecules are not fully oriented (as-spun fibers). As a result, the fibers can undergo considerable elongation during drawing. The advantage of this spinning method is that during the drawing process the chain molecules have a better opportunity to rearrange themselves into defect-free positions under the large deformation.

Figure 8 illustrates the retention of modulus of the BPDA-PFMB fibers with different draw ratios ($\lambda=5$ and 8) when heated at a constant rate (10°C/min). This indicates that high draw ratios are necessary to maintain good mechanical properties at elevated temperatures. The temperature (~200°C) where the modulus starts to decrease may suggest that segmental motion in the chain molecules begins. At 400°C and 420°C, the fibers can retain 93% and 85% of their initial moduli even after a long time period (1.5 h and 3 h, respectively), as shown in Figure 9.

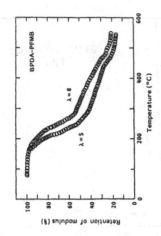

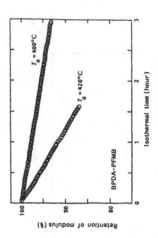

Figure 8. Retention of modulus
as a function of temperature
at λ =5 and 8.

Figure 9. Retention of modulus
as a function of time at 400°C
and 420°C.

For copolyimide fibers, we have found that with increasing
the chain rigidity by introducing the PMDA comonomers, the
tensile properties are improved. For example, a 15% increase of
these properties can be seen in the case of 70/30 copolyimide
fibers. Detailed structure characterizations and their
relationships to the properties will be reported elsewhere.

Polyimide Thin Films

Thin films were prepared spreading a 2% (w/w) m-cresol
solution on a glass plate with a doctor's knife, followed by
drying at 150°C for 5 h under reduced pressure in a vacuum oven.
The thickness of the films range from 10 to 15 μm.
In order to study the crystal and chain orientation in the
films, WAXD reflection and transmission modes were adopted.
Figures 10 and 11 shows sets of WAXD patterns for BPDA-PFMB
films annealed at 425°C obtained through both modes, as an
example[7]. With increasing annealing time, the WAXD patterns
gradually change. After a few minutes only the (hk0) crystal
planes are diffracted in the reflection mode, while the (001)
planes are diffracted in the transmission mode. This indicates
that the c-axis of the crystals initially is somewhat randomly
distributed between a parallel orientation and a perpendicular
orientation relative to the film surface. As soon as the film
is annealed, the chain molecular orientation in the crystal
changes to an in-plane orientation. This conclusion can also be
proven by comparing these two WAXD patterns obtained through
both reflection and transmission modes with those of the fiber

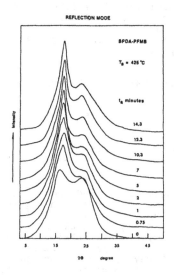

Figure 10. WAXD pattern for
BPDA-PFMB fibers during
annealing in reflection mode.

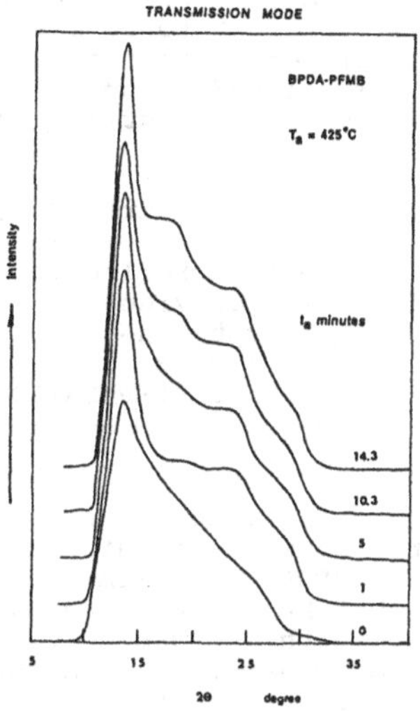

Figure 11. WAXD pattern for
the films during annealing in
transmission mode.

patterns (Figure 5) along the equational and meridian direction.
The correspondence is evident as shown in Figure 12 and 13.

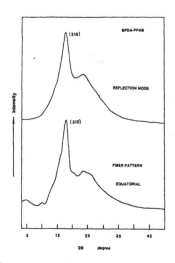

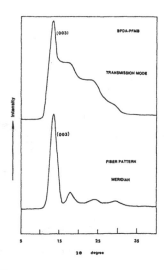

Figure 12. WAXD patterns
obtained from reflection mode
in film and fiber pattern
along the equatorial

Figure 13. WAXD patterns obtained from transmission mode
in film and fiber pattern along
the meridian direction.

The dielectric constant of the polyimide films (the thickness ranges from 20 to 30 μm) are extremely low. Based on an ASTM (D-150) test, the dielectric constant of BPDA-PFMB is 2.6 at 25°C. For a 50/50 copolyimide, it is 2.5. Table I gives some dielectric constant data for comparison.

Table I. Dielectric Constants for Various Polymers

Polytetrafluoroethylene	2.0
Polyethylene	2.25
Polystyrene	2.45
BPDA-PFMB	2.6
BPDA(50%)-PMDA(50%)-PFMB	2.5

Of equally importance is the temperature-dependence of the dielectric constant at different frequencies as shown in Figure 14 for the BPDA-PFMB films (the second heating cycles). It is evident that below 200°C the dielectric constant is almost a constant. With increasing the frequency, the dielectric constant slightly decreases. The thermal expansion coefficient reflects the dimension stability of the film. However, meaurements of these coefficients are based on the initial force applied to the film. True thermal expansion coefficients should be obtained by

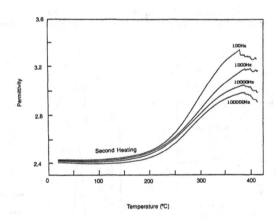

Figure 14. Dielectric constants as a function of temperature for the BPDA-PFMB film.

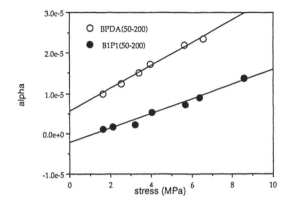

Figure 15. Thermal expansion
coefficients of the homo- and
copolyimide as a function of
the stress applied.

extrapolating the stress to zero, as shown in Figure 15 for both
BPDA-PFMB and 50/50 copolyimide in a temperature range of 50°C
to 200°C. The homopolyimide shows a slightly higher thermal
expansion coefficient (5.43×10^{-6}) than that of the copolyimide
(-2.40×10^{-6}). Both of the coefficients are very small, and that
of the copolyimide is even slightly negative.

ACKNOWLEDGEMENT

This research was supported by NASA-Langley Research Grant
(NAG-1-448), Edison Polymer Innovation Cooperation (EPIC), and
Science and Technological Center of Advanced Liquid Crystal of
Optical Materials (DMR-8920147), National Science Foundation.

References

[1] H. Lee, D. Stoffey and K. Neville, in New Linear Polymers
    (McGraw Hill, New York, 1967), pp. 183, 224.
[2] C.E. Sroog, J. Polym. Sci. Macromol. Rev., 11, 161 (1976).
[3] F.W. Harris and S.L.-C. Hsu, High Perform. Polym., 1, 1
    (1989).
[4] S.Z.D. Cheng, S.K. Lee, J.S. Barley, S.L.-C. Hsu and F.W.
    Harris, Macromolecules, 24, 1883 (1991).
[5] S.Z.D. Cheng, Z.-Q. Wu, M. Eashoo, S.L.-C. Hsu and F.W.
    Harris, Polymer, in press.
[6] F.E. Arnold, Jr., S.Z.D. Cheng, S.-F. Lau, S.L.-C. Hsu
    and F.W. Harris, Manuscript in preparation.
[7] S.Z.D. Cheng, F.E. Arnold, Jr., A.-Q. Zhang, S.L.-C. Hsu
    and F.W. Harris, Macromolecules, in press.

EFFECT OF MOISTURE ON THE PHYSICAL PROPERTIES OF POLYIMIDE FILMS

RAJEEVI SUBRAMANIAN, MICHAEL T. POTTIGER, JACQUELINE H. MORRIS, AND JOSEPH P. CURILLA , Du Pont Electronics, P.O. Box 80336, Experimental Station, Wilmington, DE, 19880-0336.

ABSTRACT

Moisture absorption and its effect on electrical properties were measured for several polyimides. A Quartz Crystal Microbalance (QCM) was used to investigate the moisture absorption in BPDA/PPD, PMDA/ODA, and BTDA//ODA/MPD polyimides. The steady-state moisture uptake in polyimides as a function of relative humidity (RH) was determined by exposing film samples to successively higher RH values ranging from 10 to 85% at 25 °C. The isothermal moisture absorption as a function of percent RH was found to be nearly linear for all of the polyimides studied. The effect of moisture on the electrical properties of a BPDA/PPD polyimide was also investigated. The relative dielectric constant at 25 °C was found to be a linear function of the moisture absorbed.

INTRODUCTION

The use of linear aromatic condensation polyimide films as interlayer dielectrics in semiconductor and packaging applications is growing due to their relatively low dielectric constants, outstanding thermal and chemical stabilities and excellent mechanical properties. Unfortunately, all polymer films absorb moisture. Moisture absorption can alter both the electrical and mechanical properties. For example, as polyimide films absorb moisture, the dielectric constant rises. Therefore, an understanding of the effects of environmental changes on these properties, especially the electrical properties, is highly desired. While numerous techniques exist for measuring the properties of thick polymer films, the majority of applications for polyimides in microelectronics involve films less than 1 mil in thickness. Measurement of the properties of thin polymeric films presents a challenge to traditional property techniques.

A direct method of measuring moisture absorption in thin, spin coated polyimide films under a controlled environment was desired. Quartz crystal microbalances (QCM) have been used extensively for measuring mass changes in-situ at an electrode and the theory of their operation has been well developed. These piezoelectric devices are very sensitive to changes in mass at the surface of the electrode and provide the capability of detecting extremely small amounts of deposited material. The use of this technique as a sensitive in-situ probe for moisture measurements in polyimide films has been described previously [1-4]. A parallel plate capacitor was used to calculate the low frequency dielectric constant of several polyimide films as a function of RH. Care was taken to insure that the films had reached equilibrium at the test conditions prior to measurement.

EXPERIMENTAL

Film Preparation

Three different polyimides were chosen for the moisture absorption study: BPDA/PPD, PMDA/ODA, and BTDA//ODA/MPD. The polyamic acid solutions consisted of 10-20 weight percent solids in NMP with number average molecular weights between 20,000-50,000 as measured by size exclusion chromatography. Sample quartz crystals were prepared by spin coating the polyamic acid solutions at the appropriate speed for 30 seconds to obtain approximately 3 micron thick cured films. The coated crystals were baked at 135 °C for 30 minutes in air in a convection oven and then heated at 2 °C/min. under flowing dry nitrogen to the final cure temperature and held for 60 minutes. Free polyimide films for the capacitance measurements were prepared in a similar fashion. The polyamic acid solutions were spin coated onto 125 mm silicon wafers containing a 1500Å thermally grown oxide layer to obtain films approximately 12 micron thick. The films were released from the wafers by dissolving the oxide layer in a 6:1 buffered hydrogen flouride solution for less than 15 minutes. Film thicknesses were measured using an AlphaStep 200 profilometer. Cured film densities were determined using a density gradient column at 25 °C.

Moisture Absorption

The experimental setup for moisture measurements involved two oscillator circuits connected to an HP computer. Each oscillator circuit consisted of a 1 inch diameter 5 MHz AT-cut quartz crystal connected to an HP 6234A power supply and an HP 5384A frequency counter. Asymmetric gold electrode patterns 2000Å thick were vapor deposited with an under layer of chromium (200Å) on the two faces of the crystal. The larger electrode with an area of 0.35 cm^2 was used as the working electrode and was coated with the polymer under study. The smaller electrode with an area of 0.18 cm^2 was the piezoelectrically active area. An AC signal is applied across the electrodes to drive the crystal at its resonant frequency. The signal is fed into the frequency counter which continuously monitors the oscillation to obtain real time studies. Two identical oscillators are used, a blank (uncoated) reference crystal and a crystal coated with the polymer under study. The reference crystal compensates for any changes that occur in the crystal itself as a function of temperature and humidity. The two crystals are enclosed in an environmental chamber where the temperature and the relative humidity can be controlled accurately.

As mass is deposited on the face of the crystal, the oscillation frequency drops. The relationship between the change in the oscillation frequency (Δf) and the change in the mass was first investigated by Sauerbrey [5] and is given by the equation:

$$\Delta m = - \left[\frac{A \left(\rho_q \, \mu_q \right)^{1/2}}{2 \, f_0^{\,2}} \right] \Delta f$$

(1)

where Δm is the change in mass; A is the piezoelectrically active area, 0.18 cm^2; Δf is the measured frequency shift; f_0 is the fundamental resonant frequency of the unloaded crystal, 5×10^6 Hz; ρ_q is the density of quartz, 2.468 g/cm^3; and μ_q is the shear modulus of quartz, 2.947×10^{11} dynes/cm^2 for a 5 MHz AT-cut quartz crystal. Dividing the change in mass by the product of the film thickness times the piezoelectrically active area yields the change in mass per unit volume. The mass uptake per unit volume is converted into the weight percent moisture absorption by dividing by the density of the polymer.

The microbalance is modeled as a composite resonator. Any additional layers deposited on the surface of the quartz crystal are assumed to behave as an extension of the crystal itself. If the acoustic impedance through the quartz and the added layers are identical, the rigid layer assumption is valid and the linear mass/frequency relationship is precise. Impedance versus phase angle plots of the coated crystals has shown that rigid layer assumption is valid for polymer films less than 6 microns thick.

All of the moisture absorption data were collected at 25 °C. A typical run consisted of preparing the crystals by drying them overnight at 150 °C in a vacuum. The crystals were then equilibrated at 0% RH (vacuum) at 25 °C for 3 hours and then the humidity was ramped up to 10, 35, 60 and finally to 85% RH, holding at each setting for 3 hours to insure that the polymers reached equilibrium, i.e. the frequency difference between the coated and uncoated crystals was a constant.

<u>Electrical Properties</u>

Capacitance measurements were used to calculate the dielectric constant. Samples for low frequency testing consisted of 60 mil diameter parallel plate capacitors. The free polyimide films were dried in a vacuum at 150 °C for a week and then gold parallel plate capacitor electrodes approximately 1/16 inch in diameter were sputter-coated onto both sides of the films. Three sets of polyimide films with the coated electrodes were equilibrated for four weeks at 0, 50 and 85% RH respectively. The capacitance of the films were measured directly on an HP 4275A LCR meter using an HP 16043B test probe while the samples were held at the same environmental conditions under which they were equilibrated. The approximate value of dielectric constant was calculated directly from the capacitance value using the equation :

$$K_{est} = \frac{C\,d}{E_0\,A} \tag{2}$$

where K_{est} is the approximate dielectric constant, C is the measured capacitance, d is the polymer thickness, E_0 is the permittivity of free space (8.85×10^{-12} F/m), and A is the area. Using ASTM D150 as a guideline, corrections for fringing capacitance effects were made using the following equation:

$$C_{edge} = P\,(0.0019K_{est} - 0.00252\ln(t) + 0.0068) \tag{3}$$

where C_{edge} is the edge capacitance, P is the perimeter of the electrode, and t is the polymer thickness. Substituting equation (3) in equation (2) yields

$$K_r = \frac{C - C_{edge}}{C_{air}} \tag{4}$$

where K_r is the relative dielectric constant and $C_{air} = (E_0\,A)/d$.

The capacitance of three sets of samples was measured at 0, 50, and finally at 85% RH. Sample A was equilibrated at 0% RH and initially measured at 0% RH and then ramped to 10, 50, and 85% RH and measured at each level after a 5 hour equilibration period. Sample B was equilibrated at 50% RH and was first measured at 50% RH and then ramped down to 0% RH, with measurements taken at each RH level after a 5 hour equilibration period. Sample C was equilibrated at 85% RH and were initially measured at 85% RH then ramped down to 50% RH and finally to 0% RH and measured at each RH level after a 5 hour equilibration period.

RESULTS

Moisture Absorption

     Figure 1 shows the moisture absorption of three polyimides obtained from the corresponding polyamic acid cured at 400 °C for 60 minutes in nitrogen. The moisture uptake is a linear function of relative humidity. This is in agreement with the previous work of Denton and co-workers [6-8] who found that moisture absorption in PMDA/ODA was a linear function of relative humidity, independent of measurement temperature. The weight percent moisture absorption in both the PMDA/ODA and BPDA/PPD polyimides did not change as a function of cure temperature as shown in Table I. Increasing the cure temperature resulted in an increase in the polymer density, however the local molecular reorganization resulting from increased imidization of the polyimide did not affect the moisture uptake on a weight percent basis. (Note that Denton and coworkers [6-8] report the moisture uptake as mass change per unit volume and indicate that the moisture uptake decreases with increasing cure temperature. Mass change per unit volume is converted to a weight percent basis by multiplying by the density of the cured polymer. The density increases with increasing cure temperature, so on a weight percent basis the moisture absorption does not change with cure temperature.) This is consistent with the work of Moylan et al. [4] who found that polyimides cured at 400 °C yielded the same moisture absorption as those cured at 450 °C.

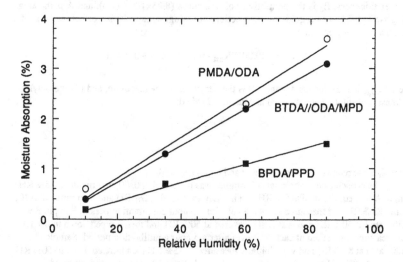

Figure 1. Steady-state moisture absorption of different polyimides cured at 400 °C.

Table I. Effect of cure temperature on moisture absorption.

Polyimide	Cure Temperature (°C)	Polymer Density (g/cc)	Moisture Absorption @ 85% RH (wt. %)
PMDA/ODA	250	1.3951	3.5 ± 0.1
PMDA/ODA	350	1.4007	3.6 ± 0.3
PMDA/ODA	400	1.4251	3.6 ± 0.3
BTDA//ODA/MPD	400	1.3900	3.1 ± 0.1
BPDA/PPD	350	1.4511	1.5 ± 0.1
BPDA/PPD	400	1.4569	1.5 ± 0.1

Electrical Properties

The capacitance measured at 1 MHz for BPDA/PPD as a function of relative humidity is plotted in Figure 2. There is a linear relationship between capacitance and moisture absorption. Equation (4) was used to calculate the relative dielectric constant at 1 MHz from the capacitance data. The data is listed in Table II.

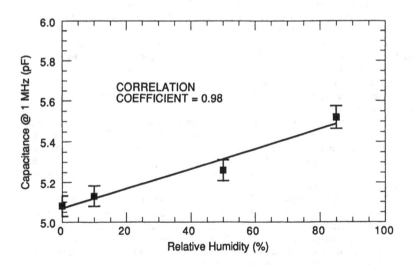

Figure 2. Capacitance versus relative humidity for BPDA/PPD cured at 400 °C.

Table II. Effect of moisture absorption on electrical properties.

Relative Humidity (%)	Moisture Uptake (wt. %)	Relative Dielectric Constant (K_r) at 1 MHz		
		Sample Set A	Sample Set B	Sample Set C
0	0	3.09		
10	0.4 ± 0.1	3.12		
50	0.9 ± 0.1	3.21	3.28	
85	1.5 ± 0.1	3.36		3.68
0		3.22	3.11	3.21

DISCUSSION

Moisture absorption in polymers is a function of several factors including temperature, relative humidity, chemical structure, polymer morphology, and sample preparation. Three modes of moisture absorption have been identified in nylon [9-10]. The first mode is absorption of water into the free volume. The amount of free volume is controlled by the polymer morphology. The second mode involves hydrogen bonding between the water molecules and the polymer. The final mode occurs when all of the free volume and hydrogen bonding sites are occupied and additional water absorption leads to plastic deformation of the polymer. These modes are not mutually exclusive. Hydrogen bonding in polyimides can occur with groups along the polymer backbone and/or with residual acid groups remaining in the polymer due to incomplete conversion from the polyamic acid to the polyimide.

The existence of two different hydrogen bonding sites has been identified in KAPTON® H using dielectric relaxation techniques from minus 190 to 0°C [11-14]. At low RH values, hydrogen bonding occurs with the stronger carbonyl bonding sites on PMDA, but at higher RH values, hydrogen bonding also occurs at the ether linkage in ODA. At room temperature, Denton, Camou, and Senturia [7] have shown that absorbed moisture behaves as free water molecules based on the increase in the dielectric constant that is comparable to the increase predicted using the Claussius-Mossotti equation. In addition, the moisture absorption was shown to be completely reversible.

Water molecules permeate into the polymer network via the free volume. The solubility limit of water in the polyimide is affected by the amount of free volume available to the water. The observed differences in room temperature moisture absorption between the three polyimides studied can therefore be attributed primarily to differences in free volume and not from any significant differences in hydrogen bonding character. Similar conclusions were reached by Moylan et al. [4].

The relative dielectric constant, K_r, is a linear function of RH. This is consistent with the findings of Denton and coworkers [6-8]. The data in Table 2 suggests the effect of moisture absorption on dielectric constant is not reversible. The dry dielectric constant for sample A is 3.09, however after ramping sample A to 85% RH and then down to 0% RH, the dielectric constant was 3.22. In addition, the dielectric constants of samples B and C measured at 0% RH were also higher than 3.09. The reason for these higher values is attributed to residual moisture in the sample. Denton et al. [7] have shown that radial diffusion in polyimides obeys a one-dimensional Fickian diffusion model. For a solid (nonpermeable) electrode, diffusion into the polymer can occur only in the radial direction. Solving Fick's second law of diffusion in cylindrical coordinates, ignoring diffusion in the z and Θ directions, with an initial uniform concentration in the polymer of zero and a constant relative humidity boundary condition, yields the following expression for the concentration of water in the polymer at the center of the electrode [15]

$$c_A(t) \approx H - H \exp\left(-\frac{5.86\, D_{AB}\, t}{R^2}\right) \qquad (5)$$

where $c_A(t)$ is the concentration of water, t is time, H is the concentration of water in the air (i.e. the relative humidity), D_{AB} is the diffusion coefficient for moisture in polyimide, and R is radius of the electrode. Using a value of 5×10^{-9} cm^2/s as the diffusion constant [8], t must be on the order of 6 days for the polyimide under an electrode of approximately 60 mils in diameter to reach 99% of the equilibrium value H. Due to the time required to reach equilibrium, care must be taken when using a parallel plate capacitor method to insure that the film between the electrodes has reached equilibrium. Another factor not taken into account in the corrections for the dielectric constant is that the gap between the electrodes increases as the polyimide swells from moisture absorption.

ABBREVIATIONS

BPDA	biphenyl dianhydride	ODA	oxydianiline
BTDA	benzophenone tetracarboxylicdianhydride	PMDA	pyromellitic dianhydride
MPD	m-phenylene diamine	PPD	p-phenylene diamine
NMP	N-methyl-2-pyrrolidone	QCM	quartz crystal microbalance

REFERENCES

1. (a) S. Bruckenstein and M. Shay, *Elecrochim. Acta*, **30**, (1985), p. 1295.
    (b) O. Melroy, K. Kanazawa, J. G. Gordon II, and D. Buttry, *Langmuir*, **2**, (1986), p. 697.
    (c) M. D. Ward, *J. Phys. Chem.*, **92**, (1988), p. 2049.
    (d) M. R. Deakin, D. A. Buttry, *Anal. Chem.*, **61** (20), (1989), p. 1147A.
    (e) K. K. Kanazawa, **New Characterization Techniques for Thin Polymer Films**, H. Tong and L. T. Nguyen, ed., John Wiley & Sons, New York, 1990, p. 125.

2. M. T. Pottiger, R. Subramanian, and M. D. Ward, "Moisture Absorption in Thin Polymer Films Using a Quartz Crystal Microbalance", presented at the Second Du Pont Symposium on High Density Interconnect Technology, Wilmington, DE, November 4-5, 1989.

3. R. Subramanian, M. T. Pottiger, and M. D. Ward, *Proc. of the Interdisciplinary Symposium on Recent Advances in Polyimides and Other High Performance Polymers* sponsored by the American Chemical Society, San Diego, CA, January 22-25, 1990.

4. C. R. Moylan, M. E. Best, and M. Ree, *J. Poly. Sci., Part B: Poly. Phys.*, **29**, (1991), p. 87.

5. G. Z. Sauerbrey, *Phys.*, **155**, 206 (1959).

6. D. D. Denton, D. R. Day, D. F. Priore, and S. D. Senturia, *J. Electr. Mat.*, **14**(2), 119 (1985).

7. D. D. Denton, J. B. Camou, and S. D. Senturia, *Proc. of the International Symposium on Moisture and Humidity*, Washington, D.C., April 15-18, 1985.

8. D. D. Denton and H Pranjoto, *Proc. of the Materials Research Society Spring Meeting*, San Diego, CA, April 1989.

9. H. W. Starkweather, Jr.,**Water in Polymers**, *ACS Symposium Series No. 127*, S. P. Rowland, ed., American Chemical Society, Washington, D.C., 1980, p. 443.

10. R. Puffr and J. Sebenda, *J. Poly. Sci. Part C*, **16**, (1976), p. 79.

11. L. Iler, W. J. Koros, D. K. Yang, and R. Yui, **Polyimides**, Vol. 1, K. L. Mittal, ed., Plenum Press, New York, 1984, p. 443.

12. G. Xu, C. C. Gryte, A. S. Nowick, S. Z. Li, Y. S. Pak, and G. S. Greenbaum, *J. Appl. Phys.*, **66** (11), (1989), p. 5290.

13. J. Melcher, Y. Daben, and G. Arlt, *IEEE Trans. on Elec. Insul.*, **24** (1), (1989), p. 31.

14. KAPTON® is a registered trademark of E. I. du Pont de Nemours & Co.

15. W. E. Boyce and R. C. DiPrima, **Elementary Differential Equations and Boundary Value Problems**, 3rd edition, John Wiley and Sons, New York, 1977.

BISMALEIMIDE/ALLYLNADIC-IMIDE BLENDS YIELD LOW DIELECTRIC CONSTANT AND LOW WATER UPTAKE

J-PH ANSERMET* AND A. KRAMER**
CIBA-GEIGY Ltd., 1701 Fribourg, Switzerland
* Materials Research
** Plastics Research

ABSTRACT

The bismaleimide resin Matrimid 5292A (I) was cocured with an allylnadic-imide resin (EP 433) which contained a long aliphatic chain as backbone (II). Water uptake, swelling, and the dielectric properties (up to 300 MHz) were studied in cast plates. The dielectric constant varied from 5.4 in (I) to 3.2 in (II) at water saturation, compared to 3.1 in (I) to 2.7 in (II) in the dry state. The glass transition temperature stayed above 200 °C at less than 80 mol% of (II).

INTRODUCTION

The reliability of the dielectric properties of polymeric materials with respect to water absorption has become a key requirement of electronics manufacturers.[1] Water absorption can give rise to : a strong dispersion in composites which distorts signals,[2] a rise of the dielectric constant which interferes with the design of interconnects of controlled impedance,[3,4] and losses which round off and attenuate digital signals.[5]
   The increase of the dielectric constant which results from water absorption has been studied by many groups.[6-9] It can be accounted for by the dipole moment of the water molecules, using the Clausius-Mosotti relationship to calculate the overall polarizability.
   In this context, we probed the effect of adding to a thermosetting polyimide a monomer which contained a long aliphatic chain.

EXPERIMENTAL

Blends based on the two monomers shown below were considered. One is a commercial Ciba-Geigy product (Matrimid 5292A). The other is an exploratory product (EP 433) which belongs to a series of polyimides which had been developed for high-temperature matrix resins.[10,11]

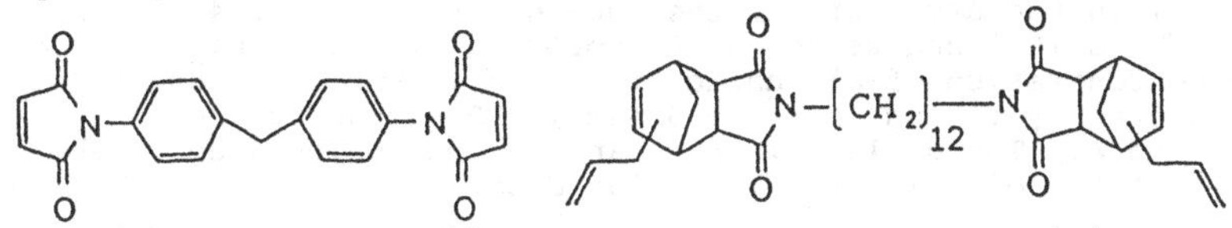

Matrimid 5292A (I)   EP 433 (II)

The samples were cast in plates of 2 mm in thickness according to the curing conditions:

Mat. Res. Soc. Symp. Proc. Vol. 227. ©1991 Materials Research Society

- Matrimid A : 10 min. 170°C, 1 h 180°C, 2 h 200°C, 6 h 280°C;
- blends : 10 min. 170°C, 3 h. 180°C, 3 h. 220°C, 12 h 250°C;
- EP 433 : 7 h. 200°C, 4 h. 210°C, 12 h. 250°C;

where the first step was under vacuum. The homopolymers were of interest only because they challenged our models. Otherwise, Matrimid A alone is very brittle, and EP 433 has a low glass transition temperature of about 150°C.

The samples were initially dried at 100°C in vacuum for 4 days, then exposed to 85% relative humidity at 85°C,[12a,b] or immersed in boiling water.

Dielectric measurements in the frequency range of 1 MHz to 1000 MHz were carried out on a home made coaxial probehead with the same samples which were used on a commercial probehead[12c] at lower frequencies. The impedance of the probeheads were measured with meters appropriate to both frequency bands.[12d] The relative accuracy of the measurements is estimated at 5%.

The swelling was measured as 3 times the change in length due to water absorption of square samples, 10 cm on a side.

DSC, TGA, DMA, TMA measurements were carried with a standard commercial equipment.[12e]

RESULTS

The basic results are presented in Figure 1 and 2. Figure 1 illustrates the stability of the material at exposures of up to 1000 hours at 85/85. The dielectric constant in the dried state and at saturation are reported in Figure 2. The lines in Figure 2 are the predictions of simple models described in the next section.

The glass transition is given in Figure 2 by a range of temperatures : the lower one corresponds to the onset of the fall of the elastic modulus, the higher one to the maximum of the loss in DMA measurements.

The coefficient of thermal expansion was of the order of 70 ± 10 ppm/K in all samples. TGA showed a sharp mass loss fall-off starting at about 420°C, with only about 2% loss at lower temperatures. DSC of the curing reactions showed three peaks, two of which correspond to the homopolymerization of each component of the blends, as reported earlier.[10,11]

INTERPRETATION

The fact that the dielectric constant decreases from 3.1 *to 2.7 in the dry state as the concentration of EP 433 varies from 0% to 100% can be accounted for by a simple model based on the concentration of aliphatic groups, c , in the blends. c is calculated by taking it to be 0.44% in EP 433 and 0% in Matrimid A. The dielectric constant is then calculated on the basis of the Clausius-Mosotti relationship :[13]

$$\frac{(\epsilon-1)}{(\epsilon+2)} = c \frac{(\epsilon_a-1)}{(\epsilon_a+2)} + (1-c) \frac{(\epsilon_i-1)}{(\epsilon_i+2)} \tag{1}$$

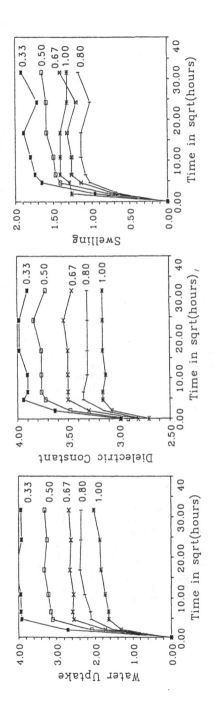

Figure 1

Water uptake (weight%), dielectric constant at 1 kHz, swelling
(%increase of volume relative to initial volume), as a function of
time of exposure to 85% relative humidity at 85°C. The lines are
guides to the eye. Molar concentration of EP 433 : (□) 33%, (■)
50%, (×) 67%, (+) 80%, (*) 100%. The diffusion constant of water
absorption at 85°C is estimated to be of 1.1 10⁻¹¹ m²/s.

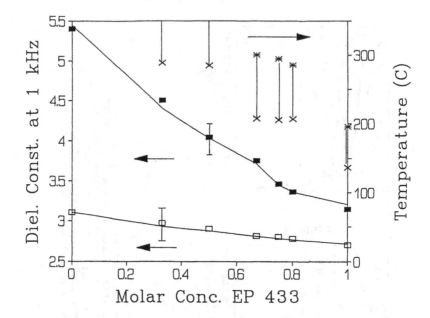

Figure 2. Dielectric constant at 1 kHz, in the dried
state (□), at saturation in boiling water (■). Glass
transition characterized by a temperature range :
the onset of the fall of the elastic modulus (x) and
the maximum of the mechanical loss (*). The lines are
predictions of our models (see text). The upper line
is calculated at the position of the data points,
using the measured water absorption.

where ϵ_a is taken to be the dielectric constant of poly-
propylene, (2.2), and ϵ_1 is taken to be 3.1 . The prediction
of this model is shown on Figure 2.

In order to account for the rise of the dielectric
constant with water uptake, we use Onsager's formula to predict
the effect of the water dipoles on the dielectric constant. Our
data include water uptake up to 6% and could be interpreted
more consistently with Onsager's formula than with the
Clausius-Mosotti formula which is usually used.[6,7] By taking
the fully relaxed dielectric constant, ϵ_m, to be that of the
sample with a water uptake m (weight%) and the unrelaxed
dielectric constant to be that in the dry state, ϵ_o, Onsager's
formula becomes :[13]

$$\frac{(\epsilon_m - \epsilon_o)\,(2\epsilon_m + \epsilon_o)}{\epsilon_m\,(\epsilon_o + 2)^2} - \frac{131\,f\,\rho\,m}{T} \tag{2}$$

where f is the polarizability of the water molecules relative
to its free state value $\alpha = \mu^2/(3kT)$ (3.2 10^{-23} pF cm^2 at 293
K), and ρ is the density of the sample. Figure 3 shows typical
dielectric spectra, which indicate that the dielectric constant
at 1 kHz can be taken to be the fully relaxed dielectric
constant. Several states of adsorbed water may be present, as
was evidenced by others in epoxy.[14,15] When all of the
dielectric data at 1 kHz are plotted in the normalized form
suggested by Eqt. 2, the points fall on a staright line with a
slope f of 0.74. In Figure 2, this model was used to calculate
the dielectric constant which was expected for the measured
amount of absorbed water.

Water uptake at saturation and the swelling at saturation
are reported in Figure 4 as a function of our estimate of the
aliphatic concentration. The swelling appears simply proportional to the concentration of hydrophylic groups in the
blends. The difference between the volume of absorbed water and
the swelling, which are both steady on a time scale of 1000
hours, is a measure of the free volume of the blends.[16] Hence
the effect of EP 433 appears to be to reduce the free volume of
the blends.

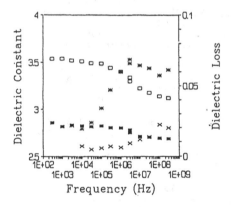

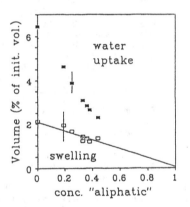

Figure 3. Sample with 67%
EP 433 : dielectric
constant in the dried
state(■), at saturation
at 85/85 (□), dielectric
loss in the dried state
(x), at saturation (*)

Figure 4. Swelling (■)
and water uptake (□)
expressed as a volume of
water relative to the
initial volume of the
sample, versus the
estimated aliphatic
concentration (see text).

CONCLUSION

Water absorption and its effect on the dielectric constant were measured in a series of blends of two cocured polyimides. As one of the imides comprised a long aliphatic chain, the data were presented in terms of the influence of the molecular structure on moisture resistance.

From a practical standpoint, these blends offer a compromise between low dielectric constant at equilibrium with moisture and high T_g. For example, T_g above 200°C could be achieved with a dielectric constant which varied from 2.8 in the dry state to 3.3 upon saturation at 85% relative humidity at 85°C.

ACKNOWLEDGMENTS

The authors are thankful to Mrs C. Irrgang for the thermomechanical characterization and to Mr. E. Baeriswyl and Mr. C. Vonlanthen for technical assistance.

REFERENCES

1. e.g. E. Wiesner (Hewlett Packard), Proc. NEPCON WEST 1990
2. W. Doeling et al., in Microelectronics Interconnects and Packaging, edited by J. Lyman, (McGraw Hill 1980), p. 140
3. B.K. Gilbert, in Packaging and Interconnection of GaAs Digital Integrated Circuits, VLSI Electronics Microstructure Science 11, (Academic Press 1985)
4. G.J. Doyle, B.J. Sheehan, Circuit Manufacturing, Dec. 1987, p. 42-46
5. A.C. Cangellaris et al., Proc. NEPCON WEST 1990, p. 215
6. D.D. Denton, Moisture Transport in Polyimide Films in Integrated Circuits, PhD Thesis, MIT, 1987
7. P.D. Aldrich, S.K. Thurow, M.J. McKennon, M.E. Lyssy, Polymer 28, 2289 (1987)
8. A.J. Beuhler et al., ACS Proc. Pol. Sci. Eng. 1988, 339
9. D. Denton, C.N. Ho, IEEE Trans. Instr. Meas. 39(3), 508 (1990)
10. A. Kramer, R. Schmid, in Advances in Polymer Blends and Alloys Technology, volume III, (Technomic Pub. to be published)
11. A. Kramer et al. , SAMPE Anaheim 1990
12. a.Weiss Technik GmbH, D-6301 Reiskirchen Germany
    b.Rotronic DV-2, Instrument Corp., Huntington N.Y. 11743
    c.DETA Polymer Laboratories Inc., Amherst, MA 01002, USA
    d.Hewlett Packard 4191A and 4192A
    e.DuPont TA9900 (DSC,DMA,TMA), Mettler TA2000C (TGA), Perkin Elmer TAS7 (TMA)
13. A.R. Blythe, Electrical Properties of Polymers, (Cambridge Uni. Press, 1979)
14. I.D. Maxwell, R.A. Pethrick, J. Appl. Polym. Sci. Technol. 28, 2363 (1983)
15. J.D. Reed et al., J. Appl.Pol.Sci.Technol. 31,1771(1986)
16. M.J. Adamson, Proc. Annu. Program Rev/Workshop 5th(1987)

CHARACTERIZATION OF THE BINDING OF INORGANIC POLYMERS TO OXIDE SURFACES BY NMR AND NMR/MAS SPECTROSCOPY

SARAH D. BURTON, WILLIAM D. SAMUELS, GREGORY J. EXARHOS, AND JOHN C. LINEHAN, Pacific Northwest Laboratory, Richland, WA 99352.

Abstract

Solution and solid state Magic Angle Spinning (MAS) Nuclear Magnetic Resonance (NMR) spectroscopy have been used to monitor the ^{31}P signal of cyclic phosphazenes and associated linear polymers in solution and of these materials bound to alumina in stabilized dispersions. The differences between the simple solution experiments and the suspensions are being studied to determine the viability of NMR techniques to probe the chemical interactions between inorganic polymer dispersants and alumina particles. It has been observed in colloidal suspensions that the adsorption of phosphazenes onto an aluminum oxide surface causes a broadening of the ^{31}P signal in solution NMR. This broadening is dependant on the amount of solid to polymer concentration, the amount of solids loading, the solvent polarity and the phosphazene substituent under investigation [1]. Conversely, the solid MAS experiments show a narrowing of the ^{31}P signal upon adsorption of the phosphazene to alumina.

Introduction

Polymeric dispersants used in ceramic processing historically are organic polymers containing polar substituents, such as a carboxylic acid [2]. One of the problems associated with using organic polymers is incomplete polymer burnout during final densification (sintering). Our laboratory has been investigating inorganic polymers, polyphosphazenes and polysiloxanes, as dispersing agents. It is known that polyphosphazenes and polysiloxanes both undergo thermal cracking to small rings and low molecular weight oligomers as well as cleavage of the side groups and further incorporation of the polymer backbone into the ceramic matrix upon sintering [3]. One area of our research is to understand the chemistry of the interaction between the inorganic polymer and the oxide surface. Localized interactions between alumina colloids and phosphazenes or other polar molecules (solvent) are assumed to involve hydrogen bonding between hydroxy groups on the surface of the ceramic and electonegative or basic groups on the polymer or solvent [1]. In the case of phosphazenes the lone pair of electrons on each nitrogen can serve as a base for hydrogen bonding.

Reported herein are the results of NMR studies which were designed to document changes in the ^{31}P signal in the phosphazenes according to the conditions and methodology of the experiment. Attractions between surface hydroxy groups on the alumina and the polymer backbone should perturb the electron density about the phosphorus center and therefore induce a shift in the ^{31}P NMR signal. Adsorption of a phosphazene onto the solid surface also may increase its rigidity and therefore lengthen the ^{31}P spin relaxation time. Finally, the phosphazene may adsorb onto the ceramic surface in a variety of configurations generating a distribution of phosphorus sites in the sample (Figure 1).

Mat. Res. Soc. Symp. Proc. Vol. 227. ©1991 Materials Research Society

162

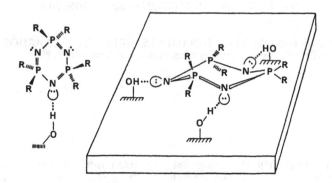

Figure 1. Adsorption of phosphazenes on an alumina surface.

Experimental

The octachlorocyclotetraphosphazene and hexachlorocyclotriphosphazene were purchased from Nippon Soda Co., Ltd. of Tokyo Japan and used as received. All other phosphazenes were prepared as described in the literature [4]. The aluminum oxide was obtained from Sumitoma, grade AKP-30.

^{31}P NMR experiments were carried out using a Varian VXR 300 MHz NMR spectrometer operating at a frequency of 121.1 MHz. MAS measurements were performed using a Doty Scientific, Inc. high-speed solid probe. The rotors used were 5 mm zirconia cylinders with vespel end caps. Experiments were carried out using 74 KHz dipolar decoupling gated on during FID acquisition. Acquisition times of 149 ms with a 59.88 KHz spectral window were employed. The rotors were spun at 7.3 +/-.2 KHz using nitrogen as both bearing and drive gasses. If the phosphazene had a melting point below 80 °C, low temperature probe procedures were employed. In cross-polarization magic angle spinning experiments, a recycle time of 10 s was necessary. The Hartmann-Hahn matching field was arrayed to determine the optimal contact time for the cross-polarization experiments. For Bloch decay acquisitions, 16-200 transients were collected with a recycle time of 60 s. All solution state ^{31}P chemical shifts are referenced to 85% phosphoric acid as an external standard. Solid state MAS chemical shifts were referenced to triphenylphosphine (-3.1 ppm) as an external standard.

Solvent suspensions were prepared by making a 0.02 N solution for each phosphazene and adding, by weight, 20% Al_2O_3 0.3 micron particles. Solid suspension samples were prepared by evaporating the solvent from the solution suspensions and evacuating the powder overnight. The coated alumina was finely ground using a mortar and pestle.

Results and Discussion

Solution ^{31}P NMR

In experiments previously reported, the observed ^{31}P signal for MEEP (bismethoxyethoxyethoxypolyphosphazene) broadened as an aqueous solution was loaded with Al_2O_3 ceramic particles [1]. Peak broadening and a noticeable chemical shift were reported for $CDCl_3$ solutions of TFET (hexatrifluoroethoxycyclotriphosphazene) bound to alumina. Broadening of the ^{31}P resonance also has been observed for other alumina loaded phosphazene solutions but it is not clear whether the line shape differences are due to the adsorption of the phosphazene onto the alumina particle or if the line shape is an artifact caused by the amount of solid present in the NMR sample (Figure 2).

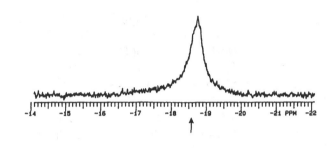

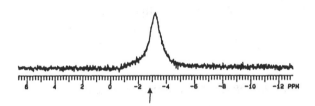

Figure 2. ^{31}P NMR of 20% alumina suspensions prepared in chloroform. The arrows indicate where the ^{31}P signal for nonadsorbed phosphorus containing moieties appear. a) Ph$_3$POOH, b) [NP(OCH$_2$CF$_3$)$_2$]$_4$

Solid State MAS Spectroscopy

Two types of solid samples were studied by MAS NMR spectroscopy: (i) the neat phosphazene; and (ii) phosphazene coated aluminum oxide. The objective of these studies is to determine whether the measurement can distinguish between adsorbed and nonadsorbed phosphazene. Both types of species may be present. Tables 1 and 2 summarize the solid and solution NMR data.

Polycrystalline samples which contain hydrogens in the phosphorus substituents are suitable for a cross-polarization experiment. This provides for a shorter relaxation time resulting in more rapid accumulation of data. For amorphous samples or samples which do not contain hydrogen in the phosphorus substituent, a Bloch decay experiment was used.   Figure 3 shows several of the MAS spectra taken for pure phosphazene materials. All solid state spectra show characteristic spinning side bands. It is important to note, the shape of the [NPCl$_2$]$_4$ spectra. This line shape is attributed to the Cl-P coupling interaction and a similar shape is exhibited by the clorotrimer.[5]

Chemical shifts for the MAS solid spectra appear down field with respect to the corresponding solution ^{31}P signals (Figure 2). The magnitude of the shift varies for each

Table 1. Solid MAS and solution ^{31}P data for cyclotriphosphazenes.

(X)$_6$	Solid PPM	Solution PPM(DCCl$_3$)
Cl	24. (m)	21.9
F	8.	13.8
NCS	28.6(d)	25.3
OCH$_3$	25.1	23.9
SCH$_3$	53.4	43.6
NHCH$_3$	23.8	23.2(D$_2$O)
N(CH$_3$)$_2$	28.1	25.3
NHNH$_2$	26.5	27.7(D$_2$O)
OPh	14.6	9.0
OCH$_2$CF$_3$	21.6	17.8
N◗	23.2	18.2

Table 2. Solid MAS and solution ^{31}P data for cyclotetraphosphazenes.

(X)$_8$	SOLID PPM	SOLUTION PPM(CDCl$_3$)
OCH$_3$	8.4	9.0
OCH$_2$CF$_3$	4.3	2.5
Cl	6.0(m)	-7.4
N(CH$_3$)$_2$	16.9	9.6

sample (Tables 1 and 2). Solid samples of phosphazenes coated on alumina exhibited an upfield shift of 0.5-10 ppm, (Figure 4). Broader lines are expected in the MAS experiments due to the multiple crystal sites possible in each solid. If only a segment of the phosphazene molecule is bound to an alumina particle, an increase in the distribution of phosphorus sites is expected leading to line broadening. Our results contrasted with this prediction. In all of these cases the linewidth decreased significantly. Figure 4 shows the spectrum of pure [NP(NMe$_2$)$_2$]$_4$ which has a line width of 267 Hz and the spectrum of a 20% alumina loaded sample which exhibits a linewidth of 90 Hz.

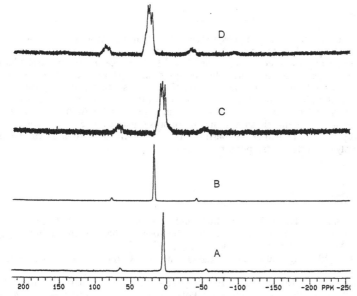

Figure 3. ^{31}P solid MAS spectra of phosphazenes. a) $[NP(OCH_2CF_3)_2]_4$, b) $[NP(NMe_2)_2]_4$, c) $[NPCl_2]_4$, d) $[NPCl_2]_3$

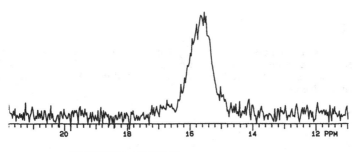

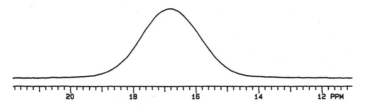

Figure 4. Expanded trace of pure and alumina loaded $[NP(NMe_2)_2]_4$ MAS spectra.

Conclusion

The ^{31}P results for measured phosphazenes show that in both solution and solid state there is a small upfield shift associated with phosphazenes bound to the alumina. Solution samples consistently displayed a broadening of the ^{31}P resonance upon cumulative addition of alumina to the suspension but solid loaded samples showed a narrowing of the resonance. Presently this narrowing of the ^{31}P signal is attributed to a diminished phosphazene-phosphazene interaction. Of interest in this research is to clearly observe both adsorbed and nonadsorbed phosphazene in ceramic suspensions and determine whether the amount of upfield shift in the phosphorus signal is related to the substitution of the phosphazene. This work demonstrates that ^{31}P NMR measurements can be used to probe polymer-surface adsorption phenomena.

REFERENCES:

1. GJ Exarhos, KF Ferris, DM Friedrich, and WD Samuels, "^{31}P NMR Studies of the Aqueous Colloidal Suspensions Stabilized by Polyphosphazene Absorption." Journal of the American Ceramic Society, 1988, 71(9):C406-7.
2. IA AKsay, FF Lang, and BI Davis, Journal American Ceramic Society, 1983, 66:C190-192.
3. WD Samuels, BJ Tarasevich, GJ Exarhos, and BD McVeety, "Characterization of the Products of the Pyrolysis of Inorganic Polymers Used as Dispersants"., 92th Annual Meeting of the American Ceramic Society, Dallas, Texas, April 22-26, 1990.
4. All synthesis for the Phosphazenes are referenced in Appendix II of Phosphorus-Nitrogen Compounds, HR Allcock, Academic Press, New York and London, 1972.
5. Crosby, R. C.; Haw, J. F., Macromolecules, 1987, 20, 2326.

ACKNOWLEDGEMENT:

Pacific Northwest Laboratory is operated by Battelle Memorial Institute for the US Department of Energy under contract DE-AC06-76 RLO 1830. This work has been supported by the Materials Science Division of the Office of Basic Energy Sciences, U.S. Department of Energy.

PHYSICAL CHARACTERIZATION OF MICROELECTRONIC POLYMERIC THIN FILMS

STEPHEN D. SENTURIA, SUSAN C. NOE, AND JEFFREY Y. PAN
Microsystems Technology Laboratories, Department of Electrical Engineering
and Computer Science, Massachusetts Institute of Technology, Cambridge, MA,
02139, USA

ABSTRACT

The measurement of the mechanical properties and adhesion of polymeric
thin films and coatings poses a number of technical problems. Elastic and
viscoelastic properties, residual stress, adhesion, the effects of extended
cure, and the effects of adsorbed moisture and process reagents are all
critical. A particular challenge is to develop measurement methods which
can be used with actual samples, preferably non-destructively. This paper
examines a number of methods which have been developed to make these measurements, with emphasis on methods which are sensitive enough to look at
the effects of process variation and the effects of moisture exposure.
Suspended-membrane methods for measuring elastic and viscoelastic properties, residual stress, and adhesion are combined with optical methods for
determining index of refraction and birefringence to yield a family of
techniques for performing physical characterization. Recent results on the
effects of extended cure and moisture uptake on elastic properties,
residual stress, and optical properties will be presented.

INTRODUCTION

The use of polymeric thin films in microelectronic structures has increased significantly during the past decade. Applications include
passivation coatings, interlevel dielectrics, die attach adhesives, and
flex-circuit substrates (see, for example, the recent conference proceedings cited in references [1-7]). Requirements on thermal and chemical
stability have favored polyimides for these applications, although other
polymer types also show promise of suitability. Polyimides, which can be
processed in soluble form and are then converted to an intractable form
with a thermal cure, have many attractive properties, such as planarization, controllable etch, low dielectric constant, low electrical conductivity, and good mechanical toughness.
    There are two principal issues that affect the decision to use poly-
mers in such applications: (1) whether the desired structure can be successfully fabricated at high yield; and (2) whether the resulting structure
will be reliable under actual use conditions. The fabrication issues depend in part on the constraints on dimensions; here, polyimides have proven
very effective in providing planarizing coatings as thin as 1 μm and as
thick as 50 - 70 μm, and provide good etch control in RIE mode down to 1 μm
feature sizes for the thinner films. Via-filling techniques using both
subtractive and additive processes have been reported. Residual stress,
however, particularly in thicker films, can create undesirable substrate
bending and delamination or crack formation at the edges and corners of
patterned features. Finally, successful fabrication depends on achieving
excellent interlayer adhesion at a variety of interfaces, either intrin-
sically, or with the use of adhesion promoters.
    The reliability of the polyimide in actual use brings up additional
issues: moisture absorption, and the effect of moisture on the electrical
properties, mechanical properties, adhesion, and corrosion resistance of
the structure. Because many of the end-use applications involve computer
and telecommunications technologies, where long-term reliability, possibly
under non-hermetic conditions, is required, it is extremely important to be
able to assess structural durability and the effects of moisture on that
durability. Furthermore, because residual stresses are such an intrinsic

part of the polyimide technology, such assessments should be done, to as great an extent as possible, on actual structures using *in-situ* measurement methods, where the residual stress is maintained.

A final issue concerns the fact that in advanced interconnect structures, where polymers are used as interlevel dielectrics with four or five levels of patterned metallization, the polymer is functioning within a three-dimensional structure, even though during deposition of any individual layer, it is still a film. In effect, the interconnect is a fiber-matrix composite constrained on a relatively rigid substrate. In-plane strains are limited by the substrate constraint, and normal strains may be constrained by relatively rigid metallization. Under thermal cycling or exposure to solvents which might produce swelling, complex loading can develop in such structures. Prediction of stress distributions in real structures requires an understanding both of the detailed geometry and also of the complete elastic (and possibly viscoelastic) behavior of the polymer, including properties normal to the plane of the film. This last subject is far from complete at this time.

BRIEF REVIEW OF PREVIOUS WORK

Residual Stress and Elastic Properties

Because of various combinations of (a) solvent loss, (b) loss of reaction products during cure, and (c) thermal mismatch to the substrate, polymer coatings typically are in a state of residual tensile biaxial (in-plane) stress. This stress does various things: First, it exerts a mechanical bending moment on the substrate, which gives rise to substrate curvature. Second, the stress can provide a driving force for coating delamination at the edge of a patterned feature. Third, stress redistribution at the corners of patterned features can create stress concentrations, leading to plastic deformation, or to fracture and crack formation. Fourth, since the stress is intrinsic to the structure, long-term creep and related viscoelastic effects may occur.

Substrate bending, which can be a serious liability during fabrication, can be used advantageously to measure residual stress. Measurement of the radius of curvature can be done either mechanically [8], optically [9], or with X-ray diffraction [10]. These methods all depend on the assumption that the coating is sufficiently thin that it does not contribute significantly to the bending stiffness of the substrate. In such a case, the elastic modulus of the coating does not affect the substrate curvature.

One problem with substrate curvature measurements is that for relatively low-stress or very thin coatings, the total bending moment may be so small that the radius of curvature cannot be reliably measured. An alternative approach has been to thin the substrate significantly, typically in the form of a cantilever beam either of quartz or silicon, and measure the tip deflection of the cantilever that results from the stress in the coating [11]. The advantage of this approach is the increased sensitivity that results from the thinned substrate. The corresponding disadvantage is that with a thin substrate, the elastic modulus of the coating can contribute significantly to the overall bending stiffness; hence, it is necessary to know the coating modulus in order to determine the stress.

A direct approach to this problem is to remove a portion of the substrate altogether, creating a suspended membrane. Suspended membranes can be used for simultaneous determination of both stress and elastic modulus. The most widely used *in-situ* method is to measure the load-deflection behavior of a suspended membrane (schematically illustrated at the bottom of Figure 1) [12-21]. In general, for a membrane with thickness t very much less than its diameter or edge-length 2a, the bending stiffness of the membrane can be neglected, leading to a load-deflection relationship in the following form:

$$p = C_1 \frac{\sigma t d}{a^2} + C_2 \ f(v) \left(\frac{E}{1-v} \right) \frac{t d^3}{a^4} \qquad (1)$$

where p is the pressure applied across the membrane, d is the deflection at the center of the membrane, σ is the residual stress, $E/(1-v)$ is the biaxial modulus, where E is the in-plane Young's modulus and v is the in-plane Poisson ratio, $f(v)$ is a geometry-dependent weak function of the Possion ratio with a magnitude of order unity, and C_1 and C_2 are dimensionless constants. If d is measured as a function of p, and p/d is plotted against d^2, the result is a straight line with a slope proportional to $f(v)$ times the biaxial modulus and an intercept proportional to the residual stress. The quantities C_1, C_2, and $f(v)$ have been determined from finite-element analysis for various geometries [17-21]. Some typical results from the use of the suspended membrane method are given later.

Because of the different Poisson-ratio dependence of the load-deflection behavior of membranes of different shapes, it is, in principle, possible to determine the Poisson ratio from a comparison of differently shaped membranes. To date, this has not yielded acceptable precision (problems of precision and accuracy are discussed later). However, by combining uniaxial measurements of free films to obtain E with membrane measurements to obtain $E/(1-v)$, values of the in-plane Possion ratio have been obtained [20].

There are several other ways of measuring the Poisson ratio. Bauer and Farris [22] have measured the pressure dependence of the stress of a uniaxially loaded polyimide film. This yields the average of the in-plane and out-of-plane Poisson ratio. However, because the polyimide must be removed from it's substrate, information on residual stress and possible elastic nonlinearities may be lost.

Recently, Maden and Farris have explored resonant methods to study suspended membranes [23]. These methods are very sensitive to residual stress. Furthermore, by cutting the suspended membrane into a ribbon shape to relieve the biaxial stress, and then using resonant methods to determine the uniaxial stress in the ribbon, the shape-dependent change of stress yields the Poisson ratio directly.

Load-deflection studies on circular suspended polyimide membranes have been used to measure yield, creep, and creep recovery [15,20,21]. The essential issue is that because of the departure of the actual deflected shape from a hemispherical cap, the principal stress achieves its maximum value at the center of a deflected circular membrane. Hence, yield and/or creep initiates at the membrane center, and proceeds radially. This has been dramatically demonstrated by Maseeh for a glassy polyimide (Dupont 2525) usually thought of as a brittle material. A circular membrane (under residual tensile stress) was loaded until creep began. After unloading, there remained a circumferential region which became flat, due to residual stress, but the central region which had undergone apparent plastic deformation due to creep, remained in a dome shape. In some cases, gradual creep recovery took place, with the sample returning to its original flat shape over a period of several days. Maseeh has explored the possibility of extracting the creep compliance from these studies [21]. The problem is significantly complicated by the non-uniform loading of the structure, but he was able to represent the behavior with a non-linear viscoleastic model with a power-law time-dependence.

Adhesion

The importance of adhesion in microelectronic reliability cannot be overstated. Methods for measurement of adhesive bond strength between coatings and substrates, and for examining the effects of surface treatment, coating chemistry, deposition conditions, cure, the use of adhesion promoters, and the effects of environmental exposure, are essential. The biggest problem faced in such measurements is that the coatings used in microelectronics are both thin and well-adhered. Therefore, it is important that the methods used for bond-strength measurement be suitable in the limit of thin films and large bond-strengths. For example, the scratch test, in which the residual tensile stress in the coating can cause a delamination at the edge of a scratch or cut, depends on having the combination of residual stress times film thickness large enough to provide the elastic driving force for debonding [24]. When the debond energy is large, as it should be in a microelectronic application, very thick films are required to create enough driving force to achieve debonding -- hence the scratch test is not a satisfactory method for studying the adhesion of practical microelectronic coatings.

The other adhesion measurement methods (reviewed in [25-27]) rely on some form of peeling of the coating from the substrate. Two of these tests are the 90° peel test and the blister test [28-30]. In these tests, a debonded portion of the coating must deliver the debonding force to the interface between the still-attached coating and the substrate. For well adhered films, this force must be large, and as the thickness of the film is reduced, this large force can create a stress in the already-debonded film which exceeds its yield or fracture point. This is called the tensile strength limit of a peel test [31].

A variant of the blister test, called the island blister test [31-33], has been developed to overcome the tensile strength limit. A suspended membrane is formed which is still attached to the substrate at a small site in the center (the island), If the bonded site is made sufficiently small, applying pressure to the surrounding suspended membrane can initiate debonding without exceeding the tensile strength of the film. After debond, the two fracture surfaces can be analyzed for morphology with SEM and for surface chemistry with XPS and Auger analysis.

PRECISION AND ACCURACY

The subject of achievable precision and accuracy of the various measurement methods has not been addressed in a systematic fashion. Since an entire strategy for measurement may be determined by these limits, it is important to examine these issues. It is not possible to do a comprehensive evalution; however, our experience in the measurement of in-plane mechanical properties, whether by uniaxial loading of free films or biaxial loading of membranes, is that the intrinsic precision of the measurement on one sample is about 5%, the modelling accuracy can be reduced to below 3% by careful FEM work, but the knowledge of the sample geometry still limits overall accuracy to a value closer to 10%. Consider, for example, the load-deflection example of Eq. 1. When load-deflection data are fitted to the form of Eq. 1, the quantities $C_1 \frac{\sigma t}{a^2}$ and $C_2 \, f(v) \left(\frac{E}{1-v} \right) \frac{t}{a^4}$ can each be experimentally determined to better than 5%, but because of the power-law dependence on edge-length (or radius) a, as little as a 2% error in edge-length creates an 8% error in the extracted biaxial modulus, even if C_2 and $f(v)$ are known perfectly. The same problem appears in the resonant method of stress determination. The intrinsic precision in determining an experimental number which is a function of both stress and geometry can be very excellent; however, small errors in geometry can dominate the accuracy of the extracted stress.

Why is this important? Typically, one seeks to compare results on different samples in order to study effects of deposition, dilution, cure, environmental history, etc. If limitations on geometry prevent the accuracy from approaching the required precision, then sample-to-sample comparisons lose much of their benefit. Therefore, it is highly desirable to discover measurement methods which can continually monitor the critical sample geometry. A related issue is the achievable precision which can be obtained in measuring properties normal to the plane of the film, such as thermal expansion and elastic modulus. These measurements depend on an accurate knowledge of film thickness. This concern over geometric accuracy is one motivation for considering optical methods of film characterization, as explained in the following section.

OPTICAL CHARACTERIZATION WITH SLAB WAVEGUIDES

A polymer film on a substrate with lower index of refraction (such as an oxidized silicon wafer) supports optical slab-waveguide modes. For sufficiently thick films, several different modes can propagate, each with a different propagation constant. The set of modes, and their propagation constants, depend on the polarization of the wave, either in the plane of the film (TE), or normal to the plane (TM). For a given film and polarization, the number of modes and their propagation constants depend on a combination of the film thickness and the index of refraction measured in the direction of the polarization. If the film propagates at least two modes, it is possible to extract both the index and thickness by measuring the propagation constants of the modes. Prism coupling is a convenient way to access the propagating modes in a slab waveguide (see Fig. 1). A commercial instrument for performing this kind of measurement is available from Metricon.

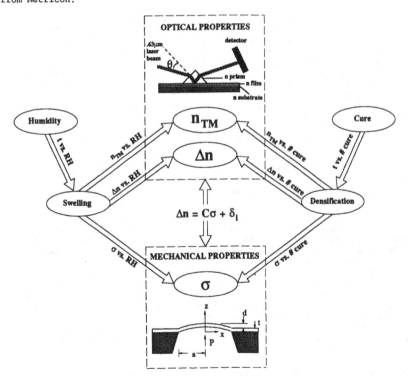

Figure 1. Schematic connections between optical and suspended-membrane measurements of index, birefringence, thickness, stress and biaxial modulus versus moisture and extended cure.

We have used a Metricon PC-2000 instrument to measure the in-plane index of refraction n_{TE}, the normal index of refraction n_{TM}, and the thickness t for several different polyimides subjected to various cure cycles. We also examined the birefringence, which is $n_{TE} - n_{TM}$. These results were correlated with measurements of residual stress and biaxial modulus using suspended membrane methods. The effect of varying ambient moisture was also examined. Details of these experiments are being reported separately [34]. The key ideas are presented here, with reference to Figure 1, which shows the experimental connections between the various quantities.

The material used to illustrate the method is Dupont Pyralin 2555, which is a polyimide made from benzophenone-tetracarboxylic dianhydride (BTDA) and an 80%-20% mixture of oxydianline (ODA) and metaphenylenediamine (MPDA). Samples were spin coated onto silicon wafers using adhesion promoter, and were then cured at 400 °C for 90 min in nitrogen. Extended cure was obtained by repeating the cure for up to 10 times. Film thickness was about 6.2 μm. For the samples used in optical studies, the silicon wafers had been previously oxidized to create a 1.3 μm thermal oxide film. This greatly enhances wave propagation in the polymer slab waveguide. Without this oxide, damping of the modes by the silicon can be so severe that the measurement of the TM modes can become impossible.

Some samples were used to make suspended membranes for determination of residual stress and biaxial modulus at room temperature, as a function of ambient moisture (for each individual sample), and as a function of extended cure (for different samples). Other samples were used for determination of index and thickness at room temperature, both as a function of moisture (for each individual sample), and as a function of extended cure (in some cases for different samples, in other cases for the same sample, reinserted in the cure furnace following measurement). In the latter case, where the same sample was used repeatedly for measurement after each cure cycle, we did achieve the desirable state of avoiding any influence of sample-to-sample variation.

Several interesting observations can be made from typical results, aided by a model of the form

$$n_{TE} - n_{TM} = C\sigma + \delta_i \qquad (2)$$

where σ is the in-plane biaxial stress, C is the stress-optic coefficient, and δ_i is a residual birefringence due to the anisotropic morphology which is typically present in a spin-coated film cured on a substrate [35-42].

Every polymer we have examined shows some difference between n_{TE} and n_{TM}. For a given sample, both n_{TE} and n_{TM} increase with moisture (consistent with the absorption of a polar molecule), but the difference between the indices, the birefringence, decreases. The BTDA-ODA/MPDA polyimide has a relatively weak birefringence, but it is readily resolvable. Figure 2 shows the relative-humidity dependence of the birefringence for a sample which has been cured once. Figure 3 shows the corresponding residual stress. We attribute the decrease in the birefringence with increased moisture to swelling of the polymer, which reduces the in-plane residual stress, and hence reduces the birefringence through the stress-optic coefficient. By cross-plotting the data from Figures 2 and 3, an estimate of the stress-optic coefficient of 270 Brewsters is obtained. In addition, the increase in thickness due to moisture uptake was observed, but was barely resolvable, at 0.02 μm.

The effect of extended cure is similar. Both n_{TE} and n_{TM} increase with cure, suggesting densification, but the n_{TE} increases more than n_{TM}. This is shown in Figure 4. The model of Eq. 2 suggests that the increased birefringence should correspond to increased stress, but because of the small changes observed, and the fact that unlike the optical data, different samples were required for the membrane measurements at each cure condition, the between-sample variation was too large to permit direct observation of the predicted stress change. However, in a different polymer, Pyralin 2556, which is a diluted version of 2555, experiments were successful in correlating stress increase with birefringence increase for extended cures. The details are presented in [34]. Finally, we were able to observe a thickness decrease due to extended cure, but again, it was just at the limit of resolution, 0.02 μm.

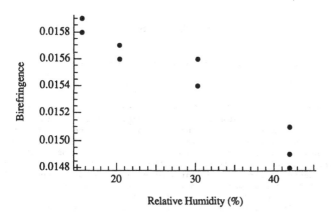

Figure 2. Birefringence versus moisture.

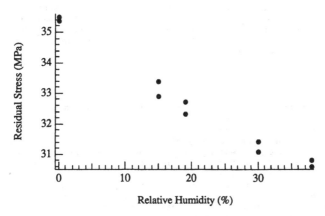

Figure 3. Residual stress versus moisture.

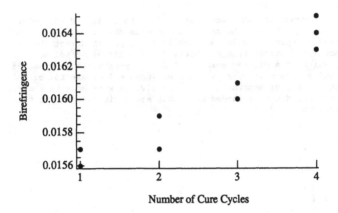

Figure 4. Birefringence versus extended cure.

CONCLUSION

There are many methods available for measuring mechanical properties of polymer films used in microelectronics. Residual stress, in-plane elastic constants, viscoelastic behavior, and adhesion can all be determined. Optical properties measured with slab waveguides offer additional insight into polymer behavior, first, by providing direct measurement of thickness and birefringence in situ, and, second, by affording highly precise access to the subtle effects of cure and moisture, effects which may be too small to be observed within the sample-to-sample accuracy of other methods. A graphic illustration is the small but steady increase in index, birefringence, and stress which results from extended cure of polyimide, never reaching a stable "full-cure" state.

ACKNOWLEDGEMENT

The authors wish to acknowledge support from the Semiconductor Research Corporation, Dupont, Raychem, 3M Company, and the U.S. Department of Justice (Contract J-FBI-88-067). The authors also wish to thank David Volfson for help with samples, and Albert Young for assistance in setting up the optical measurements facility.

REFERENCES

1. Proc. First Int'l. Conf. on Polyimides, 1982, published as: K. L. Mittal, ed., *Polyimides - Synthesis, Characterization, and Applications* (Plenum Press, New York, 1984).
2. Proc. Second Int'l Conf. on Polyimides, 1985, published as : W. D. Weber and M. R. Gupta, eds., *Recent Advances in Polyimide Science and Technology* (Mid-Hudson Section, Society of Plastics Engineers, Poughkeepsie, New York,1987).
3. Proc. Third Int'l Conf. on Polyimides, 1988, published as: C. Feger, M. M. Khojasteh, and J. E McGrath, eds., *Polyimides: Materials, Chemistry and Characterization* (Elsevier, New York, 1989).
4. M. J. Bowden and S. Richard Turner, eds., *Polymers for High Technology: Electronics and Photonics*, ACS Symposium Volume 346, (American Chemical Society, Washington, D.C., 1987).

5. R. Jaccodine, K. A. Jackson, and R. C. Sundahl, eds., *Electronic Packaging Materials Science III*, Materials Research Society Symposium Proceedings, Volume 108, (Materials Research Society, Pittsburgh, 1988).

6. J. Lupinski and R. S. Moore, eds. *Polymeric Materials for Electronics Packaging and Interconnection*, ACS Symposium Volume 407, (American Chemical Society, Washington, D.C., 1989).

7. Proc. PME'89, published as: *Polymers for Microelectronics -- Science and Technology* (Kodansha, Tokyo, 1990).

8. K. Crowe, M.S. Thesis, Dept. of Mechanical Engineering, Massachusetts Institute of Technology, Cambridge, MA, 1985 (unpublished).

9. StressGauge, Ionic Systems, Inc.

10. P. Geldermans, C. Goldsmith, and F. Bendetti, in Ref [1], p. 695.

11. R. W. Hoffman, in C. H. S. Dupey and A. Cachard, eds. *Physics of Nonmetallic Thin Films*, NATO Advanced Study Institutes, Plenum Press, Vol B-14, p. 273 (1976).

12. J. W. Beams, in C. A. Neugebauer, ed., *Structure and Properties of Thin Films*, p. 183 (1959).

13. E. I. Bromley, J. N. Randal, D. C. Flanders, and R. W. Mountain, J. Vac. Sci. Tech. B, $\underline{1}$, 1364, (1983).

14. M. G. Allen, M. Mehregany, R. T. Howe, and S. D. Senturia, Appl. Phys. Letters, $\underline{51}$, 241 (1987).

15. F. Maseeh, M. A. Schmidt, M. G. Allen, and S. D. Senturia, Technical Digest, 1988 IEEE Solid-State Sensor and Actuator Workshop, Hilton, p. 84.

16. O. Tabata, K. Kawahata, S. Sugiyama, H. Inagaki, and I. Igarashi, Technical Digest of the 7th Sensor Symposium, Tokyo, 1988, IEE, Tokyo, p. 173.

17. P. Lin and S. D. Senturia, Mat. Res. Soc. Symp. Proc. Vol. 188, p.41, (1990).

18. J. Pan, P. Lin, F. Maseeh, and S. D. Senturia, Technical Digest, IEEE Solid-State Sensor and Actuator Workshop, Hilton Head, (IEEE, New York, 1990), p. 70.

19. J. Y. Pan and S. D. Senturia, Technical Digest, SPE ANTEC'91, Montreal, May 1991, in press.

20. F. Maseeh and S. D. Senturia, in Ref [3], p.575.

21. F. Maseeh, Ph.D. Thesis, Dept. of Civil Engineering, Massachusetts Institute of Technology, Cambridge, MA, 1990, (unpublished).

22. C. Bauer and R. Farris, in Ref [3], p 549.

23. M. Maden and R. Farris, Technical Digest, SPE ANTEC'91, Montreal, May 1991, in press

24. A. G. Evans and J. W. Hutchineson, Int'l. J. Solids and Structures, $\underline{20}$, 451 (1986).

25. K. L. Mittal, Electrocomponent Sci. and Tech., $\underline{3}$, 21 (1976).

26. A. J. Kinloch, J. Mat. Sci., $\underline{15}$, 2141 (1980).

27. S. Wu, *Polymer Interfaces and Adhesion*, (Marcel Dekker, 1982).

28. H. Dannenberg, J. Appl. Polymer Sci., $\underline{5}$, 125 (1961).

29. J. Hinkley, J. Adhesion, $\underline{16}$, 115 (1983).

30. M. G. Allen and S. D. Senturia, J. Adhesion, $\underline{25}$, 303 (1988).

31. M. G. Allen and S. D. Senturia, J. Adhesion, $\underline{29}$, 219 (1989).

32. M. G. Allen and S. D. Senturia, Proc. ACS Division of Polymeric Materials Science and Engineering, $\underline{56}$, 735 (1987).

33. M. G. Allen, P. Nagarkar, and S. D. Senturia, in Ref [3], p.705.

34. S. C. Noe, J. Y. Pan, and S. D. Senturia, Technical Digest, SPE ANTEC '91, Montreal, May, 1991, in press.

35. T. P.Russell, H. Gugger, and J. D. Swalen, J. Polymer Sci: Phys., $\underline{21}$, 1745 (1983).

36. N. Takahashi, D. Yoon, and W. Parrish, Macromolecules, $\underline{17}$, 2583 (1984).

37. C. A. Pryde, J. Polymer Sci.: Chem., $\underline{27}$, 711 (1989).

38. C. A. Pryde, SPE ANTEC '90, p. 439.
39. N. E. Schlotter and J. F. Rabolt, J. Phys. Chem., 88, 2062 (1984).
40. W. M. Prest, and D. J. Luca, J. Appl. Phys., 50, 6067 (1979).
41. W. M. Prest, and D. J. Luca, J. Appl. Phys., 51, 5170 (1980).
42. I. Savatinova, S. Tonchev, R. Todorov, E. Venkova, E. Liarokapis,and
    E. Anastassakis, J. Appl. Phys., 67, 2051 (1990).

THE DETERMINATION OF THE RESIDUAL STRESS STATE AND ITS EFFECT ON PEEL TESTS IN POLYIMIDE COATINGS.

RICHARD J. FARRIS, MICHELE A. MADEN, JAY GOLDFARB, KUN TONG
Polymer Science and Engineering Department, University of Massachusetts, Amherst, MA 01003.

ABSTRACT

The determination of residual stresses in thin polymer coatings is crucial in the prediction of failure due to delamination and cracking. This is especially important in the electronics industry where thin polymer dielectric layers are used. The dielectric layers are often under three-dimensional constraints and have holes through which metal interconnects run. These act as locations of high stress concentration. Three complementary methods have been developed which will provide a better base for understanding the delamination and performance of thin coatings.

These techniques include, a holographic interferometry method for determining the residual stresses in coatings (~10 μm thick); a deformation calorimetry technique for establishing an energy balance for peel tests; a set of experiments for measuring the nine elasticity coefficients for orthotropic films. The combination of these methods allows a comparison between the residual stresses present in coatings and the contribution of those residual stresses to the work needed for peeling. The knowledge of the stress state and elasticity coefficients is essential for determining the elastic energy in a coating. The total elastic energy which can be stored in a stressed coating increases with the coating thickness. Stressed coatings exceeding a critical thickness will spontaneously delaminate if the elastic energy in the coating is greater than the energy required for debonding. The elastic energy in the coating per unit area at the critical thickness is a measure of the interfacial strength.

INTRODUCTION

Elastic energy in a coating resulting from residual stresses will decrease the effective adhesion of the coating to its substrate.[1, 2] The peel test is a simple mechanical test which is extensively used to measure the adhesion of flexible films to substrates. In the absence of energy dissipation due to plasticity or viscoelasticity, the energy required to peel the film from its substrate is a direct measure of the adhesion. However, for thin films which are strongly bonded to rigid substrates, energy dissipation due to plastic deformation occurs and the peel energy will significantly exceed the true adhesion. Most of the energy expended in peeling is consumed by dissipative processes.[3, 4] The effect of elastic energy in a coating is critical to the adhesion and performance of the coating. However, when the peel energy contains a large dissipative contribution, the elastic energy in the film will be small compared to the peel energy. Thus, the detrimental effect of residual stress upon adhesion will not be detectable by peel testing since its magnitude may be smaller than the precision of the peel energy measurements. While the elastic energy may be small compared to the peel energy, it is significant when compared to the true adhesion and will result in spontaneous delamination of films which exceed a critical thickness or films which are subject to local three dimensional constraints imposed by device geometries. Spontaneous delamination requires less energy than peeling because it occurs with less energy dissipation.

The stresses in thin films and the nine orthotropic elasticity coefficients have been accurately measured using a holographic interferometry technique and a deformation calorimeter was used to measure the work and heat of peeling. This paper describes how knowledge of the mechanical state of a coating may be used in combination with knowledge of how the energy expended in peeling is consumed to better understand the adhesion and performance of thin coatings.

THEORY

The holographic technique makes use of classical vibration theory in an innovative way to determine the biaxial stresses in coatings [5]. This technique applies measured membrane vibration behavior to the solution of an engineering stress measurement problem. This method for determining the state of stress in a membrane begins with the most fundamental equation of motion. The vibration of a circular membrane is described by the differential equation:

$$\frac{\partial^2 u}{\partial r^2} + \frac{1}{r}\frac{\partial u}{\partial r} + \frac{1}{r^2}\frac{\partial^2 u}{\partial \theta 2} = k^2 \frac{\partial^2 u}{\partial t^2} \tag{1}$$

where,

$u(r, \theta, t)$ = out of plane deflection (m)
θ = tangential cylindrical coordinate
r = radial coordinate(m)
k^2 = ρ/σ
ρ = mass per unit volume (kg/m^3)
σ = biaxial stress (N/m^2)
t = time (seconds)

Applying the boundary condition for a membrane fixed at the outer edge, $u(r = R) = 0$ (where R is the outer radius), the solution to this equation is obtained in terms of Bessel functions of the first kind, $J_n(\omega_{ni}kr)$ which determine the vibrational modes of a circular membrane. The derivation of this analysis has been presented in a previous paper [6]. It is known that the stress is related to the resonant frequencies and the zeros of the Bessel functions as follows:

$$\sigma = \rho(R\omega_{ni}/Z_{ni})^2 \tag{2}$$

where Z_{ni} is the argument of the Bessel function $J_n = 0$.

The order, n, and the zero, i, of the Bessel function are determined from the vibration patterns observed at resonant frequencies. The patterns are seen using real-time holographic interferometry when the membrane is excited sinusoidally at one of its resonant frequencies. Vibrations of the zeroth order have no nodes (regions of zero displacement) and produce a circular vibration pattern. Vibrations of the first order have one node of vibration, etc..

Energy Balance for Peeling

Residual tensile stresses in adhesive layers reduce their effective adhesion [1]. Residual stress imparts elastic energy to the bonded adhesive layer which is released upon delamination. Thus, the elastic energy reduces the external work required to remove an adhesive layer from a substrate. The total elastic energy in the adhesive layer is

$$U_{ELASTIC} = \frac{1}{2} V_0 (\sigma_{ij} e_{ij}) \qquad (3)$$

Where σ_{ij} is the stress tensor, e_{ij} is the Cauchy strain tensor, V_0 is the volume of the adhesive layer, and $i, j = 1, 2, 3$. When a coating is applied to a substrate in a liquid state and solidified, the resulting film is isotropic and under equilateral biaxial tension. There are two non-zero stress components; $\sigma_{xx} = \sigma_{yy} = \sigma$. For a coating of thickness t, the elastic energy per unit area is

$$\Delta U_{ELASTIC} = \frac{t\sigma^2}{E} (1-\upsilon) \qquad (4)$$

Where E is the Young's modulus, and υ is Poisson's ratio. The work done in separating adhesively bonded layers, ΔW, can be equated to the difference between the internal energy change, ΔU, of the structure and the heat flowing from the structure, ΔQ.

$$\Delta W = \Delta U - \Delta Q \qquad (5)$$

Neglecting the breaking and formation of chemical bonds, the internal energy change of a body undergoing fracture can be partitioned as

$$\Delta U = \Delta U_{SURFACE} + \Delta U_{ELASTIC} + \Delta U_{STORED} \qquad (6)$$

The change in surface energy is equivalent to the thermodynamic work of adhesion, W_A, which is the energy required to separate the coating and substrate reversibly.

$$\Delta U_{SURFACE} = W_A \qquad (7)$$

If the peeled film is unloaded after peeling, the change in elastic energy, $\Delta U_{ELASTIC}$, is equal to the stored strain energy in the bonded adhesive layer, which is released upon delamination. The total energy consumed in separating the bonded layers is the sum of the work done by the external force of the testing instrument and the elastic energy in the adhesive layer. Some of the work expended in plastic deformation is dissipated as heat. Materials which undergo ideal plastic deformation dissipate all of the energy consumed by plastic deformation as heat. Real materials, which undergo physical changes during plastic deformation may store a large fraction of this energy, ΔU_{STORED} and only dissipate part of it.

The energy balance per unit area for separating an adhesive layer in equilateral biaxial tension from a rigid substrate is

$$\Delta W = \frac{t\sigma^2}{E} (1-\upsilon) + \Delta U_{SURFACE} + \Delta U_{STORED} - \Delta Q \qquad (8)$$

Where ΔW, ΔU and ΔQ are the work, internal energy change and heat dissipated per unit area peeled.

Constitutive Behavior of Orthotropic Films

During the processing of free-standing thin films, different material properties are developed in different directions in the film. For example, in making some commercial polymer films, the film is held on each side by tenterhooks and is held taut as the film is pulled from the front, orienting the film to different degrees across its transverse direction. This causes the film to have different properties in orthogonal directions. The out-of-plane properties will be quite different from those in the in-plane transverse and machine directions.

The principal directions of stress can be found using the real time holographic interferometry technique. It has been observed [6] that for an anisotropic material certain modes of vibration will appear at two distinct frequencies with their nodal line(s) at 90° to each other. The directions parallel to the nodes of vibration indicate the principal directions of stress in the film. A square washer mounted with its sides parallel to the principal directions allows the magnitudes of the principal stresses to be determined. Once the principal directions in the plane of the film are known, all nine elasticity coefficients may be determined using standard tensile tests and other methods developed in our labs. The in-plane Young's moduli, E_{11} and E_{22} are measured by doing tensile tests on samples cut in the principal 1 and 2 directions. G_{12} is determined by doing a tensile test on a sample cut at 45° to the principal directions, which gives E_{xx}, and then using a transformation equation (9) to solve for G_{12}.

$$\frac{1}{E_{xx}} = \frac{\cos^4\theta}{E_{11}} + \frac{\sin^4\theta}{E_{22}} + \frac{1}{4}\left[\frac{1}{G_{12}} - \frac{2\upsilon_{12}}{E_{11}}\right]\sin^2 2\theta \tag{9}$$

The Poisson's ratio in two different in-plane directions can be determined from the square membrane samples. A narrow ribbon is cut in each of the principal directions from two square membranes of known principal stress from the same sheet of film. The one dimensional stress is measured in each direction using holographic interferometry to observe the vibrations, and the vibrating string equation below to determine the stress in the 11 and 22 directions.

$$\sigma_{1D} = \frac{\rho\omega^2 L^2}{n^2\pi^2} \tag{10}$$

Here, σ_{1D} is the one dimensional stress, ρ is the density, ω is the angular frequency of vibration, L is the length of the ribbon, and n is the order of vibration.

For a planar anisotropic film, two independent Poisson's ratio's exist, υ_{12} and υ_{21}. They can be determined using the general form of Hooke's law [7, 8] and knowledge of one- and two- dimensional stresses. The Poisson's ratio in the out of plane (33) direction is the most difficult quantity to obtain for thin films, as it is difficult to measure the minute changes in thickness which occur as the film is stretched. It is proposed (based on work done by Bauer [9]) that dilatometric measurements on thin film samples can be used to measure this value.

A ribbon is held at constant length in the 11 direction and a hydrostatic pressure is applied. The change in stress with change in pressure is plotted, and the out of plane Poisson's ratio, υ_{13} can be determined from the slope. By making the same measurement on another sample cut in the 22 direction, υ_{23} can be similarly obtained.

In order to fully characterize the material, a torsion method must be adapted to measure the two remaining shear moduli, G_{31}, and G_{23}. Using a ribbon with a rectangular cross section, we propose to do weighted torsion experiments as described by Allen [10]. He extended the analysis for isotropic fibers to that for an orthotropic fiber using cylindrical coordinates. The analysis can be done for an orthotropic ribbon in rectangular coordinates. Knowledge of orthotropic elasticity coefficients will enable the modelling of delamination using finite element analysis.

EXPERIMENTAL

Residual Stress Measurements

The sample preparation for the holographic experiments was developed making use of the fact that far from the edges of a coating on a substrate there are no stresses that act between the substrate and the coating. Therefore, if the coating is constrained (in this case by a ring mounted with epoxy) a portion of the substrate or the entire substrate may be removed without affecting the original state of stress in the coating.

The substrate is steel sheet coated with a thin layer of tin. A polyamic acid solution is spin coated onto the substrate, then cured to 225 °C. The final thickness of these films is about 10 μm. Steel washers are mounted with epoxy on top of the polyimide (pyromellitic dianhydride - oxydianiline or PMDA-ODA). The sample is placed in a mercury bath which dissolves the tin layer between the polyimide and the steel. The polyimide film (its original state of stress intact) attached to the washer can be easily lifted from the plate [11].

The membranes (polyimide films on washers) are placed in a fixture rigidly mounted inside a vacuum chamber. The entire chamber is connected to a piezoelectric transducer driven by a frequency generator. An image of the static object is recorded on a thermoplastic holographic plate using a holographic camera. Using the frequency generator, the frequency of vibration of the vacuum chamber and sample is increased steadily until a resonant frequency of the membrane is reached. This point is evident from the appearance of a vibration pattern superimposed on the object. The pattern is observed through a video camera connected to a monitor. The resonant frequency, along with the density and the radius of the membrane are then used to calculate the residual biaxial stress in the film.

Orthotropic Elasticity Coefficients

The membrane samples for the orthotropic measurements are made by mounting a washer to a free-standing film(Upilex R by Ube Corp., Kapton 30H film by DuPont or oriented poly(vinyl alcohol)) with epoxy and curing at 200°C (65°C for the poly(vinyl alcohol samples). For orthotropic films, differences in coefficients of thermal expansion in orthogonal directions, cause the film to self load in a state of unequal biaxial stress.

The holographic measurements done on orthotropic films are similar to those described above for spin coated films. The principal directions are determined by observing the nodes of the vibration patterns using holographic interferometry. A circular sample is used to determine the principal directions. Then a square sample is made with its edges parallel to the principal directions to determine the two principal stresses. This is done by mounting a second washer with a square cutout in the middle, inside the circular membrane sample. Using the same square samples, a ribbon is cut in each of the principal directions for measurement of the Poisson's ratio. Using the original freestanding film sample, and knowledge of the principal directions, samples can be cut in the principal directions for modulus and out of plane Poisson's ratio measurements.

RESULTS

Kendall observed a decrease in the peel strength of a rubber film, stretched and bonded to glass, equal to the elastic strain energy in the film [2]. In this case, the peel energy is comparable in magnitude to the stored elastic energy. We have investigated the peeling of polyimide films spin coated onto chromic acid etched aluminum substrates and thermally cured at 360°C. The estimated elastic energy in the film per unit area is

$$U_{ELASTIC} = \frac{tE(\Delta\alpha\Delta T)^2}{1-\upsilon} \qquad (11)$$

where $\Delta\alpha$ is the difference in the thermal expansion coefficients of the coating and the substrate and ΔT is the temperature change. The elastic energy in a film can be related to a biaxial stress in the film using equation (4). We have found that the residual stresses in spin coated polyimide films can be measured to a high degree of precision using the holographic technique. For samples cured to 400°C we measured approximately thirty resonant frequencies all of which yielded redundant measures of stress of 25 +/-0.5 MPa.

At 25 °C, the estimated thermal stress in a polyimide film cured at 360 °C on aluminum is 21 (MPa). The estimated stress is in good agreement with the value measured in the film cured to 400 °C. The thermal expansion coefficients of aluminum and brass are similar, 20×10^{-6} (°C^{-1}), and the thermal expansion coefficient of the spin coated polyimide is 40×10^{-6} (°C^{-1}). While the difference in expansion coefficients is relatively small, the thermal stress buildup during the cool-down of polyimide films cured at elevated temperatures is substantial because the polymer does not exhibit a glass transition and thus has a high modulus, 3 (GPa), throughout the temperature range of post curing cool-down. The strain energy in a 63 μm film corresponding to the estimated stress values is 12 (J/m^2). The measured 180° peel energy of a 63 μm polyimide film from aluminum at a peel rate of 2.71 (cm/min) is 660 (J/m^2). The elastic energy stored in the coating as residual stress therefore represents a small fraction of the peel energy.

Figure 1 shows the work, heat and internal energy change of peeling polyimide films of varying thickness from aluminum. The decrease in work with increasing film thickness, displayed in figure 1, is an order of magnitude larger than the variation in internal elastic energy with film thickness. Most of the energy expended in peeling the polymer film goes into plastic deformation. Approximately one half of the work of peeling is dissipated as heat. The internal energy change is the result of stored energy in the peeled polymer [12]. Thus, almost all of the peel energy is consumed by plastic deformation. The stored elastic energy is small compared to the total peel energy. However, the stored elastic energy is large compared to the energy required to reversibly create the surfaces exposed by delamination.

Spontaneous delamination was observed for coatings thicker than 120 µm. The strain energy in these films is estimated to be 23 J/m². Spontaneous delamination requires less energy than peeling because there is much less plastic deformation than with peeling. The effect of film thickness on peel energy is dominated by plastic deformation. The greater rigidity of thicker films leads to a larger bend radius and lower strains near the point of detachment in peeling, thus, reducing plastic deformation.

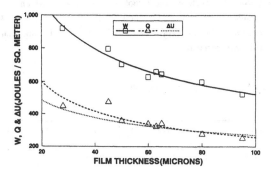

THE THERMODYNAMIC WORK, (W), HEAT, (Q), AND INTERNAL ENERGY CHANGE,(ΔU), WERE MEASURED DURING 180° PEELING IN THE DEFORMATION CALORIMETER. THE PEEL RATE WAS 2.71(cm/sec).

Figure 1. Work and heat vs. film thickness for a polyimide film peeled from aluminum sheet.

The results from measurements made on orthotropic polyvinyl alcohol (draw ratio 3.6:1), Upilex R (TM Ube, Corp.) and Kapton (TM DuPont) material are shown below in Table 1. The Young's moduli measured in our lab on Upilex R (12.5 µm thick) are 20% higher than those reported in the manufacturer's literature (measured in the machine direction on 25 µm thick films). There are several possible reasons for this, a.) our measurements were made in the principal directions (which are not necessarily the machine and transverse directions; b.) The films we used for our measurements were half as thick as the manufacturer's samples (possibly causing a higher orientation in the plane).

The Kapton and Upilex films do not exhibit a high degree of anisotropy in tensile properties, however significant differences in Poisson's ratios in the plane are seen. Commercially made films such as Kapton and Upilex possess varying degrees of orientation across the width of the film as it comes off the production line. The widths of these films may be as much as 1.5 m in the transverse direction. Any variation in the location from which the film samples were taken will affect the angle of orientation slightly. The three sets of Upilex data reported are from measurements done on three separate Upilex films. As the position in the film line from which they were taken is unknown, this must be considered as a possible source of error.

The orientation in the polyvinyl alcohol films is much higher, however, a corresponding degree of anisotropy of properties is not observed. The differences between the two polyvinyl alcohol samples may be in part because the washers used for mounting were of different materials with different thermal expansion properties. Conformation of the angles of the orthotropic axes will be done using a DuPont thermo-mechanical analysis instrument to measure the coefficients of thermal expansion of the films at 30° intervals with respect to the machine and transverse directions of the films.

A check for symmetry of the compliance matrix is used to confirm that a material is indeed orthotropic, so that its properties are fully described by nine elasticity coefficients. The off diagonal elements for this matrix for each sample is listed under table 1. The polyvinyl alcohol measurements seem to indicate that these materials are not perfectly orthotropic based on the wide difference in these elements. Further material testing will be done with these materials to explore this hypothesis. Constants which remain to be determined are out of plane Poisson's ratios (υ_{23} and υ_{13}), the Young's modulus (E_{33}) and the out of plane shear moduli (G_{23} and G_{13}), which will be measured using dilatometry and torsion pendulum experiments respectively.

Table 1.   Orthotropic Material Property Measurements

Sample	Angle	E_1 (GPa)	E_2 (GPa)	υ_{12}	υ_{21}	G_{12} (GPa)
PVOH[a]	35°	4.8	4.2	0.42	0.63	3.5
PVOH[b]	25°	6.0	4.4	0.26	0.58	1.6
Upilex R[c]	22°	4.0	4.2	0.40	0.63	1.6
Upilex R[d]	11°	4.0	3.8	0.32	0.25	1.1
Upilex R[e]	20°	4.0	3.9	0.36	0.23	1.7
Kapton 30H[f]	20°	3.8	4.5	0.39	0.52	1.4

a: Samples mounted on aluminum washers; $-\upsilon_{12}/E_{11} = 0.09$; $-\upsilon_{21}/E_{22} = 0.15$
b: Samples mounted on brass washers; $-\upsilon_{12}/E_{11} = 0.04$; $-\upsilon_{21}/E_{22} = 0.13$
c: $-\upsilon_{12}/E_{11} = 0.10$; $-\upsilon_{21}/E_{22} = 0.15$
d: $-\upsilon_{12}/E_{11} = 0.08$; $-\upsilon_{21}/E_{22} = 0.07$
e: $-\upsilon_{12}/E_{11} = 0.09$; $-\upsilon_{21}/E_{22} = 0.06$
f: $-\upsilon_{12}/E_{11} = 0.10$; $-\upsilon_{21}/E_{22} = 0.11$

CONCLUSION

We have shown that using real time holographic interferometry we can fully characterize the state of residual stress in a film. Using environmental chambers, the effects of humidity and temperature can also be analyzed [6]. Using the holographic technique to determine the principal stress directions also allows the determination of the orthotropic material properties of thin films.

The residual stress measurement technique is an important component for determining the energy balance for peeling. A tremendous amount of plastic deformation is done when peeling a thin coating, well bonded to a rigid substrate. The magnitude of the stored elastic energy of the residual stress is small compared to the total peel energy. However, the magnitude of residual shrinkage stresses in polymeric coatings is significant compared to the adhesion between the coating and substrate and can lead to spontaneous delamination. Spontaneous delamination occurs with little plastic deformation compared to peeling and, thus, the work of adhesion determined from delamination methods is a good measure of the adhesion strength.

REFERENCES

1. S. G. Croll, J. Coat. Tech., 52 (1980).
2. K. Kendall, J. Phys. D: Appl. Phys., 6, 1782 (1973).
3. K. S. Kim, J. Eng. Mat. & Tech., 110, 266 (1988).
4. A. N. Gent and G. R. Hamed, Polymer Eng. Sci., 17, 462 (1977).
5. M. A. Maden, and R. J. Farris in Electronic and Packaging Materials Science, (Mater. Res. Soc. Proc. ,Pittsburgh, PA 1989).
6. M. A. Maden, K. Tong, and R. J. Farris in Thin Films: Stresses and Mechanical Properties II, (Mater. Res. Soc. Proc.,Pittsburgh, PA 1990)
7. L. E. Malvern, Introduction to the Mechanics of a Continuous Medium, Chapter 6, (Prentice Hall, Inc., New Jersey, 1969).
8. S. W. Tsai, and H. T. Hahn, Introduction to Composite Materials, (Technomic Publishing Co., Connecticut, 1980).
9. C. L. Bauer, PhD thesis, University of Massachusetts, 1987.
10. S. R. Allen, Polymer, 29(6), 1091, (1988).
11. Yaseen,M., and Ashton, H.E., J. Coat. Tech., 49(629), 50.
12. J. L. Goldfarb and R. J. Farris, "Calorimetric Measurements of the Heat Generated by the Peel Adhesion Test", *to be published*, (1991).

INTERNAL STRESS DEVELOPMENT IN SPIN COATED POLYIMIDE FILMS

MICHAEL T. POTTIGER AND JOHN COBURN
Du Pont Electronics, P. O. Box 80336, Wilmington, DE 19880-0336

ABSTRACT

The effect of processing on the development of internal stresses in spin coated polyimide films was investigated. The internal stresses are a result of the coefficient of thermal expansion (CTE) mismatch between the polymer and the substrate. Birefringence and CTE were used to characterize the in-plane molecular orientation. In-plane orientation was shown to be sensitive to processing conditions. Increasing the spin speed results in higher in-plane orientation as observed by an increase in birefringence and a corresponding decrease in CTE. Heating rate during cure was observed to have a significant effect on in-plane orientation. Faster heating rates during cure resulted in a lower birefringence. The lower birefringence is attributed to relaxation effects that can occur during a rapid cure. The decrease in orientation was accompanied by an increase in internal stress.

INTRODUCTION

Residual stresses in solvent based polymer coatings arise from differences between the in-plane expansion or shrinkage of the thin polymer film and the substrate. The factors that control these residual stresses include: the in-plane coefficient of thermal expansion (CTE) of the polymer film, polymer film morphology such as crystallinity, and exposure to absorbants (e.g., moisture). The in-plane CTE of the polymer film is determined primarily by the degree of in-plane molecular orientation induced during processing. Internal stresses developed as a result of film shrinkage lead to higher in-plane orientation. The increase in the in-plane orientation during processing is partially countered by molecular relaxation effects. The extent to which relaxation can occur is influenced by the processing conditions. The degree of in-plane orientation in the final film is therefore a function of process history.

The use of polyimide films by the microelectronics industry as interlayer dielectrics and stress buffers in both semiconductor and packaging applications is growing. Thin polyimide films are generally prepared by spin or spray coating a polyamic acid (PAA) solution onto a substrate. The coated film is subjected to a variety of heat treatments to both remove the solvent from the coating and complete the conversion of the PAA into the polyimide. Both of these processes result in volumetric film shrinkage. Internal stresses can develop in the film if the shrinkage occurs after the film has reached a critical point where it can no longer relax. Failure of the coated structure may occur if the residual stresses exceed either the adhesion strength between the layers or the cohesive strength of the materials.

A number of investigators have studied the effects of processing on internal stress development in pyromellitic dianhydride-oxydianiline (PMDA/ODA) films prepared from polyamic acids dissolved in N-methyl-pyrollidone (NMP). Russell et al. [1] compared chemical versus thermal conversion processes and found that chemically converted films had higher in-plane orientation. Geldermans et al. [2] measured the in-plane stresses generated during cure and found that no intrinsic stress was generated during the conversion process. Residual stresses in the fully cured film resulted from the CTE mismatch between the polyimide and the substrate during cooling. Elsner [3] found that the state of stress in fully cured spin coated films was independent of the soft-bake conditions and nearly independent of thickness. Han et al. [4] investigated the effect of heating rate during cure and found that the state of stress in fully cured spin coated films increased with increasing heating rate. Bauer and Farris [5] investigated the development of stresses in polyamic acids during solvent removal. Films were prepared by dipping rubber strips into a PAA solution. During heating, stresses did not develop in the film until a critical concentration was reached, at which point the material gelled. Above the critical concentration, further solvent evaporation resulted in film shrinkage and stresses developed in the plane. The internal stresses were found not to be a function of the initial solution concentration, but the concentration when the material could first sustain a load. In this paper, a physical explanation in terms of molecular orientation is given for the observed effects that processing has on the in-plane CTE and in turn the internal in-plane stress.

Shrinkage occurs during both solvent evaporation and the conversion of the PAA into the polyimide. A spin coated polyimide film is constrained in the plane by the substrate, the bulk of the shrinkage therefore occurs in the thickness direction. The residual stresses that develop during this gel film collapse induce in-plane molecular orientation resulting in anisotropy in the physical and mechanical properties in the film [6]. The amount of orientation that is developed is influenced by a number of factors including: polyimide chain stiffness, the extent of conversion prior to solvent removal, polymer/solvent and polymer/polymer interactions, and stresses introduced during the coating process. Since the in-plane CTE and modulus of the polyimide are controlled by the degree of in-plane orientation, the stress level that develops in the film as the polyimide is cooled to room temperature is governed by the degree of in-plane molecular orientation.

EXPERIMENTAL

Film Preparation

PMDA/ODA polyamic acid solutions consisting of 15.0, 13.6 and 12.5 weight percent solids dissolved in N-methyl-2-pyrrolidone (NMP) were used in this study. The lower solids solutions were obtained by diluting the 15% solids solution with NMP. Films were prepared by spin coating the PAA solutions onto 125 mm diameter silicon wafers coated with a 1500Å thermally grown oxide layer. To study the effect of spin speed on in-plane orientation, films were spin coated at different speeds ranging from 1000 to 5000 rpm for 30 seconds. These films were soft-baked at 135 °C and fully cured in nitrogen using a standard curing process of heating the wafer at 2 °C/minute to 200 °C, holding at 200 °C for 30 minutes, heating at 2 °C/minute to 350 °C and holding at 350 °C for an additional 30 minutes.

To study the effect of soft-bake and cure conditions on the development of in-plane orientation and stress, films were spin coated at 3000 rpm for 30 seconds. The coated films were soft-baked either in a convention oven for 30 minutes at 55, 120 or 150 °C in air, or on a hot plate for 6 minutes at 60, 90 or 120 °C. These films were then cured using one of two heating methods as shown in Table 1. Samples cured by the slow cure method were heated at 2 °C/minute from room temperature to 350 °C and held at 350 °C for one hour. The samples cured by the rapid cure method were placed in an oven preheated to 350°C and held at 350 °C for one hour.

Table 1. Cure Conditions.

	Heating Rate	Final Cure
Slow Cure	Heat from room temperature at 2°C/min to 350 °C in nitrogen	Hold for 60 min. at 350 °C in nitrogen
Rapid Cure	Place soft-baked wafer in oven preheated to 350 °C	Hold for 60 min. at 350 °C in nitrogen

Orientation Measurements

Optical techniques are valuable for characterizing in-plane orientation. An increase in birefringence, the difference between the in-plane and the vertical refractive indices, is indicative of higher in-plane molecular orientation. The refractive indices were measured using an Abbe refractometer. The measurements were made using a sodium D lamp in transmission. Under cross polarizers, no preferred in-plane orientation was observed in any of the films, i.e. the films were completely isotropic in the plane. Therefore, the in-plane refractive index is constant, independent of in-plane direction. The in-plane refractive index was measured using a polarizer on the eyepiece with light polarized in the horizontal in-plane direction. The vertical refractive index was obtained by rotating the polarizer on the eyepiece 90° to polarize light vertical to the in-plane direction.

Coefficient of Thermal Expansion Measurements

The in-plane CTE was measured on fully cured film using a Du Pont 2940 Thermal Mechanical Analyzer. The film was heated at 10 °C/min under a constant load of 0.050 N. The CTE was determined from the slope of the linear expansion versus temperature curve between 100 and 250 °C.

Wafer Deformation Measurements

The wafer deformation was determined by measuring the radius of curvature of the wafer using a Flexus Thin Film Stress Measuring Apparatus (TFSMA) Model 2-300 manufactured by Flexus, Inc. of Sunnyvale, California. An "effective" radius of curvature, R, was calculated as follows:

$$R = (R_1 R_2)/(R_1 - R_2) \tag{1}$$

where R_1 and R_2 are the radius of curvature before and after film deposition. The relationship between wafer deformation and the "effective" radius of curvature, as shown in Figure 1, is:

$$h = R - \sqrt{R^2 - L^2/4} \tag{2}$$

where L is the wafer diameter, h is the height of the deformation of the wafer, and R is the radius of curvature.

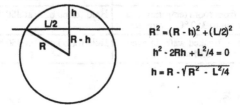

$$R^2 = (R - h)^2 + (L/2)^2$$

$$h^2 - 2Rh + L^2/4 = 0$$

$$h = R - \sqrt{R^2 - L^2/4}$$

Figure 1. Relationship between radius of curvature, R, and wafer deformation, h.

RESULTS

Orientation Measurements

The birefringence of the fully cured films obtained using the slow cure is shown in Figure 2 as a function of spin speed. The birefringence increases with spin speed. The increase in birefringence indicates that the in-plane orientation in the fully cured film is influenced by the spin speed used to apply the PAA solution. The increase in birefringence is accompanied by a decrease in the in-plane CTE as shown in Figure 3. The CTE varied from 22 to 32 ppm, while the birefringence varies from 0.07 to 0.09.

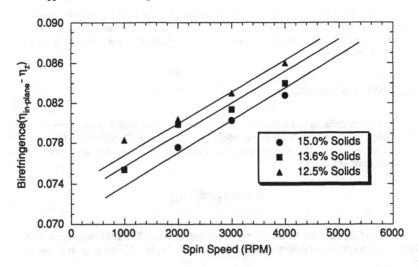

Figure 2. The effect of spin speed on in-plane orientation.

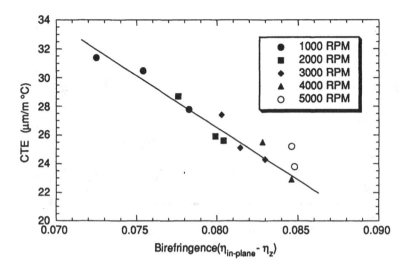

Figure 3. The relationship between CTE and in-plane orientation.

The in-plane orientation is insensitive to the soft-bake conditions used in this study (Figure 4). The heating rate during cure exhibits a much greater effect on the in-plane orientation than the soft-bake conditions. The birefringence of films cured by heating at 2 °C/minute from room temperature to 350 °C is higher than the birefringence of the films cured by being placed directly into an oven preheated to 350 °C. The birefringence was lower for all films that were rapidly heated to 350°C, with the exception of the film soft-baked at 150 °C. The difference in the films soft-baked at 150 °C and fast cured is attributed to the onset of imidization during soft-bake. A higher imide level was observed by IR measurements on the films soft-baked at 150 °C.

Wafer Deformation Measurements

The wafer deformation versus film thickness for the slow and rapid cures are shown in Figures 5 and 6 respectively. The stress level in the film is insensitive to both the initial concentration of the PAA solutions and the soft-bake conditions. The wafer deformation is about 60% higher for the films cured using the rapid cure compared to the films cured using the slow cure. Assuming that the majority of the stress in the films is a result of the CTE mismatch between the film and the substrate, the films cured using the slow cure would be expected to have lower CTE's. This is consistent with the orientation measurements.

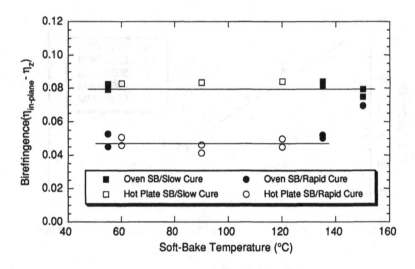

Figure 4. The effect of soft-bake and curing conditions on in-plane orientation.

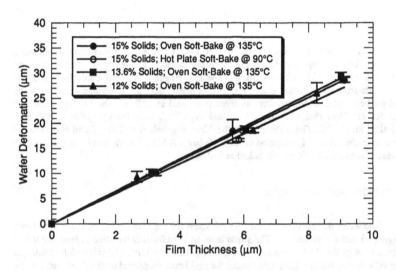

Figure 5. Wafer deformation for polyimides cured at 2 °C/minute.

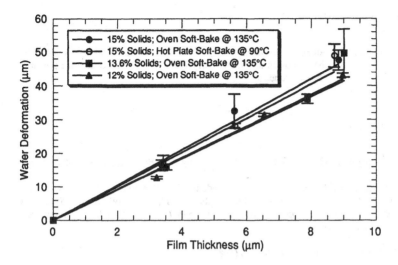

Figure 6. Wafer deformation for polyimides rapidly cured.

DISCUSSION

The sensitivity of in-plane orientation to heating rate during cure is attributed to the effect of heating rate on the "effective glass transition" of the film [2]. The ability of the partially converted polyamic acid to relax depends on the solvent concentration, the extent of conversion and the temperature. For films cured at slow heating rates, the effective glass transition temperature of the film rises faster than the film temperature. For films rapidly heated to 350 °C, the effective glass transition temperature is below the film temperature during much of the curing process. Since the film temperature is above the glass transition temperature during most of the cure, molecular relaxation can occur, resulting in a loss of in-plane orientation. Less in-plane orientation results in a higher in-plane CTE. The CTE of the PMDA/ODA polyimide is significantly higher than the CTE of silicon, which is approximately 3 ppm. A higher PMDA/ODA CTE results in a larger CTE mismatch between the polyimide and the substrate. The larger mismatch is expected to lead to higher in-plane stress as was observed in these studies.

Although the heating rate has an effect on the measured birefringence, the heating rate may also affect the film in other ways. First, the conversion reaction is complex. The formation of amic acid from the dianhydride and diamine is a reversible reaction. The relative rate for amic acid formation compared to the rate of the reverse reaction is much greater at room temperatures. At temperatures around 100 to 200 °C, the rates for the reverse reaction is significant [7]. At still higher temperatures, the rate of formation of the imide from the amic acid increases, favoring the forward reaction of amic acid formation. Second, PMDA-ODA is known to crystallize under certain processing conditions [8]. The impact that crystallinity may have not only on the film stress, but on other film properties such as toughness, makes further studies on the effect of these cure conditions on film morphology worth pursuing.

194

ACKNOWLEDGEMENTS

The authors would like to thank the help of Richard Farris, Anand Jagota, and Steve Mazur for numerous helpful discussions, and Larry Watson and Doris Oranzi for preparing samples and running the experiments.

REFERENCES

1. T. P. Russell, H. Gugger, and J. D. Swalen, *J. Poly. Sci.: Poly. Phys.*, **21**, 1745 (1983).

2. P. Geldermans, C. Goldsmith, and F. Bedetti, "Measurement of Stresses Generated During Curing and in Cured Polyimide Films", in *Polyimides: Synthesis, Characterization and Applications*, Vol. 2, pp. 695-711, K. L. Mittal, ed. Plenum Press, New York (1984).

3. G. Elsner, *J. Appl. Poly. Sci.*, **34**, 815 (1987).

4. B. Han, C. Gryte, H. Tong, and C. Feger, *Proc. of the ANTEC '88 Meeting*, November, 1988.

5. C. L. Bauer and R. J. Farris, *Poly. Eng. and Sci.*, **28**(10), 688 (1988).

6. Y. Inoue and Y. Kobatake, *Kolloid-Zeitschrift*, **159**(1), 18 (1959).

7. R. A. Dine-Hart and W. W. Wright, *J. Appl. Poly. Sci.*, **11**, 609 (1967).

8. K. H. Gardner, J. R. Edman, J. E. Freida, S. C. Freilich, and L. E. Manring, *Proc. of the Interdisciplinary Symposium on Recent Advances in Polyimides and Other High Performance Polymers* sponsored by the American Chemical Society, San Diego, CA, January 22-25, 1990.

STRESS MEASUREMENT OF BPDA/PDA POLYIMIDES FROM POLY(AMIC ACID) AND
PHOTOSENSITIVE POLYIMIDE PRECURSORS ON SILICON

HIDESHI NOMURA, MASUICHI EGUCHI, KATSUHIRO NIWA, AND MASAYA ASANO
Electronic and Imaging Materials Research Laboratories, Toray Industries,
Inc., 2-1 Sonoyama 3-chome, Otsu, Shiga 520, Japan

ABSTRACT

The stresses in three kinds of BPDA/PDA polyimides made from a poly
(amic acid) and two types of photosensitive polyimide precursors on silicon
were measured in situ in curing and cooling cycle. The photosensitive
polyimide precursors contained, respectively, the acid-base linkage and
ester linkage.
The stresses in the polyimides from the photosensitive polyimide
precursors after the thermal cycle were larger than the stress in the
polyimide from the poly(amic acid). Since the photosensitive groups in the
precursors acted as plasticizers, the thermal expansion coefficients became
larger with the degrees of in-plane orientation of molecular chains. Thus,
these thermal stresses were larger.
The stress in the polyimide from the precursor containing the ester
linkage after the thermal cycle was larger than that in the polyimide from
the precursor containing the acid-base linkage. Since the volatile product
evaporated at 400 °C, the intrinsic stress occured in the polyimide from the
precursor containing the ester linkage. Thus, the stress after the thermal
cycle became the sum of the thermal stress and intrinsic stress. On the
other hand, the stress in the polyimide from the precursor containing the
acid-base linkage, as similar to that in the polyimide from the poly(amic
acid), was equal to the thermal stress.

INTRODUCTION

Polyimides are becoming more important for the microelectronics
industry, because of their attractive properties such as high thermal and
chemical stability, high mechanical strength, low dielectric constant and
easy film fabrication [1]. There is a primary area of application for
polyimides in the form of thin films as interlayer dielectrics in high
density interconnects used for multi-chip modules [2, 3]. In the
interconnect having a multilayer wiring structure, the polyimide film is
incorporated together with the other two basic components of the inorganic
substrate such as glass or ceramic and metallic conductor such as copper or
aluminum. Since there is a mismatch in the thermal and mechanical
properties between the polyimide and other materials, the use of the
polyimide in this area encounters the mechanical problem. The mismatch in
the thermal expansion coefficient (TEC) generates a thermal stress in the
multilayer wiring structure because the heating process is requiered for
the polyimide film fabrication [4, 5]. The stress can lead to film
cracking or interlayer delamination and can induce defect formation in the
substrate. In order to minimize the stress, it is desired that TEC of
polyimide matches that of the inorganic substrate. Previously, it was
reported that the polyimides with rod-like molecular skeletons had the TECs
as low as those of general inorganic materials [6].
On the other hand, the recent high density interconnection technology
has involved the use of photosensitive polyimide precursors which can be
patterned directly [3]. In this technology the number of processing steps

are reduced and use of harmful reagents like hydrazine hydrate can be avoided by eliminating the need for photoresists. Thus, the technology results in increasing throughput and decreasing cost for manufacturing.

In this study in situ stress measurements for three types of polyimide films deposited on silicon substrates were performed during curing and cooling cycle. Though these polyimide films were composed of the same rod-like molecular skeleton, they were formed from the different polyimide precursors which were the common poly(amic acid), and two kinds of photosensitive polyimide precursors containing the acid-base linkage [7] or ester linkage [8, 9]. After the stress measurements, the TECs of the polyimide films were measured. Furthermore, using the thermogravimetric (TG) analysis and mass-spectroscopy, the cure behaviors of the precursors were investigated. We shall show what effects the photosensitive groups in the precursors have on the thermal and mechanical properties of the polyimide films.

EXPERIMENTAL

Materials
The polyimide with rod-like molecular skeleton used in this study was based on the 3,3', 4,4'-biphenyltetracarboxylic dianhydride (BPDA) and p-phenylene diamine (PDA). The polyimide films were formed from three kinds of polyimide precursors which were the poly(amic acid), poly(amic acid) complexed with dimethylaminoethyl methacrylate (DMM) by acid-base linkage and poly(amic acid) combined with 2-hydroxyethyl methacrylate(HEMA) by ester linkage. The chemical structures of these precursors are shown in Fig. 1.

The BPDA/PDA precursor in the state of solution was prepared by mixing BPDA (1 mol) and PDA (1 mol) in N-methyl-2-pyrrolidone (NMP) at room temperature. By adding DMM (2 mol) and a small amount of photoinitiators and sensitizers to this solution at room temperature, the photosensitive BPDA/PDA/DMM precursor solution was obtained [7]. The photosensitive BPDA/PDA/HEMA precursor solution was synthesized as follows [8]. BPDA (1 mol) was converted into the corresponding diester in NMP at 50 °C by adding HEMA (2 mol) with triethylamine and hydroquinone. Next, the reaction with thionyl chloride (2 mol) to form the acid chloride was carried out at 0 °C using pyridine. Then, PDA (1 mol) was added at the same temperature. After precipitation with water, washing by ethanol and drying in vacuum, BPDA/PDA/HEMA was dissolved in NMP with a small amount of photoinitiators and sensitizers.

The substrate was a (1 0 0)-oriented silicon wafer 100 mm in diameter and 525±15μm in thickness. In order to get proper adhesion between the silicon wafer and polyimide film, a dilute solution of an aminosilane was spin coated onto the wafer and was heated at 200 °C for an hour in a nitrogen atmosphere. After this pretreatment the correct thickness of wafer was measured using a micrometer.

Stress measurment
In order to determine the stress in the film on the substrate during curing and cooling cycle, a thin film stress measuring apparatus (F2300, FleXus) was used. This apparatus measured the changes in the radius of curvature of the substrate caused by the presence of the stressed thin film on its surface using the reflection of a laser beam from the curved surface [10]. When the film has a uniform thickness, much less than substrate thickness, and uniform, isotropic stress in the film plane, the stress in the film can be calculated using [11],

$$\sigma_f = E_s t_s^2 / 6(1-\nu_s) t_f R, \qquad (1)$$

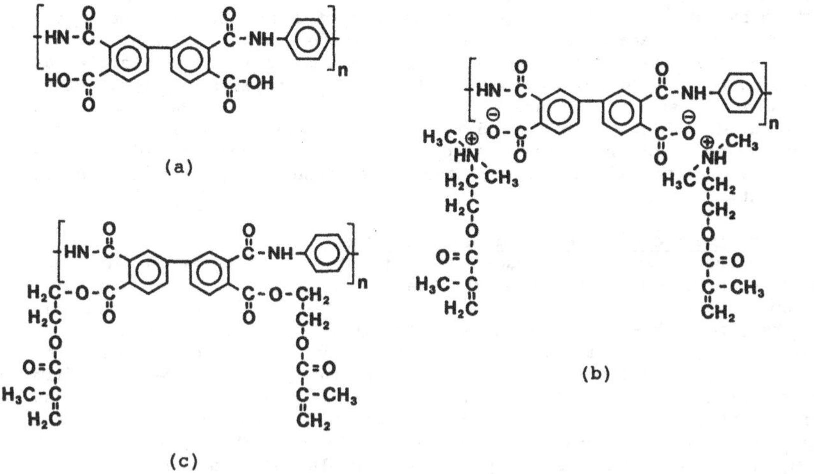

Fig. 1. Chemical structures of polyimide precursors: (a) BPDA/PDA, (b) BPDA/PDA/DMM and (c) BPDA/PDA/HEMA.

where σ_f is the stress in the film, E_s the Young's modulus of the substrate, ν_s the Poisson's ratio of the substrate, t_s and t_f the thickness of the substrate and film, respectively, and R the net change in radius of curvature of the substrate due to the film. For the (1 0 0)-oriented silicon substrate, $E_s/(1- \nu_s)$ called the biaxial elastic modulus is 1.805×10^{11} MPa [12].

The procedure of in situ stress measurement in this paper is given as follows. A polyimide precursor solution was spin coated onto the silicon wafer of which radius of curvature had been measured beforehand. The spin coating was performed with a rotation speed for the film thickness after curing to become 10 μm. By heating on a hot plate at 80 °C for 30 min, the polyimide precursor film was formed on the wafer. The photosensitive polyimide precursor was radiated by the ultraviolet light at an exposure energy of 500 mJ/cm^2 using a parallel light mask aligner (PLA-501F, Canon). The wafer deposited the polyimide precursor was placed in the stress measuring apparatus. After heating at 80 °C for 30 min in a nitrogen atmosphere to remove water absorbed in the film, the wafer in nitrogen atmosphere was heated up to 400 °C at a rate of 5 °C/min and then it was maintained at 400 °C for 30 min, and was cooled down to 25 °C. During this thermal cycle, the stress in the film was measured.

Thermal expansion coefficient measurement

For the polyimide film spin coated on the silicon wafer, it is known that there is the anisotropy of thermal expansion in the lateral and vertical directions [13]. Though the lateral thermal expansion of the film relates to the thermal stress, the vertical one does not. Thus, in this paper the only lateral TECs of the polyimide films after the thermal cycle in the stress measurement were measured using a thermomechanical analyzer (TMA8141BS, Rigaku).

The TEC measurement was performed using the following procedure. By soaking in a 3:1 mixture (by weight) of hydrofluoric acid and nitric acid, the cured polyimide film was stripped off from the silicon wafer. After washing by water, the film sample in the form of a cylinder was set in the thermomechanical analyzer, and then it was loaded with a constant weight. In order to remove absorbed water and any residual strain, the sample was heated up to 400 °C and cooled down to 25 °C in a nitrogen atmosphere. After this pretreatment the dimensional change in length of the sample was measured at a heating rate of 5 °C/min in nitrogen atmosphere. In this paper the value of the dimensional change from 25 to 400 °C divided by the temperature difference was used as the TEC because the TEC was temperature dependent.

Thermogravimetric analysis and mass-spectroscopy

For the polyimide precursor films, the behaviors of weight loss and evaporation of volatiles during the heating process were measured using a system consisting of a thermogravimeter (TGC-30, Shimadzu), gas-chromatography-mass-spectrometer (GCMS-QP1000, Shimadzu) and thermal analyzer (DT-30, Shimadzu). The formulation used to perform the measurement is given as follows. Initially, the polyimide precursor film was formed on a silicon wafer using the same procedure as that of stress measurement. The ultraviolet light exposure was carried out for the photosensitive precursors. After stripping the film off the wafer, about 20 mg of the piece of film was put on the sample pan, and then was set in the sample unit of the measurement system with 50 ml/min of helium flow. In order to evaporate water absorbed in the film, the temperature in the unit was held at 80 °C for 30 min. Then, the temperature was ramped up at rate of 5 °C/min to 450 °C. During this heating process the TG curve and mass-chromatogram were obtained.

RESULTS AND DISCUSIION

Figure 2 illustrates the stresses in three types of films on silicon wafers during curing and cooling cycle. The stress values were caluculated using the cured film thicknesses of 10μm. In the BPDA/PDA film before curing, 41 MPa of initial stress at 80 °C was induced by the volume shrinkage [5]. On the other hand, there were only 5 MPa and 0 MPa of initial stresses for the films of BPDA/PDA/DMM and BPDA/PDA/HEMA, respectively. It is thought this difference is caused by the fact that the photosensitive groups in the films act as plasticizers.

Owing to the thermal expansion and decrease of elastic modulus, the stress in the BPDA/PDA film decreased with increasing temperature and became 0 MPa at about 200 °C as shown in Fig. 2(a). It is thought that the film transformed into the rubber-like state at this temperature because the stress remained about 0 MPa from 200 to 400 °C. On the other hand, the BPDA/PDA/DMM film transformed into the rubber-like state at about 120 °C, and the BPDA/PDA/HEMA film had already transformed into the rubber-like state at 80 °C.

As shown in Fig. 2(c), there was a peak of stress in the film of BPDA/PDA/HEMA at about 200 °C. It indicates that a peak of evaporation amount of volatiles was at this temperature. Thus, the stress was thought to be induced by the volume shrinkage, although it was relaxed with increasing temperature. Similarly, there was a small peak of stress in the film of BPDA/PDA/DMM at about 200 °C as seen in Fig. 2(b). On the other hand, no peak of stress in the film of BPDA/PDA film was shown in Fig. 2(a).

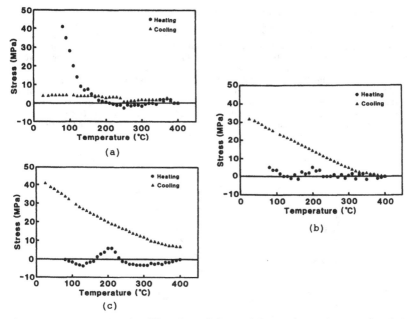

Fig. 2.   Stresses in the films formed from (a) BPDA/PDA, (b) BPDA/PDA/DMM
and (c) BPDA/PDA/HEMA on the silicon wafers during curing and cooling cycle
in a nitrogen atmosphere at a heating rate of 5 °C/min.   The film
thicknesses after curing are 10 μm.

When the temperature reached 400 °C, the stresses in the all films
became 0 MPa. However, 7 MPa of intrinsic stress occured in the film of
BPDA/PDA/HEMA after holding for 30 min at 400 °C, whereas no stress occured
in the films of BPDA/PDA and BPDA/PDA/DMM.

During the cooling process the thermal stress originated from the TEC
mismatch between the polyimide and silicon [4, 5].   The formula used to
describe the thermal stress is given as [11]

$$\sigma_t = E_f(\alpha_f - \alpha_s)(T_b - T)/(1 - \nu_f), \tag{2}$$

where σ_t is the thermal stress, E_f the Young's modulus of the film, α_f and
α_s the TEC of the film and substrate, respectively, T_b the final film
baking temperature, T the temperature at which the stress is measured, and
ν_f the Poisson's ratio of the film. The stress in the film of BPDA/PDA
during the cooling process scarcely varied, whereas the stress in the film
of BPDA/PDA/DMM and BPDA/PDA/HEMA increased monotonously with decreasing
temperature.

The relationship between the dimensional change in the film length and
temperature for three types of cured films is shown in Fig. 3.
Furthermore, the TECs are listed with the stresses in Table I. The initial
stress in the film of BPDA/PDA was much larger than that in the film of the
photosensitive precursor.   Thus, it is thought that the degree of in-plane
orientation in the molecular chain of BPDA/PDA became much larger than that

of the photosensitive precursor [14]. Since the TEC of the cured polyimide film decreased with increasing degree of in-plane orientation in the molecular chain [6], the TEC of the polyimide film formed from BPDA/PDA was much lower than that of the film formed from the photosensitive precursor. Moreover, the TEC of the film formed from BPDA/PDA was nearly equal to that of silicon. Thus, the thermal stress became very small.

On the other hand, there was a small initial stress in the film of BPDA/PDA/DMM, whereas there was not an initial stress in the film of BPDA/PDA/HEMA. Since a small degree of in-plane orientation originated in the molecular chain of BPDA/PDA/DMM. the TEC of the film formed from BPDA/PDA/DMM was lower than that of the film formed from BPDA/PDA/HEMA.

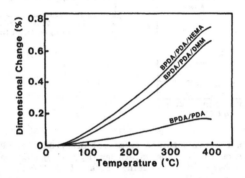

Fig. 3   Thermal expansion behaviors of the polyimide films formed from BPDA/PDA, BPDA/PDA/DMM and BPDA/PDA/HEMA at a heating rate of 5 °C/min in a nitrogen atmosphere.

Table I   Stresses on silicon substrates and thermal expansion coefficients for three types of films in a nitrogen atmosphere.

	BPDA/PDA	BPDA/PDA/DMM	BPDA/PDA/HEMA
Initial Stress (MPa) at 80 °C	41	5	0
Intrinsic Stress (MPa) at 400 °C	0	0	7
Final Stress (MPa) at 25 °C	4	32	41
Thermal Expansion Coefficient (ppm/ °C) from 25 to 400 °C	4.2	17.6	19.9

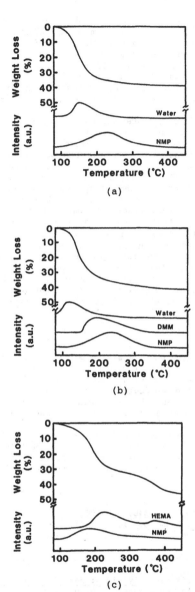

Fig. 4   Behaviors of weight loss and volatile evaporation during the heating process at a rate of 5 °C/min: (a) BPDA/PDA, (b) BPDA/PDA/DMM and (c) BPDA/PDA/HEMA.

Thus, the thermal stress in the film formed from BPDA/PDA/DMM became smaller than that in the film formed from BPDA/PDA/HEMA. Furthermore, since there was not an intrinsic stress in the polyimide film formed from BPDA/PDA/DMM, the final stress was equal to the thermal stress, as similar to the case in the film formed from BPDA/PDA. On the other hand, the final stress in the film formed from BPDA/PDA/HEMA was the sum of the intrinsic stress and thermal stress. Thus, the final stress in the film formed from BPDA/PDA/DMM became smaller than that in the film formed from BPDA/PDA/HEMA.

Figure 4 shows the TG curves and masschromatograms of the polyimide precursor films during the heating process. The weight loss of the BPDA/PDA/DMM proceeded at the lowest temperature. Furthermore, the peak temperature of water evaporation in BPDA/PDA/DMM was lower than that in BPDA/PDA. It reflects the fact that the imidization began at the lower temperature because of the DMM action as a catalyzer [15]. Here the DMM fragment was not detected at the same temperature as water evaporated because the boiling point of DMM was much higher than the temperature of the water evaporation from the precursor with imidization. The behavior of NMP evaporation in BPDA/PDA/DMM was much the same as that in BPDA/PDA.

The weight loss of BPDA/PDA/HEMA proceeded in two steps whereas that of BPDA/PDA and BPDA/PDA/DMM did in one step. It reflects the fact that the imidization of the photosensitive precursor containing the ester linkage proceeds in two steps [9]. The weight loss in the first step is thought to be related to the stress peak at about 200 °C during the curing process as shown in Fig. 2(c). Furthermore, the weight loss in the second step, which did not finish at 400 °C, is thought to cause the intrinsic stress.

CONCLUSIONS

(1) The initial stress in the photosensitive polyimide precursor film was smaller than that in the non-photosensitive precursor film because the photosensitive group acted as a plasticizer. Since the degree of the in-plane orientation of the molecular chain decreased with decreasing initial stress, the TEC of the polyimide film formed from the photosensitive precursor became larger. Therefore, the final stress became larger.
(2) The final stress in the polyimide film formed from BPDA/PDA/HEMA was larger than that in the polyimide film formed from BPDA/PDA/DMM. It is caused by the fact that the intrinsic stress occured in the film formed from BPDA/PDA/HEMA at the final film baking temperature, 400 °C.

REFERENCES

[1] S.D.Senturia, ACS Symp. Ser. 346, 428 1987.
[2] A.W.Wilson, Thin Solid Films 83, 145 (1981).
[3] K.Moriya, T. Ohsaki, and K. Katsura, Electron Components Conf. 34th, 821 (1984).
[4] C.Goldsmith, P.Geldermans, F.Bedetti, and G.A.Walker, J.Vac. Sci. Technol. A1, 407 (1983).
[5] C.L.Bauer and R.J.Farris, in Polyimides: Materials, Chemistry and Characterization, edited by C.Feger, M.M.Khojasteh, and J.E.McGrath (Elsevier Science Publishers B. V., Amsterdam, 1989), p.549.
[6] S. Numata, S. Oohara, K.Fujisaki, J.Imaizumi, and N.Kinjo, J. Appl. Polym. Sci. 31, 101 (1986).
[7] N.Yoda and H.Hiramoto, J.Macromol.Sci.-Chem. A21, 1641 (1984).
[8] R.Rubner, H.Ahne, E.Kuhn, and G.Kolodziej, Photogr. Sci. Eng. 23, 303 (1979).
[9] H.Ahne, W.-D.Domke, R.Rubner, and M.Schreyer, ACS Symp.Ser. 346, 457 (1987).

[10] J.T.Pan and I.Blech, J.Appl.Phys. 55, 2874 (1984).
[11] R.W.Hoffman, in Physics of Thin Films, edited by G.Hass and R.E.Thun (Academic Press, New York, 1966), Vol. 3, p.211.
[12] W.A.Brantley, J.Appl.Phys. 44, 534 (1973).
[13] G.Elsner, J.Kempf, J.W.Bartha, and H.H.Wagner, Thin Solid Films 185, 189 (1990).
[14] T.P.Russell, H.Gugger, and J.D. Swalen, J.Polym.Sci., Polym. Phys.Ed. 21, 1745 (1983).
[15] R.W.Snyder and P.C.Painter, Polym. Mater. Sci. Eng. 59, 57 (1988).

PROPERTIES OF A PHOTOIMAGEABLE THIN POLYIMIDE FILM II.

TAISHIH MAW AND RICHARD E. HOPLA
OCG Microelectronic Materials, Ardsley, NY 10502

ABSTRACT

The polyimide synthesized from benzophenonetetracarboxylic dianhydride and alkyl-substituted diamines is inherently photosensitive at ≤365 nm, and a solvent soluble, negative-acting system can be formulated from the fully-imidized resin. The lithographic, thermal, mechanical, and electrical properties of the polyimide films have been characterized. This polyimide film shows good thermal, mechanical, and electrical properties, and a 1:1 aspect ratio is consistently achieved on 10 μm thick films. The thermal properties of the films were determined using TGA and TMA methods. The decomposition temperature was 527°C, the weight loss of the cured film at 350°C in nitrogen was 0.04 %/hour and the thermal expansion coefficient was 37 ppm/°C. The dielectric constant and dissipation factor of the film were 3.0 and 0.003 respectively at 4% humidity. The effects of hard-bake time, hard-bake temperature, nitrogen purge rate during heat treatment, and humidity on the thermal and electrical properties of the thin film were also examined, and are presented here. The rate of weight loss of the cured film increases when the rate of nitrogen purge decreases or when the cure temperature increases. Longer heat treatments resulted in a slight decrease in the CTE and an increase in the T_g. The electrical properties of the films are dependent both on the humidity during measurement and on the hard-bake temperature.

INTRODUCTION

Polyimides possess high thermal stability, good chemical resistance, low dielectric constants and excellent planarization capabilities. The combination of these characteristics makes them useful for passivation, alpha particle barriers, stress buffers and interlayer dielectrics in ICs as well as in multilayer thin-film high density interconnect packages[1].

In 1985, J. Pfeifer and O. Rohde of Ciba-Geigy reported the synthesis of a class of solvent soluble polyimides which are based on a benzophenone tetracarboxylic dianhydride (BTDA) and ortho-alkylated diamine polymer backbone, and are inherently photosensitive. They fulfill all the requirements of a storage-stable, high purity, non-shrinking photoimageable polyimide system [2]. These materials are marketed under the trade-name of PROBIMIDE 400 formulations.

The lithographic, mechanical, thermal, and electrical properties of this polyimide thin films have been reported[2-5], and the impact of the processing and cure conditions on the mechanical properties have been examined[5].

This report describes the impact of the processing and cure conditions on the thermal and electrical properties.

EXPERIMENTAL

The polyimide used in this study was PROBIMIDE 414 (87-707), which is a solution of Probimide 400 polyimide in γ-butyrolactone. The polyimide solution is commercially available from OCG Microelectronic Materials, and was used as received.

Four inch silicon test wafers were coated with Probimide 414 by spin coating 3 ml polyimide solution at spin speed of 1.9 Krpm on a MTI MultiFab wafer track line. The coated wafers were then soft-baked (3 min @ 110°C on hot plate and 30 min @ 110°C in a convection oven under nitrogen), exposed (broad-band, 1200 mJ/cm² measured at 365 nm with an OAI Exposure Analyzer, Model 356.), developed and hard-baked for 2 hours at 350°C under nitrogen in a Heraeus hot plate oven. This gave a 10.5 μm thick film after soft bake and 9.5 μm after hard-bake as measured by a Tencor Instruments Alpha-Step 200 Thin Film Measurement System. Polyimide films on wafers were diced into 1/8 inch wide strips using a programmed dicing saw by Micro Automation, Inc., Model 1006, and then removed from the wafers by treatment with buffered ammonium hydrogen fluoride solution at ambient temperature. Strips were rinsed with high-purity DI water and air dried.

The glass transition temperature (T_g) and the coefficient of thermal expansion (CTE) of these hard-baked films were measured by a Perkin-Elmer TMA-7 Thermomechanical Analyzer with a 50 ml/min flow rate of nitrogen as purge gas. Zinc strips (CTE(literature) = 35 PPM/°C) in width of 1/8 inch were used to check the overall calibration of the TMA (CTE(observed) = 38 PPM/°C).

The onsets of the thermo-decomposition temperatures of the soft-baked, exposed, developed and hard-baked Probimide 414 films were determined using a Perkin-Elmer TGA-7 Thermogravimetric Analyzer with a heating rate of 10°C/min and a nitrogen purge rate of 50 ml/min. The rate of weight loss was also determined by TGA. Soft-baked, exposed and developed films were heated at a rate of 3.6 °C/min from 50°C to the hard-bake temperature (350°C or 400°C), and held at this temperature for 14 hours, which is similar to the heating cycle used for processing in a Heraeus hot-plate oven, and the rate of weight loss was obtained from the weight loss-time curve. The rate of nitrogen purge in these studies was 0-100 ml/min.

Thermal diffusivity, specific heat and thermal conductivity of hard-baked Probimide 414 films were determined by R. Gardner from Sinku-Riko, Inc. The ac calorimetric method which uses a thermal diffusivity meter (Model Pit-R1) was used to determine the above thermal properties.

The dielectric constant and dissipation factor of the above hard-baked Probimide 414 films were determined with a Hewlett Packard 4277A LCZ meter. Surface resistivity, volume resistivity and breakdown voltage of Probimide 400 films were determined by A. Agarwal from SRI International.

RESULT AND DISCUSSION

The Effect of Processing Parameters on the Thermal Properties:

1. Glass Transition Temperature

Thin films of PROBIMIDE 414 were soft-baked, flood exposed, developed and then hard-baked from 0.5 to 14 hours at 350 and 400°C under nitrogen atmosphere. The thin films were removed from wafer, the glass transition temperatures of the sample films were then determined by TMA. The results are given in Figure 1. The increase of hard-bake time from 0.5 to 14 hours resulted in an increase in the T_g. The increase in T_g is small at 350°C (e.g. 10°C), but much larger at 400°C.

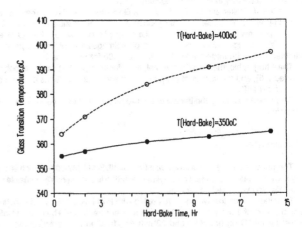

Figure 1. T_g of hard-baked Probimide 414 films.

2. Coefficient of Thermal Expansion:

CTE values of the above sample films were also determined by TMA, and the results are given in Figure 2. As the hard-bake time increases, a decrease in the CTE value was observed initially, and then reached a constant value of 37 ppm/°C. The higher value of CTE for the short-baked film is believed to be due to the presence of lower molecular weight components in the film.

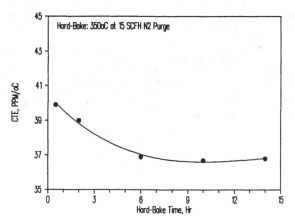

Figure 2. CTE of hard-baked
Problmide 414 films

3. Rate of Weight Loss:

Thin films of Probimide 414 were soft-baked, exposed, developed and then peeled off from the wafers. The thermal stability of the films was analyzed by TGA. The procedures for the hard-bake process were used here. Sample films were purged with nitrogen for thirty minutes at room temperature, heated up to the hard-bake temperature at a rate of $3.6°C/min$, and then held at the hard-bake temperature for 14 hours under nitrogen atmosphere. The rate of weight loss was obtained from the slope of the weight-time curves. The results are shown in Figure 3 and 4. The rate of weight loss was high in the first half hour due to the loss of the residual solvent, and reached a constant level after more than two hours at the hard-bake temperature. Similar changes in the rate of weight loss were also found for higher hard-bake temperature.

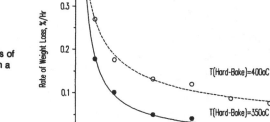

Figure 3. TGA analysis of
Probimide 414 films in a
nitrogen atmosphere

Probimide 400 is a pre-imidized polyimide resin in γ-butyrolactone. This study shows that it takes more than 0.5 hour at 350°C to remove all the residual low molecular weight components from the sample films. The presence of these volatile components may contribute to the higher CTE and lower T_g for the polyimide films.

When the films were heated at the hard-bake temperature under various nitrogen purge rates, the rate of weight loss after two hours heat treatment were given in Figure 4. As the nitrogen purge rate decreases, there was an increase in the rate of weight loss observed, which reached a very high value under ambient conditions. Our earlier studies in the mechanical properties of the hard-baked films also showed similar sensitivity to the curing environment, where the mechanical properties were highly depended on the curing environment. Improved mechanal properties were obtained for films hard-baked at higher nitrogen purge rate.

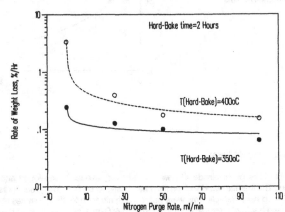

Figure 4. Effect of nitrogen purge rate on the thermal stability of Probimide 414 films

4. Thermal Decomposition:

Thin films of Probimide 414 were soft-baked, exposed, developed and then peeled off from the wafers. The thermal stability of the films was analyzed by TGA, and the results are shown in Figure 5. The polymer film also showed good thermal stability, and degradation occurs at 527°C. The thermal decomposition temperature also remained fairly constant for films which were hard-baked from 0.5 to 14 hours at 350°C.

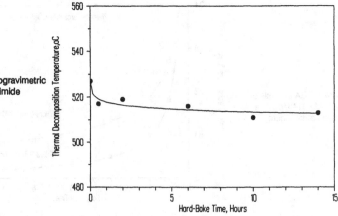

Figure 5. Thermogravimetric Analysis of Probimide 414 Films

5. Thermal Diffusivity, Specific Heat and Thermal Conductivity

The specific heat, thermal diffusivity and thermal conductivity of the hard-baked Probimide 414 films was reported to be 0.32 cal/g/°C, 0.0104 cm²/s and 41E-4 cal/cm/s/°C, these values are comparable to those of other polyimides.

6. Electrical Properties:

Thin films of PROBIMIDE 414 on high conductivity wafers were soft-baked, exposed, developed and then hard-baked at 350 and 400°C from 0.5 to 14 hours under nitrogen atmosphere. Aluminum dots were evaporated onto the hard-baked film through a shadow mask to form capacitors, and the electrical properties of these hard-baked films were analyzed at a humidity of 4% and 73%, and the results are given in Figure 6. For polyimide films which were hard-baked at 350°C from 0.5 to 10 hours, the dielectric constant of these films maintained constant at 3.0 under dry condition (% Humidity=4), but increased to a higher, constant value of 4.0 under wet condition (% Humidity=73%). All the measured dielectric constants fluctuated less than 1% over test frenquencies from 10 kHz to 1 MHz. The dissipation factor is relatively constant between 0.003 to 0.006 under dry conditions, and reaches a higher values between 0.011 to 0.013 under wet conditions.

Figure 6. Electrical properties of Probimide 414 films. (Frequency=1MHz)

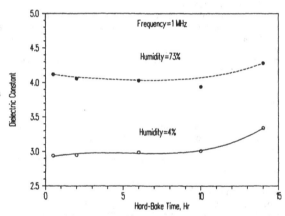

Surface and volume resistivity of Probimide 408 films (film thickness=1.90μm) were reported to be 1.2E+16Ω and 8.6E+16Ω-cm. The dielectric strength of Probimide 408 films and Probimide 412 films (film thickness=9.25μm) were found to be 3.34E+6 and 2.53E+6 volts/cm. The electrical properties of these Probimide films are comparable to those of other polyimides.

CONCLUSION

Probimide 400 materials are a class of solvent soluble polyimides which are based on a benzophenone tetracarboxylic dianhydride (BTDA) and ortho-alkylated diamine polymer backbone, and are inherently photosensitive.

The lithographic and mechanical properties of this polyimide have been reported earlier. Patterns with a 1:1 aspect ratio can be obtained from this photosensitive polyimide with vertical developed profiles. The cured film possesses very good mechanical properties. The mechanical properties of the polyimide films does not depend on the level of exposure energy (from 0 to 2 J/cm²). Excellent retention of the mechanical properties was obtained for thin films which were hard-baked at 350°C at a nitrogen purge rate of 15 SCFH. However, reduced mechanical properties were obtained for films which were hard-baked under low flow rates at either ambient or reduced pressure.

Unlike conventional polyamic acid coatings, the solvent content in the sample films was approximately 5-6% after the soft-bake process, which reduces to 1-2% at even higher soft-bake temperatures (e.g. 170°C). The solvent content in the film increases to 15-20% during the development process. There is much less change in both film thickness and feature dimensions after hard-bake compared to polyamic acids, and a vertical developed profile results.

Thermal properties of the cured Probimide 414 thin films have been analyzed by TMA and TGA. As the hard-bake time increases, an increase in the glass transition temperature is observed. The increase in Tg was small at 350°C hard-bake temperature, and was higher at 400°C hard-bake temperature. The CTE of the hard-bake films reaches to a constant value of 37 ppm/°C at a hard-bake time of more than four hours; films hard-baked at a shorter times showed slightly higher values of CTE. The onset of the decomposition temperature is 527°C. Although Probimide 400 resin solution is a pre-imidized polyimide resin in γ-butyrolactone, where most of the solvent (e.g. >95%) is removed from the polyimide film during the soft-bake process, it takes heat treatment at 350°C for more than two hours to remove the residual low residue volatile components from the polyimide film. The presence of these volatile components in the film will result lower T_g, higher rate of weight loss and higher CTE.

For films which were hard-baked at 350°C, the rate of weight loss under isothermal heat treatment at 350°C was low (e.g. 0.04 ppm/°C) under high nitrogen purge conditions. However, the rate of weight loss increases sharply under ambient air. The loss of thermal stability is believed to be due to thermal oxidative degradation of the films.

Excellent retention of the electrical properties is obtained for thin films which are hard-baked at 350°C at a nitrogen purge rate of 15 SCFH. The hard-baked films possess a low dielectric constant of 3.0 in a dry environment, however, the dielectric constant increases to 4.0 as the relative humidity increases to 73%.

These results show that this material has excellent mechanical, thermal and electrical properties for use in multi-chip modules.

REFERENCES

1. K. Kimbara, A. Dohya and T. Watari in Advanced Electronic Packaging Materials, edited by A.T. Barfknecht, J.P. Partridge, C.J. Chen, and C, Li (Mater. Res. Soc. Proc. 167, Pittsburgh, PA 1990) pp. 33-42.
2. J. Pfeifer and O. Rohde, Recent Advances in Polyimide Science and Technology, (SPE Inc, New York, 1987), pp. 336-350.
3. R. Mo, T. Maw, A. Roza, K. Stefanisko and R. Hopla, Proc 7th Int'l IEEE VLSI Multilevel Interconnect Conference, 390 (1990).
4. M. Ree & K.-J. R. Chen, Polymer Preprints, 31 (2), 594 (1990).
5. T. Maw & R. Hopla, presented at the 1990 MRS Fall Meeting, Boston, MA 1990 (to be published).

STRUCTURE AND PROPERTIES OF BPDA-PDA POLYIMIDE FROM ITS POLY(AMIC ACID) PRECURSOR COMPLEXED WITH AN AMINOALKYL METHACRYLATE

MOONHOR REE*, THOMAS L. NUNES, K.-J. REX CHEN, AND GEORGE CZORNYJ
IBM General Technology Division, Hopewell Junction, NY 12533

ABSTRACT

BPDA-PDA poly(amic acid) precursor was functionalized through its carboxylic acid groups being linked with a crosslinkable aminoalkyl methacrylate, 2-(dimethylamino)ethyl methacrylate (DMAEM), by acid/base complexation. BPDA-PDA polyimide films, which were thermally imidized from the precursors complexed with various amounts of DMAEM, were characterized by means of wide angle x-ray diffraction, stress-strain analysis, and residual stress analysis. The structure and properties of the BPDA-PDA polyimide film were dependent upon the history of the precursor, that is, the complexation of the poly(amic acid) precursor with DMAEM. The molecular packing order was enhanced with the history of DMAEM loading while the molecular order along the chain axis was disrupted. Overall, physical properties, such as mechanical properties and residual stress, were degraded with DMAEM loading. The moisture induced stress relaxation behavior was sensitive to the history of DMAEM loading, whereas the creep induced stress relaxation was varied little due to its high T_g. These properties are understood in terms of structure/property relationships, as well as microvoids, which were possibly generated by outgassing the bulky DMAEM pendent groups during thermal imidization.

INTRODUCTION

Since Kerwin and Goldrick [1] reported a photosensitive polyimide (PSPI) in 1971, several PSPIs, including both negative [2–5] and positive acting types [6–9] , have been developed. PSPIs have recently gained great attention in the microelectronics industry due to the simple fabrication process based on their direct patternability. In general, PSPI is a polyimide precursor type [1–4, 6–8] or a preimidized soluble polyimide type [5,6,8,9]. The former type of PSPI is possible from any polyimide precursor, whereas the latter type is very much limited to a few soluble polyimides.

For a precursor based PSPI, photochemically crosslinkable groups, such as acrylate and methacrylate derivatives, are commonly attached to the carboxylic acid groups of a poly(amic acid) precursor through covalent bonds or acid/base salt formation. In general, the functionalized precursor is further formulated with a photosensitizer package to be selectively active to a certain wavelength of UV-light. At thermal imidizing, all the photosensitive groups are debonded from the precursor backbone and finally outgassed. The photosensitive groups may affect the structure and properties of the final polyimide, although they are outgassed during thermal imidization. To examine the effct of an attached photosensitive group on the final structure and properties of a polyimide, poly(p-phenylene biphenyltetracarboximide) (BPDA-PDA) polyimide was chosen in the present study as a model system. The BPDA-PDA poly(amic acid) was functionalized with 2-(dimethylamino)ethyl methacrylate (DMAEM) through acid/base salt formation (see Fig.1). BPDA-PDA polyimide films, which were thermally imidized from the precursors complexed with various amounts of DMAEM, were characterized by means of wide angle x-ray diffraction (WAXD), residual stress and stress relaxation analysis, and stress-strain analysis.

EXPERIMENTAL

BPDA-PDA poly(amic acid) (ca. 40K $\overline{M}_w$ and 13.5 wt%) solution in N-methyl-pyrrolidone (NMP) and 2-(dimethylamino)ethyl methacrylate (DMAEM) were used in the present study as a polyimide precursor and a crosslinkable monomer, respectively. The crosslinkable DMAEM was attached to carboxylic groups of BPDA-PDA precursor through acid/amine salt formation. DMAEM was added to the BPDA-PDA precursor solution and mixed at room temperature for one day, using a roller mixer. The seven different precursor solutions were prepared by varying DMAEM loading; 0, 7, 15, 25, 50, 75, and 100 mole% equivalent

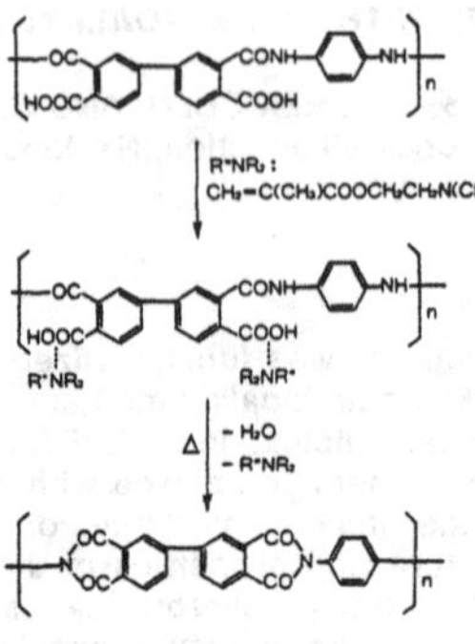

Fig.1. Schematic diagram of the chemical structures of BPDA-PDA poly(amic acid), its complex with DMAEM and the resulting polyimide.

to the carboxylic group of the precursor. The functionalized precursor solutions were stored in a refregirator at ca. 5°C. The refrigerated precursor solutions were warmed for ca. 5 hours at room temperature before use.

The precursor solutions were spin-coated on Si(100) wafers of 82.5 mm diameter, which were precleaned by ashing in a Plasmaline 515 asher of Tegal Corporation (300 watts/5 min and 535 cc/min oxygen flow). Then, the spin condition was 500 rpm/15 sec plus 1,600 rpm ~ 2,300 rpm/20 sec to give ca. 10 μm final cured film thickness. The spin-coated samples were softbaked at 80°C in a convection oven with nitrogen flow for 30 min, and followed by thermal imidization in a Heraeus oven through a cure process (150°C/30 min, 230°C/30 min, 300°C/30 min, and 400°C/1 hr) in nitrogen ambient. For samples to be used in stress measurements, double-side polished silicon wafers (82.5 mm diameter and ca. 400 μm thickness) of known curvature were used as substrates. In this case, a primer solution, 0.1 v% A1100 (γ-aminopropyltriethoxy silane) solution in 90 v% EtOH/10 v% water or distilled water, was spin-coated at 2,000 rpm for 20 sec on the precleaned wafers and baked at 120°C for 20 min in air. Then, the precursor solutions were spin-coated on the primer-coated wafers and soft-baked at 80°C in a convection oven with nitrogen flow for 30 min. These softbaked samples were used for dynamic residual stress measurements.

Wide angle x-ray diffraction (WAXD) measurements were performed for the cured films at room temperature in both reflection and transmission geometry. All WAXD measurements were carried out using a Rigaku horizontal diffractometer in ca. 20 mmHg vacuum. The CrK_a radiation source was operated at 35 kV and 40 mA. The 2θ scan data were collected at 0.1° step intervals for the range of 6° ~ 80°. The collecting time was 35 seconds per step. The mechanical properties were measured at room temperature using an Instron Tester (Model 1122). The grip gauge length was ca. 50 mm and the strain rate was of 1.6×10^{-2} sec^{-1}. The width of film strips, which were diced on a dicer with a circle type of blade, was 3.175 mm. Dynamic residual stress measurements were performed in nitrogen ambient during thermal imidization of the softbaked precursor films on Si(100) wafers and subsequent cooling from 400°C to room temperature, using a double He-Ne laser beam based stress analyzer (Model 2-300, Flexus Company) equipped with a hot-stage, controlled by an IBM PC/AT computer. Also, just after thermal curing, the stress relaxation was monitored at 25°C in air ambient with 50 %RH (relative humidity) or at 100°C in nitrogen ambient as a function of time.

RESULTS AND DISCUSSION

BPDA-PDA polyimide, a semi-rigid polymer, exhibits excellent properties such as low thermal expansion coefficient (TEC), good mechanical properties, high chemical resistance, and high thermal stability [10]. The poly(amic acid) precursor of this attractive polyimide was functionalized by attaching DMAEM, a photochemically crosslinkable monomer, through carboxylic acid/amine group complexation. The structure and properties of the BPDA-PDA polyimide films, which were thermally imidized from its functionalized precursors, were investigated in the present study.

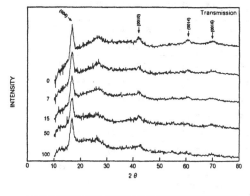

Fig.2. Transmission WAXD patterns of BPDA-PDA polyimide films prepared from the poly(amic acid) precursor with various DMAEM loadings; 0, 7, 15, 50, and 100 mole % DMAEM equivalent to the carboxylic acid groups of the precursor.

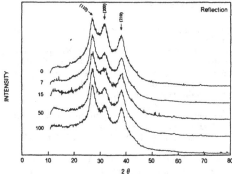

Fig.3. Reflection WAXD patterns of BPDA-PDA polyimide films prepared from the poly(amic acid) precursor with various DMAEM loadings; 0, 7, 15, 50, and 100 mole % DMAEM equivalent to the carboxylic acid groups of the precursor.

For the BPDA-PDA polyimide films, WAXD measurements were performed at room temperature in reflection and transmission geometry. The results are shown in Figs. 2 and 3 as a function of DMAEM loading. The morphological structure of thermally cured BPDA-PDA polyimide was previously investigated with WAXD measurements and structure refinement analysis by M. Ree et al. [10]. In the present study, the peaks of both reflection and transmission WAXD patterns were assigned according to their results. The transmission WAXD patterns show mainly (00ℓ) peaks; (004), (0010), (0014), and (0016). The appearance of the multiple (00ℓ) peaks indicates that the BPDA-PDA polyimide molecules in the solid film are highly oriented along the chain axis. As shown in Fig.2, the (004) peak apparently was not influenced in both peak intensity and width by DMAEM loading. However, higher ordered (00ℓ) peaks were affected as the DMAEM loading increased; (0010), (0014), and (0016). The (0010) peak became weak in intensity and wide in shape as the DMAEM loading increased. This kind of peak broadening was significant in both (0014) and (0016) peaks. The broadening of these peaks appeared even for the sample cured from the 7 mole % DMAEM loaded precursor. Furthermore, both (0014) and (0016) peaks almost disappeared for the polyimide sample from the 100 mole % DMAEM loaded precursor. These WAXD results indicate that for BPDA-PDA polyimide, the molecular order along the chain axis is disrupted by DMAEM loading and the degree of its disruption increases with increasing DMAEM loading. Fig.3 shows the reflection patterns from the BPDA-PDA polyimide films. The reflection WAXD pattern was also influenced by DMAEM loaded to the precursor. All the BPDA-PDA films showed three major peaks; (110), (200), and (210). The (110) peak intensity increased as the DMAEM loading increased. The (210) peak was little changed with DMAEM loading. For the films imidized from both 50 mole % and 100 mole % DMAEM loaded precursors, the (210) peaks are slightly sharper than that of the BPDA-PDA film from the precursor without DMAEM loading. These results indicate that the intermolecular packing order was enhanced in the BPDA-PDA polyimide films from the

Table 1. Variation of Mechanical Properties of BPDA-PDA with History
of DMAEM loading

Mole % DMAEM[*]	Modulus GPa	Yielding stress MPa	Yielding strain %	Stress @ break MPa	Strain @ break %
0	10.2 (.4)[**]	280 (2)	8.0 (.5)	570 (61)	45 (7)
7	7.2 (.2)	214 (1)	9.1 (.9)	428 (14)	66 (4)
15	7.1 (.2)	209 (0)	9.1 (.8)	405 (69)	61 (15)
25	7.1 (.3)	199 (4)	8.9 (.4)	338 (77)	49 (22)
50	6.6 (.4)	187 (6)	9.6 (1)	382 (47)	77 (12)
75	6.6 (.5)	201 (.9)	8.8 (.2)	341 (36)	66 (13)
100	6.6 (.3)	196 (5)	9.3 (2)	314 (55)	52 (22)

* Mole % of DMAEM equivalent to the carboxylic acid groups of the
poly(amic acid) precursor.
** The numbers in parentheses indicate one standard deviation.

DMAEM complexed precursors. As the DMAEM loading increased, the (200) peak weakened in intensity, but did not broaden in shape.

The WAXD results suggest that by the complex formation of the BPDA-PDA precursor with DMAEM, the intermolecular packing order in the resulting polyimide film is increased while the molecular order along the chain axis decreased. The attached bulky DMAEM groups might further disrupt the low degree of molecular order in the condensed state and finally lead to less molecular order along the chain axis in the fully cured polyimide. On the other hand, attaching bulky DMAEM group to the precursor might lower its T_g, and consequently provide more mobility to the precursor chain. During thermal imidization, the detached DMAEM groups might act as plasticizers or lubricants until outgassed. This may also increase the molecular chain mobility leading to better intermolecular packing order.

Stress-strain curves of BPDA-PDA polyimide films were measured at room temperature. Overall, the mechanical properties were significantly degraded by DMAEM attached to the precursor as shown in Table 1. Even 7 mole% DMAEM loading drastically changed the overall mechanical properties of the resulting polyimide films. The modulus dropped to between 6.6 GPa and 7.2 GPa from 10.2 GPa, depending upon the degree of DMAEM loading. The yielding stress decreased to between 187 MPa and 214 MPa from 280 MPa and the ultimate tensile strength also dropped to between 314 MPa and 428 MPa from 570 MPa. In contrast, the strain property was improved by DMAEM loading. The strain at break was enhanced to between 49 % and 77 % from 45 %.

These mechanical property changes are understood in terms of structure-property relationships. Here, the measured mechanical properties are the in-plane characteristics of BPDA-PDA films in which the polyimide molecules are highly oriented in the plane at ca.10 μm film thickness [11]. The Young's modulus generally is proportional to the molecular chain linearity. In the present study, the modulus was highest for the BPDA-PDA polyimide film from the poly(amic acid) itself and lowest for the polyimide film from the 100 mole% DMAEM loaded precursor, indicating that the overall molecular linearity in the BPDA-PDA polyimide film was disrupted by the DMAEM loaded to the precursor. This is consistent with the WAXD results. Another possible contribution to the change of the mechanical properties is from microvoids, which might be formed in the polyimide films by outgassing of the bulky DMAEM groups during thermal imidization.

The residual stress and its relaxation behavior were investigated with the aid of a Flexus stress analyzer. Here, the residual stress (σ_F) was calculated from the radii of wafer curvatures before and after polyimide film deposition, using a simple equation [12] valid under $t_F << t_S$;

$$\sigma_F = \{E_S t_S^2/6t_F(1-\nu_S)\}\{1/R_F - 1/R_\infty\} \qquad (1)$$

Here, the subscripts F and S denote polymer film and substrate, respectively. The symbols E, ν, σ, and t are Young's modulus, Poisson's ratio, stress, and thickness of each layer material. R_F and R_∞ are the radii of a substrate with and without a polymer film, respec-

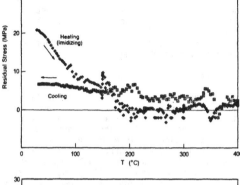

Fig.4. Residual stress versus temperature dynamically measured in nitrogen ambient during curing the softbaked BPDA-PDA poly(amic acid) precursor film on a Si(100) wafer through a 4-step cure process and subsequent cooling. The heating rate from one step to the next step was 2.5°C/min and the cooling rate was 1.0°C/min.

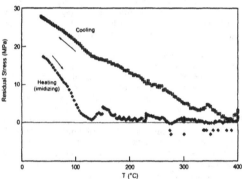

Fig.5. Residual stress versus temperature dynamically measured in nitrogen ambient during curing the softbaked 100 mole % DMAEM loaded precursor film on a Si(100) wafer through a 4-step cure process and subsequent cooling. The heating rate from one step to the next step was 2.5°C/min and the cooling rate was 1.0°C/min.

tively. For Si(100) wafer, $E_s/(1 - v_s)$ is 1.805 x 10^5 MPa [13]. During thermal curing and subsequent cooling, the stress was dynamically measured as a function of temperature over the range of 25°C to 400°C. For the BPDA-PDA poly(amic acid) precursor film softbaked at 80°C for 30 min, the stress was 20.9 MPa at room temperature. As shown in Fig.4, this residual stress gradually decreased with elevating temperature during the curing. During curing at 150°C for 30 min, the first step curing, the stress again rose mainly due to both imidization and residual solvent evaporation. After this first step cure, the stress gradually decreased again with increasing temperature, and reached zero stress level at ca. 200°C. Above 300°C, the stress increased slightly again. For the 100 mole% DMAEM loaded precursor film which was softbaked at 80°C for 30 min, its stress at room temperature was 17.4 MPa, as shown in Fig.5. This stress value is lower than that (20.9 MPa) of the softbaked BPDA-PDA poly(amic acid) precursor film. Except for relatively lower stress below 200°C, the 100 mole% DMAEM loaded precursor film showed similar stress versus temperature behavior as was observed on the precursor without DMAEM loading. During curing, other precursor films, which were complexed with DMAEM of 7 - 75 mole%, exhibited the same stress versus temperature behavior as shown in Fig.5 (see the heating curve).

However, in the cooling run after completion of thermal imidization, the stress versus temperature behavior was significantly dependent upon the amount of DMAEM loaded to the precursor. For the polyimide film from its poly(amic acid), the stress increased with decreasing temperature to 6.7 MPa at room temperature from 2.6 MPa at 400°C (see the cooling curve in Fig.4). On the other hand, the stress for the polyimide from the 100 mole% DMAEM loaded precursor drastically increased to 27.9 MPa at room temperature from 3.1 MPa at 400°C (see the cooling curve in Fig.5). The polyimide films from the precursors loaded with 7 - 75 mole% DMAEM exhibited intermediate stress between those of the polyimides from the poly(amic acid) and 100 mole% DMAEM loaded precursor as shown in Fig.6. Even though only 7 mole % carboxylic acid groups of the poly(amic acid) precursor was functionalized with DMAEM, the final stress of the resulting polyimide at room temperature increased to 19.1 MPa from 6.7 MPa. With further increasing DMAEM loading, the stress slowly increased and then leveled off. The stress at room temperature was 21.1 MPa for 15 mole%, 21.7 MPa for 25 mole%, 26.6 MPa for 50 mole%, and 27.9 MPa for 100 mole% DMAEM loaded.   For these BPDA-PDA polyimide films, the measured overall stress is dominated by thermal stress (σ_t) driven from the mismatch of thermal expansion coefficients (TECs) between polyimide film and Si wafer [14,15];

$$\sigma_t = (\alpha_F - \alpha_s) E_F (T_f - T)/(1 - v_F) \qquad (2)$$

T_f and T are the final temperature of film heat-treatment or glass transition temperature and the temperature of the curvature radius measurement, respectively. α_F and α_s are the coefficients of thermal expansion of a BPDA-PDA polyimide film and a Si wafer, respectively. For all the BPDA-PDA films, their T_g's are the same (ca. 380°C in the dynamic mechanical and thermal analysis), and Poisson's ratios are assumed to be same. Then, the change in residual stresses of BPDA-PDA polyimide films are dependent mainly upon two factors, α_F and E_F. The attached DMAEM lowered significantly the modulus of BPDA-PDA polyimide as shown in Table 1.   Despite lowered modulus (E_F) with DMAEM loading, the stress drastically increased, indicating that for BPDA-PDA polyimide film, the TEC (α_F) significantly increased by DMAEM loading.   For the BPDA-PDA polyimides from the DMAEM loaded precursors, the relatively high TECs might be due to the morphological structure with disrupted molecular orientation along the chain axis (see the WAXD results in Fig.2).   In general, the TEC of a polymer is proportional to the molecular chain linearity.

For BPDA-PDA polyimide films on Si wafers, their stress at room temperature relaxed with time in air ambient.   In general, the stress of a polymer film on a substrate may be relaxed due to its creep behavior as well as moisture uptake.   However, for all the BPDA-PDA polyimide films in the present study, this creep induced stress relaxation was less than 1 MPa at 100°C, regardless of the history of DMAEM loading.   This result might be due to the high T_g (ca. 380°C) of BPDA-PDA polyimide films.   In contrast, the stress relaxation was very sensitive to moisture.   For this measurement, the polyimide film on a Si wafer was thermally cured at 400°C through the cure process, and subsequently cooled to 25°C at 1.0°C/min rate.   The sample was suddenly exposed to air ambient (50% RH) by removing a quartz cover from the Flexus hot-stage when the sample was cooled to 25°C. Then, the stress relaxation was monitored isothermally as a function of time.   This moisture induced stress relaxation was very much dependent upon the history of DMAEM loaded to the BPDA-PDA precursor as shown in Fig.7.   The stress of the polyimide from the poly(amic acid) precursor decayed exponentially with time.   After 3 hrs, the stress relaxation still continued with time.   On the other hand, for the polyimides from the DMAEM loaded precursors, the stress decay with time significantly accelerated as the DMAEM loading increased.   The stress difference before and after relaxation for 3.0 hrs was approximately 2.8 MPa for 0 mole%, 3.7 MPa for 7 mole% DMAEM, 8.6 MPa for 25

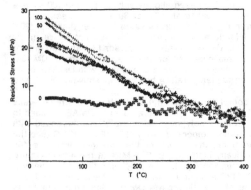

Fig.6. Residual stress variations with temperature, dynamically measured during cooling at 1.0°C/min rate after the BPDA-PDA polyimides were cured from the poly(amic acid) with various DMAEM loadings; 0, 7, 15, 25, 50, and 100 mole % DMAEM equivalent to the carboxylic acid groups of the precursor.

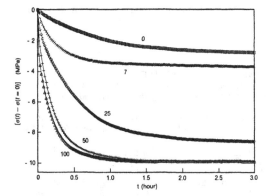

Fig.7. Residual stress relaxation with time, measured at 25°C in air ambient with 50 %RH for the BPDA-PDA polyimide films on Si(100) wafers immediately after their thermal imidization through a 4-step cure process; the DMAEM loadings were 0, 7, 25, 50, or 100 mole % equivalent to the carboxylic acid groups of the precursor.

mole% DMAEM, and 9.9 MPa for 50 mole% ~ 100 mole% DMAEM loaded precursor based polyimides. This result indicates that the polyimide film from a DMAEM complexed BPDA-PDA precursor absorbs more moisture than that from the poly(amic acid) precursor and its moisture uptake increases with increasing DMAEM loading.

In the BPDA-PDA polyimides from the precursors complexed with DMAEM, the relatively high moisture uptake is thought to be due to the contribution of the moisture uptake through voids or cavities which might be generated by degassing the linked bulky DMAEM groups during curing. Each bulky DMAEM group linked to the BPDA-PDA precursor occupies a certain volume in the condensed precursor film after drying. When the imidization takes place through the thermal cure process, the outgassing of these DMAEM groups generates cavities or voids corresponding to their occupied volume. Then, these cavities may not be healed completely because of both freezing polyimide molecules below T_g and limited molecular chain mobility with low degree of superheating above T_g due to its high T_g characteristic during thermal imidization. These microcavities are currently being studied in detail by means of small angle x-ray and light scattering [16].

CONCLUSIONS

The functionalized BPDA-PDA poly(amic acid) precursor was made by attaching crosslinkable DMAEM monomer through acid/base complexation. The structure and properties of BPDA-PDA polyimide films, which were thermally imidized from the functionalized precursors with various DMAEM loadings, were investigated with respect to the effect of DMAEM loading history by WAXD, residual stress and its relaxation analysis, and stress-strain analysis. We found that the structure and properties of BPDA-PDA polyimide film from its precursor complexed with DMAEM were quite different from those of the polyimide from its conventional poly(amic acid) precursor. The WAXD results show that the molecular packing order was improved with DMAEM loading, whereas the molecular chain order was disrupted. The mechanical properties significantly degraded with DMAEM loading. Although the Young's modulus was lowered with DMAEM loading, the overall residual stress, mainly due to the thermal stress, drastically increased, indicating that the TEC was significantly increased with DMAEM loading. The creep induced stress relaxation was changed only slightly, less than 1 MPa, with DMAEM loading. In contrast, the stress relaxation due to moisture uptake was very much dependent upon the history of DMAEM loading. The moisture uptake in the polyimide film increased as the DMAEM loading increased. These properties are understood by considering the microcavities which might be formed by degassing out the bulky DMAEM groups during curing, in addition to the structural change.

ACKNOWLEDGMENT

The authors would like to thank Dr. Dan Kirby for reviewing this paper and valuable comments.

REFERENCES

1. R.E. Kerwin and M.R. Goldrick, Polym. Eng. Sci. 2, 426 (1971).
2. R. Rubner, H. Ahne, E. Kuhn, and G. Kolodziej, Photogr. Sci. Eng. 23, 303 (1979).
3. N. Yoda and H. Hiramoto, J. Macromol. Sci.- Chem. A21, 1641 (1984) and references therein.
4. F. Kataoka, F. Shoji, I. Takemoto, I. Obara, M. Kojima, H. Yokono, and T. Isogai, in Polyimides: Synthesis, Chract. Appl., edited by K.L. Mittal (Plenum, New York, 1984), p. 933.
5. J. Pfeifer and O. Rohde, in Polyimides: Synthysis, Charact., Appl., edited by K.L. Mittal (Proc. 2nd Tech. Conf. Polyimides, SPE, Ellenville, NY, 1985), p. 130.
6. D.N. Khanna and W.H. Mueller, Polym. Eng. Sci. 29, 954 (1989).
7. S. Kubota, Y. Yamawaki, T. Moriwaki, and S. Eto, Polym. Eng. Sci. 29, 950 (1989).
8. T. Omote, K. Koseki, and T. Yamaoka, Macromolecules 23, 4788 (1990) and references therein.
9. T. Omote, H. Mochizuki, K. Koseki, and T. Yamaoka, Macromolecules 23, 4796 (1990) and references therein.
10. M. Ree, D.Y. Yoon, L. Depero, and W. Parrish, submitted for publication.
11. T.P. Russell, H. Gugger, and J.D. Swalen, J. Polym. Sci.: Polym. Phys. Ed., 21, 1745 (1983).
12. R.J. Jaccodine, and W.A. Schlegel, J. Appl. Phys. 37, 2429 (1966).
13. J.J. Wortman and R.A. Evans, J. Appl. Phys. 36, 153 (1965).
14. W.R. Hoffman, in Physics of Thin Film, edited by G. Hass and R.E. Thun (Academic, New York, 1966), Vol.3, p. 211.
15. S. Timoshenko, J. Opt. Soc. Am. 11, 233 (1925).
16. M. Ree and J.S. Lin, in preparation.

THERMAL EXPANSION AND VISCOELASTIC PROPERTIES

OF A SEMI-RIGID POLYIMIDE

THOR L. SMITH* AND CHURL S. KIM**
*IBM Research Division, Almaden Research Center, San Jose, CA 95120
**Present Address: IBM GTD, Hopewell Junction, NY 12533

ABSTRACT

Studies were made of the physical properties of the commercially available polyimide Upilex-SGA, which is prepared from biphenyl dianhydride and p-phenylene diamine. Annealing the Upilex-SGA for 2 hr under N_2 at 400°C gave a film that expanded continuously when heated at a fixed rate, in contrast to the as-received film. The linear expansion showed a change of slope at 84°C and also at 295°C, the later being T_g. The thermal coefficient of linear expansion at all temperatures was very small, even above 295°C it is 27.8 x 10^{-6}. Its stress-strain curve did not exhibit a yield point, even though its ultimate elongation is ~23%. Similar behavior is shown by the PMDA-ODA polyimide, except its ultimate elongation is ~70%. The unusual stress- strain curves exhibited by these polyimides is undoubtedly caused by their liquid-crystalline morphology. The stress-relaxation modulus was measured at 0.5% extension and 12 temperatures from 30 to 330°C. Derived isochrones showed that the 1-s tensile modulus at 20°C is 9.0 GPa, but at 330°C it is 2.0 GPa. Creep curves were also measured at a stress of 30 MPa and at 10 temperatures from 30 to 340°C. Master curves prepared from the relaxation and creep data are discussed briefly and evidence is given which show that the superposition method is not truly valid for this polyimide, which actually is not surprising.

INTRODUCTION

Interest has developed recently in the polyimide prepared from biphenyl dianhydride (BPDA) and p-phenyl diamine (PDA), whose repeat unit is shown in Fig. 1. This polyimide is available commercially from the Ube Industries in Japan. We have studied their material designated Upilex-SGA. (The GA indicates some treatment that promotes adhesion.) The structure shown in Fig. 1 can be represented as a slanted line-segment, followed by a horizontal segment, and another slanted line-segment, etc; that is, by a staircase sketch. Because of the quite high elongation exhibited by this polyimide, discussed subsequently, it is unlikely that this staircase can continue indefinitely. If so, it would be a highly rigid molecule, and so the bulk material could be stretched by only a small amount, in contrast to the relatively high extension observed. In a solution of the polyamic acid precursor, rotations can occur around the bond in the biphenyl moiety, and so it is probable that some such conformations, formed by such rotations, exist in the solid state of the polyimide.

The purpose of the present study was to determine on well annealed specimens the following: the thermal coefficient of linear expansion over a broad temperature range; the stress-strain curve at ambient temperature; and the viscoelastic properties as exemplified by the stress-relaxation modulus in tension and the creep compliance, both at temperatures from 30 to 330 or 340°C.

EXPERIMENTAL

The Upilex-SGA film studied had a thickness of 0.051 mm. When a specimen was heated continuously at 10°C/min in the tensile unit of a DMTA (Dynamic Mechanical Thermal Analyzer from the Polymer Laboratories) the displacement (Fig. 2) increased until at about 300°C contraction began. At about 360°C, expansion started again and continued until the test was terminated at about 465°C. (**Note**: The displacement obtained from the DMTA is not the expansion of the specimen precisely, but it gives a semiqualitative measure.) The obtained data indicate that the original specimen was not isotropic. After annealing the original film for 2 hr at 400°C in a nitrogen atmosphere, the film expanded continuously up to 420°C, as shown in Fig. 2. Annealed films were used in all subsequent tests. Strictly speaking, we cannot claim that the annealed film is completely isotropic, but it is more nearly so than the original film.

BIPHENYL DIANHYDRIDE +
p-PHENYLENE DIAMINE

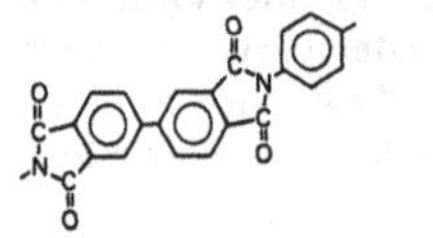

Fig. 1 The repeating structure in the BPDA-PDA Upilex-SGA.

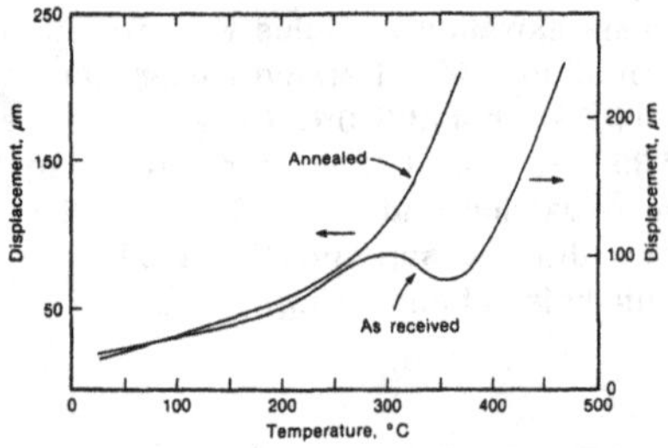

Fig. 2 Temperature dependence of the displacement of the original and the annealed films during heating at 10°C/min in a DMTA.

The thermal coefficient of linear expansion was measured by using a special method[1] with a Dynastat (Imass, Inc.) equipped with an Invar jig on specimens 7 mm wide and slightly more than 35 mm long between the compression grips. The advantage of this method, which can be applied only to flexible films, is that the specimen is not subjected to a tensile load, except occasionally to a load of 1 g for a short period while data are being taken.

The stress-relaxation modulus in tension was determined with a Dynastat at a fixed strain of 0.5% and at 14 temperatures from 30 to 330°C. The Dynastat was also used to determine the creep compliance at a stress of 30 MPa and at 10 temperatures from 30 to 340°C.

RESULTS AND DISCUSSION

Thermal Coefficient of Linear Expansion

Figure 3 shows the expansion data on a plot of log L against temperature from about 20 to 350°C; L is the length of the specimen. Data were first obtained during stepwise heating, then during stepwise cooling, and finally during heating again. The data agree closely, except for two points from the first heating. To determine the significance of the two breaks in the data, log E′ and tan δ were measured with the DMTA at 10 Hz and a heating rate of 10°C/min at temperatures up to 500°C. The results are shown in Fig. 4.

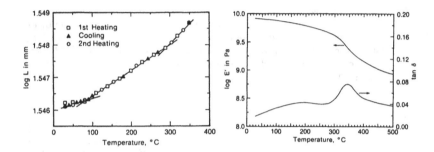

Fig. 3. Thermal expansion data.

Fig. 4. Log E′ and tan δ determined at 10°C/min with a DMTA at 10 Hz.

As the high-temperature tan δ peak in Fig. 4 begins to develop somewhat below 300°C, it follows that the break at 295°C in Fig. 3 is T_g. The origin of the break at 84°C in Fig. 3 is not definitely known, but it is considered to be a reflection of the broad shallow secondary peak in tan δ shown in Fig. 4.

The slopes of the line segments in Fig. 3 are related to the thermal coefficient of linear expansion α by

$$\alpha = \frac{d \ln L}{dT} = 2.303 \frac{d \log L}{dT}$$

The values of α in units of 10^{-6}/°C obtained from the data in Fig. 3 are: 8.5 from 30 to 84°C; 18.2 from 84 to 295°C; and 27.8 from 295 to 350°C. As T_g is 295°C, it follows that α above T_g is only slightly greater than that for aluminum foil, as shown elsewhere[1].

There apparently are no literature data to compare with the values of α given above. It is true that Numata and co-workers[2], and other studies by Numata and associates referenced in Ref.2, have carried out extensive studies of the "expansion coefficients" of numerous polyimides. But they adopted the policy: "Since the thermal expansion coefficient had a temperature dependency, the average value between 50 and 250°C was used as a representative value." They report that α for a film cured free standing, after solvent removal at 100°C, is usually larger than those for films prepared by restricting the contraction in one or two directions. For the BPDA-PDA polyimide cured free standing, their curve shows that α is about 32 x 10^{-6}/°C. at 300°C. Our value is 27.8 x 10^{-6}/°C at temperatures above 295°C. Their data[2] are shown by curves that exhibit no distinct breaks as are shown in Fig. 3.

Stress-Strain Curves and Stress-Relaxation Data

Figure 5 shows stress-strain curves determined in 6 tests at an extension rate of 5.7 min[-1] on the as-received and the annealed films of Upilex-SGA. There is only a slight difference in the curves for the two films. The tensile modulus from the initial slope is about 8.04 GPa which is essentially same as that shown in Fig. 4 at 20°C.. A value given in the literature[3] is 11 GPa.

The key factors shown by Fig. 5 are that the ultimate elongation is approximately 23% and that the curves show no well defined yield point; such is also shown by the PMDA-ODA polyimide. But the ultimate elongation for the later polyimide is ~70%. The similar shapes of the stress-strain curves for Upilex (BPDA-PDA) and PMDA-ODA undoubtedly results because the morphology of both is liquid-crystalline[4,5].

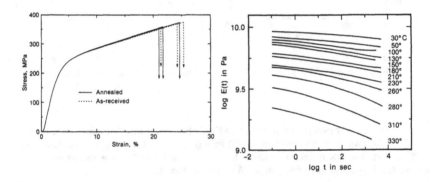

Fig. 5 Stress-strain curves determined on the as-received and annealed Upilex-SGA films.

Fig. 6 Time dependence of the stress-relaxation modulus at temperatures from 30 to 330°C.

Stress relaxation was determined at a fixed strain of 0.5% (e = 0.005) and at 14 temperatures from 30 to 330°C and over a range of time t from 0.1 to ~6000 s. The stress-relaxation modulus in simple tension is: $E(t) = \sigma(t)/e$ where $\sigma(t)$ is the time-dependent stress and e is the strain. Measurements at several strains from 0.2% to 0.7% at 30°C showed that $E(t)$ is sensibly independent of strain over the range covered and thus $E(t)$ at a strain of 0.5% appears to be in the range of linear viscoelastic behavior. The stress-relaxation data are presented in Fig. 6 on doubly logarithmic coordinates. Curves at two temperatures are omitted for clarity.

Figure 7 shows the temperature dependence of $E(t^*, T)$ where t^* is a fixed time whose values in Fig. 7 are 0.1 and 100 s. (The data in Fig. 7 are isochrones obtained from Fig. 6 and partially from curves not shown in Fig. 6. Figure 7 shows that the modulus at ~20°C and 1 s is approximately 9 GPa, which as expected is greater than 8 GPa obtained from the stress-strain curves in Fig. 5, but at 330°C, the 1-s modulus is 2.0 GPa.

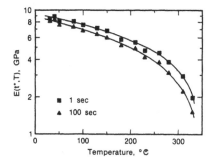

Fig. 7 Temperature dependence of 1 s and 100 s isochrones of the stress-relaxation modulus.

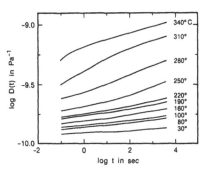

Fig. 8 Time dependence of the creep compliance D(t) from creep data at 30 MPa.

Figure 8 shows the time dependence of log D(t) given in reciprocal Pa, where D(t) is the creep compliance determined at 30 MPa. D(t) equals the constant applied stress divided by the time-dependent strain.

Master, or composite, curves were prepared from the curves in Fig. 6 as well as those in Fig. 8. In brief, an arbitrary reference curve is selected at some temperature. Then, the other curves are shifted along the log t axis (to the right and left) to obtain a master curve. The needed shift distance for each curve is the temperature function log a_T. A plot of all values of log a_T against 1/T, where T is in K, gave a straight line at all temperatures less than 240±14°C. This line gave an activation energy of 70 Kcal. At temperatures greater than about 240°C, the slopes on the plot of log a_T against 1/T changed and gave two straight lines, one from the relaxation data and the other from the creep data. The activation energy from the line from creep data was 140 Kcal and that from the relaxation data was 160 Kcal. Even though the relaxation and creep data each gave good master plots, it appears that the superposition procedure is not valid, strictly speaking, which is not surprising.

REFERENCES

1. C. S. Kim and T. L. Smith, *J. Polym. Sci.; Polym. Phys. Ed.* **28**, 2119 (1990).

2. S. Numata, S. Oohara, K. Fujisaki, J. Imaizumi, and N. Kinjo, *J. Appl. Polym. Sci.* **31**, 101 (1986).

3. S. Numata and T. Miwa, *Polymer* **30**, 1170 (1989).

4. N. Takahashi, D. Y. Yoon, and W. Parrish, *Macromolecules* **17**, 2583 (1984).

5. D. Y. Yoon, personal communication.

ELECTRICAL PROPERTIES OF THIN POLYIMIDE FILMS
PREPARED BY IONIZED CLUSTER BEAM DEPOSITION

HYUNG-JIN JUNG*, DONG-HEON LEE*, JEON-KOOK LEE*,
AND CHUNG-NAM WHANG**
* Korea Institute of Science and Technology, Division of
Ceramics, Seoul, Korea
** Yonsei University, Physics Department, Seoul, Korea

ABSTRACT

The electrical properties of gold - polyimide - silicon
structures were investigated experimentally by capacitance -
voltage (C - V) measurement. Polyimide films were deposited on
silicon substrate by using ionized cluster beam deposition
(ICBD) of PMDA-ODA followed by in-situ thermal curing in N_2
atmosphere. The resulting C - V plots show hysteresis, and it
was believed to be due to the injection of carriers. The
interface trap density is fairly low because of the clean
interface provided by ICB technique.

INTRODUCTION

Polyimide films are reliable high temperature planarizing
insulators for multilevel interconnection systems and promising
passivating layers for semiconductor devices [1]. Polyimides
as dielectric coatings exhibit many advantages over the
conventional sputtered or ceramic dielectrics, i.e., easy
processing, low cost, low defect count, and low dielectric
constant [2].
Conventional chemical processing for preparing polyimide
films is a two-step processing,i.e., preparation of appropriate
concentrations of polyamic acid solutions in N-methylpyrolidone
(NMP) solvent, and then spin coating followed by thermal curing
[3]. Unfortunately, this intermediate polymer solution creates
several processing difficulties; polyamic acids are hydrolytically unstable, solvent is retained within the film, and
pollution problems connected with the use of the solvent.
Also, the problems due to the spin coating is created such as
difficulty in film thickness and uniformity control.
In the present work, we report the electrical properties
of metal-polyimide-silicon structure using polyimides prepared
by ionized cluster beam deposition (ICBD) technique, which is
essentially dry processing.

EXPERIMENT

Sample preparation

The ICBD system ,the same as used in the previous
experiment [4], consists of ionized cluster beam sources, one
for PMDA and one for ODA monomer. The molecules ejected from

the heated crucible through the narrow nozzle form clusters due
to the adiabatic expansion process. Then, they are ionized by
the electrons from electron emitters, and accelerated by the
acceleration voltage between the crucibles and the substrate
holder.

In this set of experiments, acceleration voltage , ionization voltage, and ionization current were fixed at 800 V, 200 V,
2 mA, respectively. In-situ curing was done at 250°C for 0.5
hours. Substrates were p-type silicon wafers, and their
resistivities ranged from 1 to 10 Ohm.cm. The back electrodes
were Al-deposited by evaporation method, and the top electrodes
were dc - sputtered Au.

Measurement techniques

A commonly used method was employed for the measurement of
C - V characteristics of gold-polyimide-silicon structures.
The 1 MHz capacitance meter (Hewlett Packard, model 4280A) and
related probe stations and plotter were used. Typical sweep
rate was 0.1 V/s.

RESULTS AND DISCUSSION

FT-IR chracterization

Fig.1 shows FT-IR results of as-deposited PMDA-ODA films
before in-situ curing. Polyamic acid shows main peaks at 1650
cm^{-1} and 880 cm^{-1} due to the primary amine N-H bending
vibration by carboxylic acid anhydride.

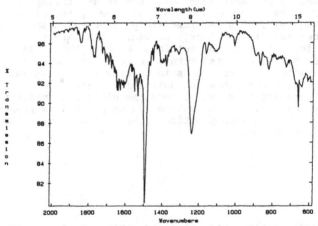

Fig. 1   FT-IR spectra of PMDA-ODA deposited by ICBD.

(before curing)

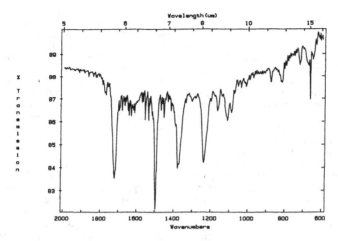

Fig. 2 FT-IR spectra of PMDA-ODA deposited by ICBD,
followed by in-situ curing.

Fig.2 shows FT-IR results of PMDA-ODA films after in-situ curing. Strong peaks at 1380 cm^{-1} and 1720 cm^{-1} are observed, and peaks at 1780 cm^{-1} and 725 cm^{-1} are also observed. Considering above results, it is confimed that polyimides are very easily deposited by ICB.

C - V hysteresis characteristics

Fig.3 shows typical C-V curves of gold-polyimde-silicon structures, where polyimdes are deposited by ICB. The resulting C-V curves show hysteresis phenomena.

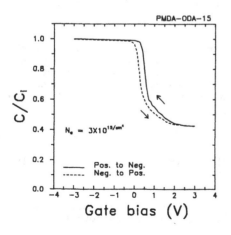

Fig. 3 Typical C-V curve of
PMDA-ODA deposited by ICBD.

The hysteresis observed in the dynamic capacitance curve indicates an evolution of charge near the silicon-polyimide interface. C - V curve will be shifted to a more positive bias for negative charges near the silicon-polyimide interface and to a more negative bias for positive charges.

Generally, there are four cases that account for the hysteresis observed in the capacitance curve, i.e. mobile ion effect [5], polarization effect [6][7], charge injection from electrodes. The experimentally observed C - V curves of polyimides are shifted in the positive direction of voltage during the return loop.

In case of C - V hysteresis due to mobile ion effect and polarization effect, capacitance curves must be shifted in the negative direction of voltage during the return loop. Therefore, the experimentally observed hysteresis curves are not due to mobile ions and polarization effect. In case of interfacial polarization , if one supposes that the native oxide thickness is 50 Å and its conductivity is about 4×10^{-6} Ohm^{-1} cm^{-1}, the sign of the accumulated charge at the polyimide-silicon interface is the same as that of the polarizing voltage applied to the metal. In that case the result is not consistent with the experimental results.

In case of metal injection, a positive voltage on the metal induces positive charges at the metal-polyimide interface, and a negative voltage induces negative charges at the interface. Therefore, in this case, the hysteresis phenomenon has the same sense as the preceding case of polarization.

Therefore, the experimentally observed hysteresis is believed to be semiconductor injection of holes and electrons. When the net potential of the metal is positive with respect to silicon, the silicon can inject electrons and, when the potential of the metal is negative, the silicon injects holes into the polyimide. But further research is required if this case is due to only semiconductor injection or combined effect of both metal and semiconductor injection. And it is very interesting that Hahn and Yoon [13] have observed hole injection in their spin casted PMDA-ODA system.

Interface state density

Interface traps are electronic states that reside at the insulator-semiconductor interface. In fact, thay are the results of dangling bonds. These energy states are located in the forbidden band gap, and they are quantified as interface trap density D_{it} with a unit of number of states/(cm^2. eV). The interface trap density D_{it} is defined as

$$D_{it} = \frac{1}{q} \left(\frac{\partial Q_{ss}}{\partial \psi_s} \right)_V \qquad \text{states/cm}^2/\text{eV}. \qquad (1)$$

where Q_{ss} is the total charge in the surface states at a given surface potential, and ψ_s is the surface potential.

The evaluation of surface state density using capacitance measurement can be achieved, for example, by the differentiation procedure, the integration procedure [9], or the temperature procedure [10]. In the differentiation procedure, which was used first by Terman [8], comparison of actual C - V curve with the ideal one gives a curve of ΔV versus V where ΔV is the voltage shift. From the above ΔV versus V curve, one can determine Q_{ss} and thus D_{it}.

The ideal curve formula we have used in this calculation is as follows [11];

$$C_s = 2C_{FBS}\left\{1 - \exp(-v_{so}) + \left(\frac{n_i}{N_A}\right)^2\left[(\exp(v_{so})-1)\frac{\Delta}{1+\Delta}+1\right]\right\} \dot{F}^{-1}(v_{so}, u_B). \qquad (2)$$

where

$$\Delta \approx \frac{F(v_{so}, u_B)}{\exp(v_{so})-1}\left\{\int_0^{v_{so}} dv\left[\frac{\exp(v_s)-\exp(-v_s)-2v_s}{F^3(v_s, u_B)}\right]-1\right\}. \qquad (3)$$

But, in most cases, a simplified high frequency MOS capacitance formula proposed by Brews was used [12].

Fig.4 shows the distribution of the interface state density of MIS structure determined experimentally according to this method. The values are fairly low, i.e., about $5\text{X}10^{10}$ states/cm^2/eV at the midgap energy, and this fact gives us confirmation that polyimide-silicon interface is very clean because the ICBD provides appropriate cluster energy so that clusters can clean the initial silicon surface and, at the same time, cannot make defects on silicon surface at the first nucleation stage.

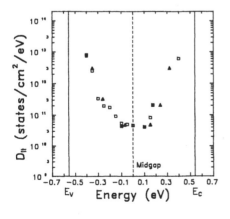

Fig. 4 Interface state density of typical polyimide films deposited on p-type (100) silicon surface by ICBD.

CONCLUSIONS

It has been shown that the polyimide films obtained by ICBD can be grown very easily on a semiconductor surface. The very low interface trap density can be attributed to the charateristics of ICBD. The hysteresis phenomena present in the C - V curve were attributed to the injection and trapping of carriers in the polyimide film. And, also, the use of polyimide films by ICBD will become attrcative for interlayer dielectrics because of small hysteresis and clean interface.

REFERENCES

1. C.E.Sroog, J.Polym.Sci.Macromal.Rev., 11, 161 (1976).
2. A.M.Wilson, in Polyimides - Synthesis, Charaterization and Application, edited by K.L.Mittal (Plenum, New York, 1984), Vol.2, p.715.
3. A.M.Wilson, Thin Solid Films, 83, 145 (1981).
4. S.J.Cho, H.S.Choe, H.G.Jang, and C.N.Whang, in Beam-Solid Interactions: Physical Phenomena, edited by J.A.Knapp, P.Borgesen, and R.A.Zuhr (Mater. Res. Soc. Proc. 157, Boston, MA, 1989) pp. 49-53.
5. B.Deal, J.Electrochem.Soc., 121, 198C (1974).
6. R.Goffaux and R.Coelho, Rev.Gen.Electr., 78 619 (1969).
7. F.A.Sewell, H.A.R.Wegewer, and E.T.Lewis, Appl.Phys. Lett., 14, 45 (1969).
8. L.M.Terman, Solid-State Electr., 5, 285 (1962).
9. C.N.Berglund, IEEE Trans. on Electron Devices, ED-13, 701 (1966).
10. P.V.Gray and D.M.Brown, Appl.Phys.Lett., 8, 31 (1966).
11. E.H.Nicollian and J.R.Brews, MOS (Metal Oxide Semiconductor) Physics and Technology (Wiley, New York, 1982) p.161.
12. J.R.Brews, Solid-State Electron., 20, 607 (1977).
13. B.R. Hahn and D.Y.Yoon , J. Appl. Phys., 65, 2766 (1989)

CHANGES IN THE CHARACTERISTICS OF THE FREE VOLUME OF GLASSY POLYMERS DURING DEFORMATION AND RELAXATION.

M.I. Tsapovestsky *, V.K. Lavrentiev * and S.A. Tishin **
*Institute of Macromolecular compounds. 199000 Leningrad Bolshoy pr.31. USSR.
**Institute of Electronics. Tashkent, USSR.

ABSTRACT

The behavior of glassy polymers was studied with the aid of mechanical tests on the basis of free volume. Three types of experiments were carried out and interpreted using modern ideas about the structure of free volume. They were: 1. The study of changes in the total free volume occurring as a result of deformation (by measurement of density) 2. The study of micro voids and their changes on deformation using small angle X-ray scattering (SAXS) and positron annihilation 3. The study of mobile regions by the measurement of dielectric loss (using an original interpretation) during deformation. This allows several parts of the total free volume to be distinguished and investigated. A model for the free volume relaxation during deformation is proposed.

INTRODUCTION

In recent years most physical phenomena observed in glassy polymers have been interpreted in terms of free volume. Physical aging [1.2,8], relaxation properties [3], and diffusion of low molecular weight additives in polymer glasses [4] are usually associated with increase or decrease in free volume and with changes in the distribution of various fractions of this volume. Such a fundamental phenomenon as the glass transition is also considered from this viewpoint [5,6,8]. Many experimental papers have been published on the analysis of various characteristics of free volume in glassy polymers. Free volume elements of diameter $d > 50$ Å are studied by probe methods, SAXS, and neutron scattering [10,11,12]. The observation of micro-holes of diameter 3-20 Å is difficult. However, Lashkov has observed the photo-chemical dissociation of anthracene covalent bonded to chain ends in PMMA. He found micro-holes with sizes 5,7 and 12 Å [13]. Positron annihilation is a more common method of observation of these micro-holes [14,15, 20,23].

Local regions of increased molecular mobility, that is, regions with a high free volume concentration, have been observed in various polymers with the aid of photon correlation spectroscopy and C^{13} NMR relaxation spectroscopy [16,17]. The presence in a polymer of micro-holes of size 4-7 Å [21-23] suggests that the concentration of these micro-holes is high in mobile regions. When the polymer is influenced by any factors, the entire structure of the free volume may change. Hence in the present paper we attempt to follow the changes in some free volume populations during mechanical tests. Free volume is observed by different methods before and during deformation at a constant rate up to failure as well as during volume relaxation after deformation and failure.

EXPERIMENTAL

The material used for most experiments was polyimide PI-1, a commercial material (an analog of Kapton H) in a film 40 μm thick. An experimental polyimide PI-2, 60 μm thick, and commercial PVC, 60-70 μm thick were also used. Mechanical deformation was carried out on a UTS-1O instrument. The measurement of the free volume parameters was performed before deformation, during deformation at a constant rate, on volume relaxation immediately after deformation or sample failure and after complete volume relaxation. 36 hours was necessary for the completion of volume relaxation after the mechanical deformation. Density was measured by the flotation method [22]. A mixture of toluene and CCl_4 was used for experiments, and the experimental temperature was $20 \pm 0.3°C$.

Small-angle X-ray scattering

The SAXS experiments were carried out in a Kratky camera with slit collimation and detection from $2\theta = 3'$ to $16'$. CuK_α radiation was used with a Ni filter. The width of the entrance slit was 30μm. Divergence of the primary beam was 2.7". As the scattered intensity is very low, the scattering curve for each sample was obtained by the adding 10-15 scans and smoothing by Chebishev polynomials. Radii of gyration of pores were calculated from scattering cures by the method of Russell [10].

Positron annihilation

In a positron annihilation experiment the interaction of positrons with the polymer is observed by recording the radiation emitted on annihilation. An experimental instrument was used which had a computer system of peak position stabilization, angular correlation, point-linear geometry, computer control of measurement program, and automatic background subtraction. The mathematical processing of spectra was carried out with the aid of PAAC Fit, Resolution and PosFit programs [28].

Dielectric losses.

The value of tan δ reflects a combination of processes occurring in the polymer. If it is assumed that dielectric losses are caused by relaxation absorption in the temperature and frequency ranges in which our measurements are carried out, then in our opinion tan δ may be used for the study of the characteristics of some fraction of polymer free volume. In fact, if these losses are related to relaxation processes, an electrically active relaxation agent changes its spatial position. The value of the dielectric loss then reflects (in addition to other characteristics) the number of dipoles changing their spatial position [18]. These dipoles can be located only in regions exhibiting high molecular mobility, i.e. in regions with a high free volume concentration. This implies that a change in tan δ indicates that the concentration of these regions changes (in the case of relaxation absorption). Dielectric losses were measured at a frequency of 1000 Hz at room temperature with the aid of an automated bridge with a computer control which has been developed. The time between measurements was 0.7 s.

RESULTS

Table 1 gives the characteristics of PI-1 before deformation. Three components could be seen in the spectrum of positron lifetime: $\tau_1 = 200 \pm 20$ ps; $\tau_2 = 385 \pm 10$ ps; $\tau_3 = 1.1 \pm 0.1$ ns. However, the intensity of the third component was low, 1%, and so it is not

Table I Free volume parameters of PI-1 measured by different methods.

positron annihilation				SAXS		Dielectric loss	Density
$\tau 1$ ns	$\tau 2$ ns	J1 %	J2 %	R_{g1} Å	R_{g2} Å	tan δ	ρ g/cm^3
0.201	0.388	21	79	160	---	0.0019	1.41

considered further. The component with a lifetime τ_2 = 385 ps is usually associated with positrons localized in microscopic voids with a mean diameter of 5-7 Å [22,23,26]. The nature of the shortest time component will be discussed later. J_1 and J_2 are the relative concentration of micro regions that correspond to the annihilation of positrons with τ_1 and τ_2. R_{g1} and R_{g2} are the radii of gyration of pores and ρ is the sample density. Since the value of tan δ is much higher than the background loses, it may be assumed that its value is related to the relaxation processes occurring in the polymer.

In the first series of experiments the polymers were investigated before deformation and after completion of volume relaxation. The time between the completion of deformation and the measurements was 36h. Figure 1 shows that sample density decreases with increasing deformation, so free volume increases. To find out which components are responsible for this increase we measured changes in the positron annihilation spectra and tan δ in PI-1. We found that after deformation and relaxation the size of micro-holes and their relative concentration do not increase as compared with those of the initial sample. After relaxation is completed, the value of tan δ is equal to that before deformation.

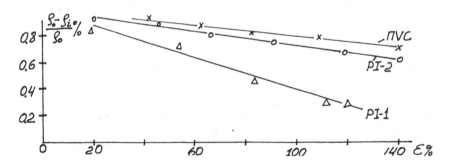

Figure 1 Fractional change in sample density vs strain after deformation. The last point represents failure. Deformation rate was 13.5cm/min.

Figure 2 shows that with increasing deformation of PI-1 , the pore size in the direction normal to the sample surface increases to 240 Å. Moreover, as a result of drawing, pores with radius 60-70 Å appear in the sample. Hence, this experimental data show that pores are the only free volume fraction leading to an increase in the total free volume.

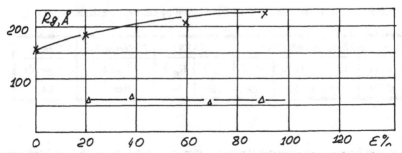

Figure 2 Radius of gyration of pores in PI-1 vs strain before and after deformation.

The next part of the paper deals with the observation of free volume characteristics during deformation and relaxation. The relative change in tan δ during deformation and immediately after relaxation (or failure) shown in Figure 3a and 3b.

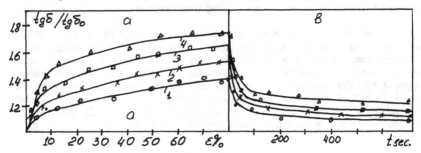

Figure 3 a) Dependence of tan δ on strain during deformation at a constant rate
　　　　 b) Dependence of tan δ on time after sample failure.

The four curves are at different rates of deformation
1, 0.0916 cm/min; 2, 0.26 cm/min; 3, 0.9 cm/min; 4, 3 cm/min.
tgδ_0 is the value before deformation; tgδ that during deformation.

The main features of tan δ will be discussed below. However, it should be noted here that the value of tan δ drastically increases during deformation and remains high throughout the process. This fact implies that the total amount of free volume in this fraction increases. It is clear that tan δ decreases during the volume relaxation; after a few seconds its value decreases by 60-70%. After 5-7 h. the value of tan δ in a deformed sample becomes equal to that before deformation.

Figure 4a and 4b give the results of measurements of positron annihilation during sample volume relaxation after deformation. After deformation was completed, the annihilation characteristics gradually attained values characteristic of the initial sample. The lifetime of the first component $\tau_1 = 210\text{-}170$ ps, strongly depends on relaxation times; the lifetime of the second component, $\tau_2 = 388\pm10$ ps, is independent of the sample state. However, considerable changes in J_2, the intensity of this component, were observed.

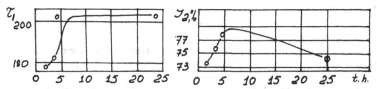

Figure 4 Characteristics of positron annihilation during relaxation of PI-1 sample deformation. Deformation rate 20 cm/min. strain 60%.

The results show two regions that differ in their rate of change. Immediately after deformation τ_1 greatly decreases, and its relaxation to the initial value takes 5-6 h. J_2 exhibits similar kinetics. However, after volume relaxation for 5-6 h. the value of J_2 is much higher than that in the initial sample. At longer times τ_1 does not change, whereas J_2 decreases. After 24 h. it becomes slightly lower than in the initial sample.

DISCUSSION

Changes in tan δ during deformation and volume relaxation

In our opinion, the increase in tan δ during deformation may be due to:

1) An increase in the number of regions with a high local free volume concentration (higher mobility). The mean relaxation time τ remains the same.

2) A constant number of these regions, but the free volume fraction present in them (in the form of micro-holes) increases. The mean characteristic relaxation time τ then decreases.

Our calculations showed that only the first case is realized. Consequently, the increase in tan δ is caused by an increase in the number of dipoles changing their position, that is, by an increase in the concentration of high mobility regions.

The curves in Figure 3a are similar to the stress-strain curves for this polymer. They exhibit three regions. In the first region the value of tan δ drastically increases up to a strain of 8-9%. This coincides with an abrupt increase in stress. In the third region the tan δ value is almost invariable and the second region is an intermediate transition region. As already mentioned, the value of tan δ corresponds to the total amount of the phase with a high local free volume concentration. However, in the first deformation region free volume is not accumulated. [2]. The authors explained this effect by saying that the regions with local high molecular mobility are formed as a result of micro-hole redistribution. The total amount of free volume in the sample does not change.

As already mentioned the presence in the polymer of holes of size 5-7 Å suggest that the regions active in dielectric losses are regions of a fluctuation increase in micro-hole concentration. At strains over 10% tan δ varies only slightly. In this region the change in stress is also slight showing that stress relaxation occurs in the polymer. Stress relaxation is a rearrangement of a region under strain and thus the formation of high mobility region. This should lead to a drastic increase in concentration of regions with a high free volume fraction. Under these conditions the quasi-stationary value of tan δ can probably imply an equilibrium between the formation of such regions and their relaxation. It is assumed that in this part of the curve the regions active in dielectric losses are formed by the generation of excess free volume. Bauwens has also expressed a similar idea[11]. In this case the disappearance of this region drastically increases micro-hole concentration in the sample, and this concentration becomes non-equilibrium. The hole gas is supersaturated and will condense.

Hence, in our opinion, in the process of deformation of a glassy polymer, free volume is generated in the form of concentration fluctuations of the hole gas and this gas condenses as a result of supersaturation. Condensation may occur either homogeneously or heterogeneously. Homogeneous condensation may lead to the formation of small new pores, whereas heterogeneous condensation results in increase of the size of available pores.

This hypothesis is confirmed by the behavior of the system in relaxation after failure (Figure 3b) The most interesting fact here is that in 5-7 h. the value of tan δ becomes the same as before deformation. This is probably related to the disappearance of local high mobility regions during volume relaxation. In the first part of the curves tan δ drops almost instantaneously. We suppose that two processes are possible here: 1. The reverse of the process observed for deformation up to strains of 9%. The strain unloading markedly increases the transition barrier, and this part of fragments becomes inactive and no longer affects tan δ. 2. The condensation of the supersaturated hole gas present in the sample.

Change in annihilation characteristics in polymer volume relaxation.

Fig.4 illustrates changes in annihilation characteristics during polymer volume relaxation after deformation occurring at a constant rate. The shortest component of the time spectrum (τ_1) is complex in nature and is provided by the contribution of at lease two annihilation channels. The short-lived states may result from the annihilation of free positrons caused by their collision with the particles of the medium in the ranges with high packing density and from out annihilation of the parapositronium atoms. We made set of experiments and it was found that these phenomena are absent in our samples.

It seems more probable that a new short-lived dynamic state is formed which is not characteristic of the initial non-deformed polymer. In our opinion, the observed effect is due to positron interaction with the regions of local increase in molecular mobility. In these regions a complete set of modes characteristic of a given volume may exist, among them low frequency modes. This may lead to local increase in the electronic density in these regions. Finally, regions of local increase in mobility with very short life-times may appear. A positron in this region will annihilate during the life-time of this region. This assumption is in good agreement with the conclusion made in the analysis of tan δ changes. The increase in τ_1 to the initial value probably results from a decrease in the concentration of this region. This is the reason for the correlation between the time dependencies of τ_1 and J_2 for the first 5-6h. (Figure 4). It is assumed that the observed decrease in the content of the long-lived component is due to a decrease in the concentration of micro-holes with d=5-7Å which diffuse out of the sample.

This study of changes in the characteristics of positron annihilation, tan δ and density during deformation and volume relaxation after deformation permits the formulation of a hypothesis of the mechanism of free volume relaxation in glassy polymer during forced highly elastic deformation. During deformation, free volume is generated in the form of local regions of unfreezing of molecular mobility. The regions of local devitrification are characterized by a fluctuating increase in concentration of these micro-holes. In the process of forced highly elastic deformation, excess regions of local devitrification are generated. Their concentration drastically increases, the equilibrium between fluctuations and hole gas shifts towards the hole gas, and these regions begin to be separated into micro-holes. In this case, the hole gas concentration supersaturates. This can lead to both homogeneous condensation of the hole gas resulting in the formation of new small pores and to heterogeneous condensation that induces an increase the number of the existing pores. After the completion of mechanical tests, the non-equilibrium system tends to equilibrium. The excess hole gas is rapidly condensed, then a much slower process takes place: the separation of regions of local devitrification into micro-holes. Finally, the slowest process occurs. This is diffusion of holes beyond the sample boundaries. Eventually, the matrix-hole gas system comes to equilibrium, and hole gas concentration becomes equal to that before the beginning of mechanical tests.

LITERATURE

I. L.C.E.Struik,Physical Aging in Amorphous Polymers and Other Materials,(Elsever Scientific Publishing CO.,Amsterdam, I978), p.270.

2. J.D.Ferry, Viscoelastic Properties of Polymers, 3rd. ed. (J.Willy and Sons,New York,I983)p.284.

3. R.H.Haward,J.Colloid and Polymer Science,258,643(I980).

4. L.G.Stuc, J.Polymer Science, Pt.B: Polymer Phys., 27, 256I (I989).

5. V.Rostiashvili, V.Irzac, B.Rozenberg, Steclovanie Polymerov,(Himia, Leningrad,I987),p.I88.

6. M.H.Cohen, P.Grest, Phis.Rev. B, 20,I077(I979); Phys.Rev. B, 2I,4II3(I980).

7. V.Bouda, Polymer Bulletin,7,639(I982); Polymer Degradation and Stability,24,3I9(I989).

8. R.Simha, T.Somcynsky, Macromolecyles,342(I979).

9. J.C.Bauwens, Polymer,2I,699(I980).

I0.T.R.Russell, Polymer Engineering and Science,24,345(I984).

II.J.H.Wendorf, Progress in Colloid and Polymer Science,66,I35(I979);Polymer,2I,533(I980).

I2.J.G.Victor, J.M.Torcelson, Macromolecules,2I,3690(I988).

I3.G.I.Lashkov, Visokomoleculiarnie soedinenia,A22,I0(I983).

I4.Metodi Positronnoi Diagnostiki i Rasshfrovki Spectrov Annigiliatsii Positronov(Nauka,Tashkent,I985),p.3I2.

I5.A.J.Hill, P.J.Jones, J.H.Ling and S.W.J.Rapsell, Polymer Science.A,26,I54I(I986).

I6.M.D.Poliks, T.Gillion, J.Scenery, Macromolecules. 23, 2764 (I986).

17.K.L.Nazi,G.Pytas, J.Polymer Science.B, Polymer Physics,24,1683,(1986).

18.A.L.Kovarsky, Dissertastia(Moskva,1989)p.169.

19.M.I.Tsapovetsky, L.A.Laius, M.I.Bessonov, M.M.Koton, Docladi A.N.USSR,256,912(1981).

20.V.I.Goldansky, Fisicheskaa Himia Positron a i Positron ia. (Nauka, Moskva, 1968),p.173.

21.P.N.West, Positron Studies of Condensed Matter. (Taylor and Francis Ltd.,1974),p.124.

22.P.Kindl, G.Reiter, Phys.Stat.Sol.,A,104,707,(1987).

23.P.Arifov, S.Vasserman, S.Tishin, Phys.Stat.Sol.,A,102,565 (1987)

24.T.I.Borisova, Dissertatsia(Leningrad, 1977)p.265.

25.P.Kirkegaard,J.Pedersen,M.Eldrup, Tech. Rep. Risa-M-2740,1989; Risa National Laboratory. DK-4000 Roskilde Denmark.-133p.

26.E.S.Dole, G.M.B.Mahbonbian-Jones, R.A.Pethrick, Polymer Communication, 26,262(1985).

DIELECTRIC RELAXATION OF SEMICRYSTALLINE POLYIMIDES

PENGTAO HUO, JEROME FRILER AND PEGGY CEBE
Department. of Materials Science and Engineering
Massachusetts Institute of Technology, Cambridge, MA 02139, U.S.A.

1. Introduction

Aromatic polyimides because of their structure process a unique property combination of outstanding thermal stability and radiation and solvent resistance. Their excellent mechanical and electrical properties make polyimides the materials of choice for a wide range of applications in the electronics industry. Polyimides have long been amorphous materials but recently, flexible hinge groups containing ether and ketone linkages have been incorporated into the polymer backbone [1]. This results in increased flexibility and development of crystallinity. The introduction of crystallinity is an effective means of improving the melt processability, however, little work has been done to determine the effects of crystallinity on the insulating characteristics of polyimides. Here we report the dielectric behavior for the semicrystalline polyimide New-TPI [2], which is synthesized by Mitsui Toatsu Chemical Co. and film processed by Foster-Miller, Inc. We have used dielectric relaxation to study New-TPI, especially its behavior near T_g.

Crystallinity has an important effect on the dielectric properties of other high performance polymers such as poly(phenylene sulfide), PPS [3]. Rigid amorphous phase, that portion of the amorphous phase still in the solid state at temperatures just above glass transition temperature [4], was found to relax little by little as the temperature was increased. In contrast, for New-TPI we find that all of the amorphous phase relaxes over a narrow temperature range near the glass transition.

2. Experimental Section

The as-received New-TPI film is a transparent amorphous film as seen by x-ray. The amorphous as-received film was dried in a Mettler hot stage at 150°C for 20 hours, then relaxed at 260°C for 20 hours. This amorphous film is called amorphous film. Semicrystalline sample was prepared by annealing the amorphous film at 300°C for 20 minutes. DSC scans were performed on a Perkin Elmer DSC-4 using a scan rate of 20°C per minute.

Mat. Res. Soc. Symp. Proc. Vol. 227. ©1991 Materials Research Society

For dielectric relaxation experiments, both amorphous film and cold crystallized semicrystalline films were coated with a gold layer of thickness around 300 Å. The capacitance and loss factor of the samples were measured by using a Hewlett Packard impedance analyzer with frequency range from 1000 Hz to 1 MHz over the temperature range from 150°C to 320°C.

3. Results and Discussion

DSC scans for amorphous and semicrystalline samples were shown in Figure 1. The glass transition temperatures are 253°C and 255°C for amorphous and semicrystalline samples, respectively. The degree of crystallinity for the semicrystalline sample, 20%, was obtained by wide angle x-ray scattering(WAXS), indicating the total amorphous phase is 80%. We also studied the heat capacity step at T_g for the semicrystalline sample. The fraction of liquid-like amorphous PPS was obtained by using the ratio of heat capacity increment of the semicrystalline sample to that of amorphous New-TPI at the glass transition temperature. The fraction of rigid amorphous phase was obtained by subtracting the fraction of liquid-like amorphous phase from that of the total amorphous phase [3]. For the semicrystalline material we studied, the fraction of liquid-like amorphous phase is about 75%. Therefore, there is only 5% rigid amorphous phase.

Dielectric relaxation results in the temperature range from 150°C to 320°C are shown in Figure 2 for amorphous New-TPI. The value of ε' is roughly constant prior to the relaxation, then increases sharply. The loss factor (tanδ) shows a maximum whose peak position shifts to higher temperature with increasing frequency. There is a slightly change in slope of ε' at about 297°C which indicates that the amorphous film is crystallized during the heating. In Figure 3, we present the ε' and tanδ as a function of frequency at several temperatures close to glass transition temperature. A peak of tanδ in frequency space shows a shift to higher frequency as a consequence of temperature increase. ε' and tanδ of semicrystalline film are shown in Figure 4. Both ε' and tanδ have quite similar behaviors, compared to amorphous sample shown in Figure 2. However, semicrystalline film has a weaker relaxation intensity. This is simply because the semicrystalline film can be viewed as a composite system in which one phase relaxes (the amorphous phase) but the other does not relax. The ε' and tanδ as a function of frequency are shown in Figure 5 for the semicrystalline New-TPI.

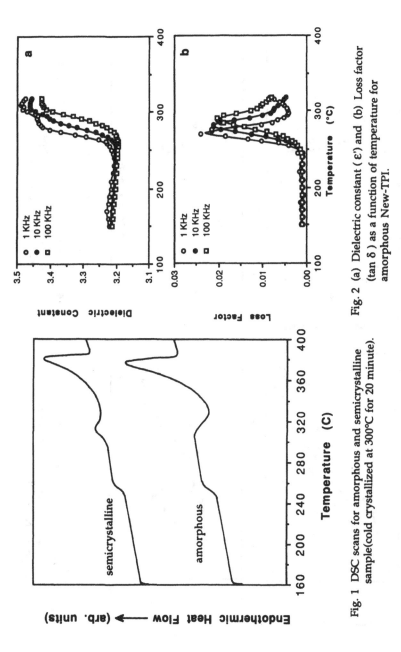

Fig. 1 DSC scans for amorphous and semicrystalline sample(cold crystallized at 300°C for 20 minute).

Fig. 2 (a) Dielectric constant (ε') and (b) Loss factor (tan δ) as a function of temperature for amorphous New-TPI.

242

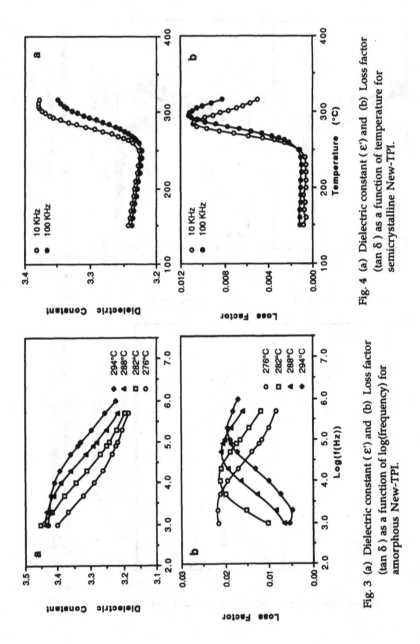

Fig. 4 (a) Dielectric constant (ε') and (b) Loss factor (tan δ) as a function of temperature for semicrystalline New-TPI.

Fig. 3 (a) Dielectric constant (ε') and (b) Loss factor (tan δ) as a function of log(frequency) for amorphous New-TPI.

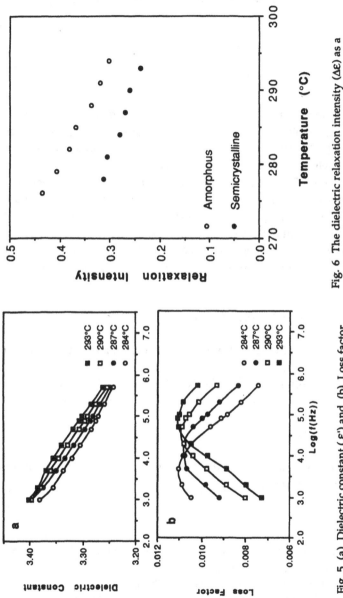

Fig. 6 The dielectric relaxation intensity (Δε) as a function of temperature for amorphous and semicrytsalline New-TPI.

Fig. 5 (a) Dielectric constant (ε') and (b) Loss factor (tan δ) as a function of log(frequency) for semicrystalline New-TPI.

Cole-Cole plots were used to get the relaxation intensity. A circle was used to fit the Cole-Cole plot because of the symmetry, which implies $a_2=1$ for :

$$\hat{\varepsilon} = \varepsilon_\infty + \frac{(\varepsilon_s - \varepsilon_\infty)}{\left(1+(i\omega\tau)^{a_1}\right)^{a_2}} \tag{2}$$

where $\hat{\varepsilon}$ is the complex dielectric function, ε_s and ε_∞ are the low and high frequency limits of the real part, ω is frequency and τ is relaxation time. The parameters a_1 and a_2 characterize the departure from the Debye equations. The intensity of relaxation, defined by $\Delta\varepsilon = \varepsilon_s - \varepsilon_\infty$ was obtained from the Cole-Cole plot $\varepsilon'' = 0$ intercepts. The relaxation intensity of the amorphous and semicrystalline samples both decrease as temperature increases, as shown in Figure 6. They also show a roughly equal slope. This is very different from other high performance polymer we have studied such as PPS [3] which shows an opposite trend. For PPS, amorphous $\Delta\varepsilon$ decreases and semicrystalline $\Delta\varepsilon$ increases. By analogy to the heat capacity increment, we define :

$$\beta(T) = (\Delta\varepsilon)_{semicryst.}/(\Delta\varepsilon)_{amorp.} \tag{3}$$

as the total amount of material that has already been relaxed at T. We found that $\beta(T)$ is about 0.75 independent of temperature from 275°C to 290°C. These results simply indicate that for New-TPI there is 5% rigid amorphous phase and it is still "rigid" up to temperatures 30°C higher than its T_g.

Acknowledgement

Research was supported by the Petroleum Research Fund, administered by the American Chemical Society.

References

1. P. M. Hergenrother, N. T. Wakelyn, and S. J. Havens, J. Polym. Sci., Polym. Chem., 25, 1093-1103 (1987).
2. Mitsui-Toatsu Chem., Inc. Technical Data Sheet/A-00
3. P. Huo and P.Cebe, Mat. Res. Soc. Symp. Proc., 215, 93(1991).
4. H. Suzuki, J. Grebowicz, and B. Wunderlich, Makromol. Chem., 186, 1109-1119 (1985).

PART III

Applications and Processing

POLYIMIDE CONSIDERATIONS IN THIN-FILM ELECTRONIC PACKAGING

W. P. Pawlowski
IBM, Systems Technology Division, 1701 North St., Endicott, NY 13760

ABSTRACT

Free-standing polyimide film is used extensively in tape automated bonding (TAB) and flexible printed circuit (Flex) applications because of its excellent chemical, mechanical, and electrical properties. Kapton˙ polyimide film has been the material of choice for many years. Recently, a new polyimide film, Upilex˙˙, has been introduced into the marketplace. The primary property differences between Kapton and Upilex result from differences in the chemical composition of each film. A review of property/product performance indicates that Upilex S is preferred with respect to dimensional stability, metal/polyimide and adhesive/polyimide adhesion, chemical inertness and glass transition temperature.

INTRODUCTION

The electronics industry is in constant pursuit of dielectrics that outperform those presently used in product fabrication. Although polyimides have a dielectric constant of about 3.5 and can absorb up to 3% by weight of water, they continue to be the material of choice. This is because their other electrical, chemical, and mechanical properties more than offset the aforementioned deficiencies. There are two primary methods of preparing thin film TAB and flexible printed circuits. The first involves subtractively etching a circuitry pattern in a continuous metal (typically copper) layer of appropriate thickness (0.5 to 2 ounce) using photoresist. This continuous metal layer is laminated to the dielectric using an adhesive, and the 3-layer structure (metal/adhesive/dielectric) is available in both roll and panel form. The adhesives used are typically dry films of epoxy or acrylic polymers. The second method involves additive plating of a metal (typically copper) onto a seed layer that was previously sputtered onto the dielectric. The seed layer typically contains a copper surface layer which acts as nucleation and growth site for the additive copper plating and a metal adhesion layer (typically nickel or chrome) between the dielectric and copper seed layers. The copper is plated to the desired thickness in the developed channels of photodefined resist. The photoresist is stripped and the seed layer is removed to isolate the circuitry. Parts prepared in this manner are referred to as 2-layer or adhesiveless. In either case, the dielectric is critical to the success of the TAB or flex circuit. This paper describes some information regarding Kapton H and Upilex S and circuits made thereof.

˙ Kapton is a registered trademark of E.I. Dupont de Nemours & Co.

˙˙ Upilex is a registered trademark of UBE Industries, Ltd., Japan.

Mat. Res. Soc. Symp. Proc. Vol. 227. ©1991 Materials Research Society

EXPERIMENTAL

Samples of Kapton H and Upilex S (2 mils thick) were cut into 0.75 X 1.25" pieces and stored at ambient until use. Reagent grade methylene chloride (MC) was used as received. The samples were weighed using a Mettler AE240 Analytical Balance prior to immersion. They were then exposed to MC at 25°C for specified intervals, wiped dry, and re-weighed on the analytical balance. The fractional weight gain of the of the two films were compared.

Test pieces of adhesiveless circuitry prepared with a chrome/copper seed layer and having the same design were built using both Kapton H and Upilex S (2 mils thick) as the dielectric. Lines (8 mils wide) were peeled on an Instron tester at 2 in./min. after specified intervals of temperature and humidity (T&H) conditioning at 85°C and 80% relative humidity.

Dimensional changes at several points in the X and Y direction over a distance of 14.5 and 10", respectively, were measured using an X/Y measuring table calibrated to +/-0.0001" at 200X magnification. All experiments were run in triplicate.

RESULTS AND DISCUSSION

Property Comparison

The chemical composition, molecular weight and method of film formation are controlling factors in the chemical and mechanical properties of the resultant polymer film. Some typical properties of Kapton H and Upilex S are shown in Table I [1].

Table I. Properties of polyimide films.

Film Type	Density (g/ml)	Moisture* Regain (%)	Tg (°C)	TCE► (ppm/°C)	Dielectric Constant	Modulus✿ (mPa)
Kapton H	1.42	2.8	360-410	20	3.5	2,944
Upilex S	1.47	1.2	500	8	3.5	8,825

*Immersion in water at 23°C for 24 hours.
►Thermal coefficient of expansion (X10⁶cm/cm/°C).
✿Tensile modulus at 25°C.

The higher density of Upilex S results primarily from replacement of the flexible oxydianiline portion of the pyromellitic dianhydride-oxydianiline (PMDA-ODA) polyimide used to prepare Kapton H with the rigid p-phenylene diamine segment of the biphenyl dianhydride-oxydianiline (BPDA-PDA, Upilex S). The moisture regain values of Kapton H and Upilex S support this. The method of film formation also attributes to the difference in density and moisture regain [1]. The thermal coefficient of expansion is significantly lower for Upilex S compared to Kapton H which should minimize dimensional changes through processing. The higher modulus of Upilex results from the increased rigidity of the polyimide backbone.

Dimensional Stability

Dimensional stability is required of TAB and Flex circuits for accurate matching of circuit to card and device. This becomes more critical as line pitches are extended to the envelope of current processing capability. In addition, dimensional changes during processing can adversely affect intra-part alignment when multiple photoresist passes are used. The intra-part dimensional changes can be corrected by compensation of sequential artwork, however, the overall product dimension must meet the desired tolerance set. Table 2 shows dimensional changes for parts prepared with Upilex S and Kapton H.

Table II. Dimensional stability of test vehicles.

Film Type	Change in X (ppm)	Change in Y (ppm)
Kapton H	-950	+310
Upilex S	-480	+140

Parts prepared with Upilex S show a similar trend but a smaller magnitude change. Lower dimensional changes for Upilex S compared to Kapton H were also found by Holzinger [2] for TAB parts prepared using the two dielectrics. The resultant dimensional tolerance process capability index was 4.21 for parts prepared with Upilex S compared to 1.52 for Kapton H indicating increased precision [2].

Metal/Polyimide and Adhesive/Polyimide Adhesion

Circuit line adhesion is critical to the reliability of TAB and Flex circuits as line lifting may result in intra- or inter-part shorting as well as increased stress at interconnects. The peel strength values of the test vehicles (8 mil lines) versus time in T&H are shown in Figure 1. Those prepared with Upilex S are less susceptible to degradation through T&H conditioning. Holzinger [2] reported a similar trend for 3 layer TAB parts after both thermal and T&H conditioning. Since the functional groups of both films are the same, it appears that peel strength is probably more adversely affected by the physical properties of the substrate as opposed to a difference in metal/polymer chemical bonding interactions. This difference may be attributed to the higher modulus, lower moisture regain (and hygroscopic expansion) and/or smaller thermally induced dimensional changes during stressing of Upilex S compared to Kapton H. Peel strength reduction may be compensated by employing overcoats (liquid or film).

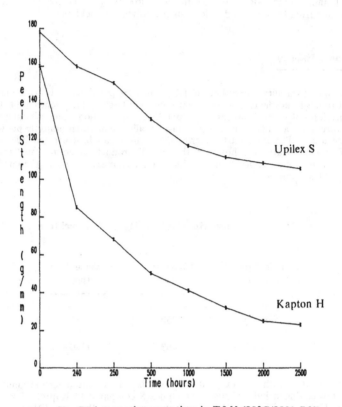

Figure 1. Peel strength versus time in T&H (80°C/80% RH)

Solvent absorption

The fractional increase in weight as a function of exposure time in MC is shown in Figure 2 for Upilex S and Kapton H [1] The equilibrium weight gain of Kapton H was about 22 weight percent after an exposure of approximately 10 minutes. Upilex S showed no detectable weight gain after 24 hours and negligible weight gain after 1200 hours of exposure. The increased rate of MC absorption for Kapton H compared to Upilex S probably results from a combination of differences in chemistry (increased rigidity and packing of Upilex S) and film preparation techniques. The decreased rate of solvent absorption for Upilex S minimizes the susceptibility of the circuit to stress induced failure resulting from chemical leaching or ionic transport during T&H conditioning. However, this can also be a hindrance as Upilex S is more difficult to chemically etch compared to Kapton H.

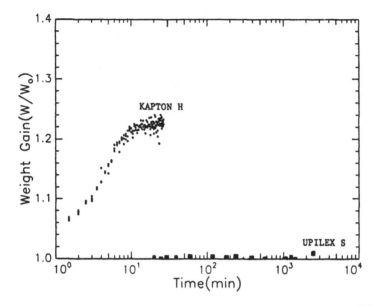

Figure 2. Fractional weight gain as a function of time in MC at 25°C.

CONCLUSIONS

The improved dimensional stability, metal/polyimide and adhesive/polyimide adhesion, solvent absorption, moisture regain and Tg of Upilex S compared to Kapton make it more desirable in thin-film applications. The emergence of several new polyimide films including Kapton E and ZT will be incorporated into future studies.

REFERENCES

1. W.P. Pawlowski, M.I. Jacobson, M.E. Teixeira, and K.G. Sakorafos, in <u>Solvent Diffusion in Selected Polyimide Films</u>, edited by A.T. Barknecht, J.P. Partridge, C.J. Chen, C.Y. Li (Mater. Res. Soc. Proc. <u>167</u>, Boston, MA 1989) pp. 147-152.

2. S. Holzinger, TAB: Mechanical, Chemical, and Thermal Considerations for Three Layer TAB Materials, Presented at EXPO SMT International, Las Vegas, NV, 1989.

REFERENCES

UV-LASER PHOTOABLATION OF THERMOSTABLE POLYMERS: POLYIMIDES, POLYPHENYLQUINOXALINE and TEFLON AF

Sylvain LAZARE*, Hiroyuki HIRAOKA**, Alain CROS***, Régis GUSTINIANI***

*Laboratoire de Photochimie et de Photophysique Moléculaire, URA 348 du CNRS, Université de Bordeaux I, 351 cours de la Libération, 33405 Talence, FRANCE
**Hong Kong University of Science and Technology, Kowloon, Hong Kong
***Groupe de Physique des Etats Condensés, URA 783 du CNRS, Faculté des Sciences de Luminy, 13288 Marseille Cédex 9, FRANCE

ABSTRACT

The UV laser ablation of three thermostable polymers, polyimide, polyphenylquinoxaline and Teflon AF of interest for microelectronics is presented. Direct absorption of the laser beam is not possible by Teflon, but a doping technique is described.

I INTRODUCTION

Photablation is the spontaneous etching that occurs when a laser pulse of high-intensity (MW/cm^2) ultraviolet light of a laser is absorbed at the surface of a material [1]. It is now developed as a widespread etching technology in microelectronics industry [2] for via holes formation in dielectric layers [3], necessary for integrated circuit fabrication and packaging. Photoablation of polyimide (PMDA-ODA) which has been used for more than a decade as a dielectric material in place of SiO_2, is the most documented case [4]. Recently, we studied the excimer laser photoablation of other polymers with good thermostability including polyphenylquinoxaline (PPQ) (see Fig.1) and Teflon AF, a new copolymer of tetrafluoroethylene and 2,2-bis(trifluoro-1,3-dioxolene) [5]. These two materials display low dielectric constants, respectively, 2.8 and 1.9, favorable for improvement of the rate of data exchange, by lowering the capacitive coupling.

Polyphenylquinoxaline, synthesized by condensation of aromatic, tetraamine and tetraacetone, displays high temperature stability (550/600°C) and good solubility in a xylene/meta-cresol mixture.

Teflon AF unlike regular polytetrafluoroethylene has a good solubility in various fluorinated solvents owing to the presence of dioxolene as comonomer. As a result it can easily be processed in essentially amorphous thin films, suitable for application in microelectronics. These two polymers, however, suffer from poor adhesion to the substrate in comparison with polyimide, but this rather important question is not discussed in this paper.

We compare the photoablation properties of these polymers and emphasize the important molecular parameters to consider prior to any application. We show that for highly absorbing polymers (see Fig.1) like PI and PPQ, direct ablation is possible at any wavelength of the excimer laser. Teflon, having virtually no absorption, must be doped with small molecules whose function is to absorb the laser radiation and to produce the volume excitation necessary to ablation.

In our next paper in this issue [6], we report more extensive data on the surface modifications introduced by ablation on PI and PPQ as measured by surface analytical techniques (XPS and contact angles). It is also shown that these modifications drastically change adhesion (metallization) and wetting of the polymer surface.

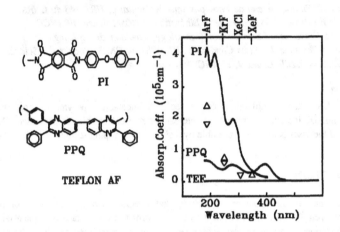

FIG.1: Absorption spectra of thin films of polyimide (PMDA-ODA), polyphenylquinoxaline PPQ and Teflon AF and ablation screening coefficient ß as defined in section III, at various wavelengths of the excimer laser for PI ▽ and PPQ △.

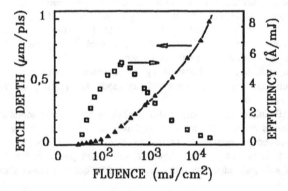

FIG. 2: Etch curves of PPQ ablated at 248 nm in air, and ablation efficiency as a function of laser fluence. Other kinetic data are given elsewhere [8].

II EXPERIMENTAL

-*Excimer lasers* are the most widely used UV lasers for photoablation. This work was done with a Lambda Physik EMG 200E. Its radiation is pulsed (25 ns) and its main wavelengths

193 nm (ArF), 248 nm (KrF), 308 nm (XeCl), 351 nm (XeF) can be used in air. A shorter wavelength 157 nm (F$_2$) will be increasingly used for materials like Teflon that do not absorb at the above wavelengths, but has to be used in vacuum or in a non absorbing medium since air strongly absorbs it.

-*The polymers* used in this study are from CEMOTA (polyimide and polyphenylquinoxaline) and from DU PONT (Teflon AF). They were spin casted from solutions: polyimide in N-methylpyrrolidone, polyphenylquinoxaline in xylene/meta-cresol and Teflon AF in perfluorinated solvents. The dopant, tris(perfluoroheptyl)triazine was purchased from ALDRICH.

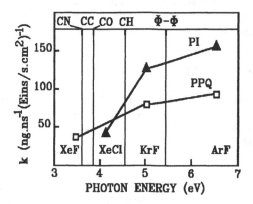

FIG.3: Ablation rate constants for PI and PPQ as a function of photon energy. Comparison can be made with the energies of the covalent bonds present in the polymers.

III PHOTOABLATION KINETICS

The ablation curves like that of PPQ at 248 nm displayed in Fig.2 have been measured with the quartz crystal microbalance technique [7], which is very sensitive and accurate. Various experimental conditions and polymers have been explored as reported elsewhere [8]. These curves can be interpreted with the moving interface model, that we want to briefly review here. As shown in Fig.2, the energetic efficiency of the UV-beam, above the ablation threshold reaches a maximum and then decreases to a lower value with increasing fluence. This phenomenon is called the screening effect and is due to the attenuation of the beam by the ablation products. In the model it is characterized by a coefficient of absorption ß, called the screening coefficient. Furthermore, we consider that the interface between products and the excited polymer moves at a speed $v = k(I_0 e^{-\beta x} - I_t)$. k is the ablation rate constant, $I_0 e^{-\beta x}$ is the screened intensity when the ablation depth is x and I_t is the ablation threshold. This model was demonstrated to reproduce the experimental data fairly well, after iterative fitting. It provides values k and ß that are used to determine the influence of the polymer structure or experimental conditions.

Values of k for PI and PPQ are displayed in Fig.3, and ß in Fig.1. These parameters are obtained for ablation in air, but background pressure has little influence on them. Rather photon energy plays a major influence. The more the energy, the higher the rate constant. This is understood by comparing photon and bond energies. If the photon exceeds the bond energies, rate constants are higher. Therefore polymer structure is also a major factor that governs rate of ablation. As we reported earlier, non-aromatic polymers ablate faster than aromatic ones which gain some stability from their numerous double bonds. Similarly, the screening effect is stronger for aromatics. ß values for PI and PPQ are displayed in Fig.1. These coefficients can be considered as high intensity absorption coefficients of the polymer which gradually changes its starting coefficient when it ablates under the action of the laser beam. It is interesting to note here that PI absorption decreases whereas that of PPQ increases, reaching approximately the same values.

IV PHOTOABLATION OF TEFLON AF

We recently demonstrated the ablation of doped Teflon AF, in a preliminary communication [9]. Ablation of regular Teflon has been attempted by ultra-short pulses, that makes coherent 2-photon absorption possible [10]. Recently sputtered Teflon thin films have been studied [11].

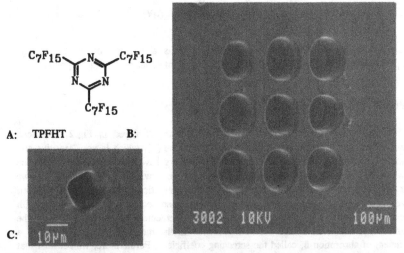

Fig.4 A: Chemical structure of the dopant molecule used for ablation of Teflon AF. B: SEM picture of a 100 μm thick film of Teflon AF doped with 50 wt% of tris(perfluoroheptyl)triazine and ablated in air through a metallic mask with one hundred pulses of 193 nm radiation of the excimer laser. C: 10 μm holes made by 193 nm ablation showing the resolution of the etching.

Teflon AF can dissolve small molecules like perfluoroheptyltriazine (TPFHT) (see Fig.4A) that are absorbing at the laser wavelengths. A 50% doping is possible owing to the good compatibilty with the polymer, and is necessary to have a reasonable absorption coefficient at 193 nm. Smaller dopant molecules with higher absorption coefficient like perfluorinated biphenyl or naphthalene were envisaged, but failed to give a nicely doped material because it evaporated away from the polymer in a few minutes owing to the lack of compatibility. Fig.4B shows a 100 μm thick film of 50% doped Teflon AF on silicon substrate after ablation with one hundredth of 193 nm pulses (1J/cm^2) through a metallic contact mask consisting of a matrix of circular holes (ϕ = 120 μm). The ablation rate is estimated to be approximately 1 μm/pulse in this case. The film of Fig.4B still contained the dopant, and it is seen that the holes are perfectly cylindrical. Removal of the doping molecules requires only a quick (5 min) heating of the film at 160°C (T$_g$ of Teflon AF 1600). Of course this annealing is accompanied with some shrinkage of the film (50%) which may or may not be isotropic depending on the conditions. Holes of Fig.4B are shown in Fig.5A after this simple dedoping procedure. It is clear that they are no longer cylindrical; however the bottom is seen to retain its original diameter. It is hoped that simulations will bring some basic understanding. Fig.4C shows 12 μm square holes that demonstrate the resolution capability of the ablation of doped Teflon AF. Again here the dedoping was shown to retain most of the shape of the hole and no extensive flow of the polymer was detected.

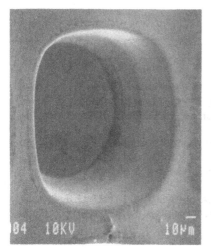

A: B:

FIG.5 A: Holes (120 μm in diameter) in Teflon AF made by UV laser ablation (193 nm) after dedoping of the material by heating at 160 °C for 5 min. B: 193 nm ablation of a 16 wt% doped Teflon film. Due to low concentration of dopant the absorption is too weak to trigger clean ablation exempt of unwanted thermal effects.

The mechanisms of the ablation step are still speculative and experiments are now directed at their understanding. For instance, a low level of doping (16 %) at the same fluence ($1J/cm^2$) and wavelength (193 nm) shows a difficult ablation, as in Fig.5B. Most of material is expelled as molten droplets, which tend to redeposit. The ablated surface is very irregular, a typical behaviour of a weakly absorbing material. This kind of slow ablation could only be due to a thermal degradation of the polymer. This shows the necessity of the high level of doping (>25 % with TPFHT) or the search for more absorbing molecules. It is also an interesting question to determine whether the chemical nature of the dopant is an important factor. For instance an unstable, gas forming, molecule might be suitable, but compatibility with the perfluorinated polymer has to be solved.

V CONCLUSION

Understanding of the ablation of high temperature polymers like polyimide and polyphenylquinoxaline can be obtained with the aid of the model of the moving interface, which describes the rate as a function of two basic parameters k and ß, respectively the rate constant and screening coefficient. Teflon AF, a soluble polymer that can be processed in thin film, is non absorbing and must be doped with TPFHT before ablation at 193 nm. Via holes have been etched in thick films (100μm) of doped (50%) Teflon with a resolution of 10 μm. Dedoping by annealing, retains the geometry of the hole but the film undergoes some shrinkage.

REFERENCES

[1] For reviews see: R. Srinivasan, Science 234 559 (1986) and R. Srinivasan, B. Braren, Chem.Rev. 89 1303 (1989); S. Lazare, V. Granier, Laser Chem. 10 25 (1989); J.H. Brannon, J.Vac.Sci.Technol. B7 1064 (1989).
[2] J.H. Brannon, US Patent 4,506,749 (1985) and F. Bachmann, Chemtronics 4 148 (1989).
[3] R.J. Jensen in Microelectronics Processing: Chemical Engineering Aspects, ACS Adv.Chem.Ser. 221 441 (1989)
[4] R.Srinivasan, B.Braren, R.W. Dreyfus, J.Appl.Phys. 61 372 (1987); J.H. Brannon, J.R. Lankard, A.I. Baise, F.Burns, J. Kaufman, J.Appl.Phys. 58 2030 (1985); W. Spies, H. Srack, Semicond.Sci.Technol. 4 486 (1989)
[5] P.R. Resnick, Polym.Preprint 31 312 (1990)
[6] H. Hiraoka, S. Lazare, A. Cros, R. Gustiniani in this issue
[7] S. Lazare, J.C. Soulignac, P.Fragnaud, Appl.Phys.Lett. 50 624 (1987)
[8] S. Lazare, V. Granier, ACS Sympos.Ser.: Polym. in Microlithography 412 Chap.25 411 (1989)
[9] H.Hiraoka, S. Lazare, Appl.Surf.Sci. 46 342 (1990)
[10] S. Kuper, M. Stuke, Microelectron.Eng. 9 475 (1989)
[11] J.H. Brannon, D. Sholl, E.Kay, Appl.Phys.A 52 160 (1991)

SPIN COATING PLANARIZATION OF INTEGRATED
CIRCUIT TOPOGRAPHY ON A ROTATING DISK

ROGER K. YONKOSKI and DAVID S. SOANE
Department of Chemical Engineering, University of California Berkeley, Berkeley CA 94720.

ABSTRACT

Polyimide is commonly used in the microelectronic industry for interconnection applications because of its ability to planarize features typically found on an IC chip. A mathematical model is developed to describe fluid flow on a rotating disk based on the principles of mass and momentum conservation. Constitutive relationships necessary for this model are proposed. Experimental data for polyimide precursor solutions are presented which enable the determination of parameters for the constitutive equations. This model is used to describe the film profiles over flat surfaces and near micron-sized features. Attention is focused on the coupling between mass transport and fluid flow as well as the effects of surface tension on film profiles over topographical features.

INTRODUCTION

Polyimides have found wide acceptance by the semiconductor industry as they exhibit low relative permittivity, high radiation resistance, excellent chemical and thermal stability, and planarizing ability. Thin films of polyimide are applied by spin coating a solution of polyimide precursor in its solvent followed by a cure step at an elevated temperature. As microcircuitry becomes more dense, it is essential that these films cover the circuitry in as planar a manner as possible in order to perform the necessary photolithography effectively. In addition, the reliability of these circuits is enhanced with a planar film through diminished chance of shorting line connections. Therefore, understanding the physics of the spin coating process near small features is of fundamental importance.

A mathematical model describing both the mass and momentum transport near an axisymmetric topographical feature on a rotating disk is being developed. The equations are based on the conservation of momentum and mass and are solved numerically using the finite element technique. Constitutive relationships based on polyimide precursor solution are coupled with this model in order to predict behavior typical of conditions occurring on wafers in the microelectronics industry The purpose of this paper is to present the progress of this work and any conclusions that we may make at this time. We will focus on the solutions to spin coating on a flat substrate as this is the logical first step to solving the planarization problem. These solutions will then be used as a boundary condition for the more complete problem which includes substrate topography.

MODELING APPROACH

The model topography is envisioned to be an axisymmetric profile far from the center of the disk. Because we are choosing an axisymmetric step we may then solve a swirling flow problem which is essentially two-dimensional with a constant azimuthal velocity. In addition, the length scale of the area around the topographical feature is small compared to the distance of the feature from the center of the disk. Therefore, we can use Cartesian coordinates to solve the problem near the feature and take the centrifugal acceleration as a constant body force. Further, we assume that far from the topographical feature the fluid behaves as if no feature is present. This allows us to take as boundary conditions to the planarization problem the flow and concentration profiles which would occur on a flat substrate. Figure 1 illustrates this proposed procedure where inflow conditions are those predicted by the solution of the spin coating problem. As outflow conditions we assume that the fluid will regain this flat plate profile sufficiently downstream from the disturbance. This modeling technique is similar to the approach for flow past a constriction in a pipe. Hence, the first step in solving the complete problem is to solve the spin coating flow of a polymeric material on a rotating disk.

Mat. Res. Soc. Symp. Proc. Vol. 227. ©1991 Materials Research Society

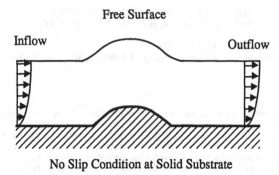

Free Surface

Inflow   Outflow

No Slip Condition at Solid Substrate

Figure 1. The approach taken to solve the planarization
problem of an axisymmetric topographical feature. Inflow
and outflow are taken to be solutions of the spin coating
problem with a flat substrate.

SPIN COATING INTRODUCTION

Spin coating of resist and polyimide films onto silicon wafers is an important step in the
fabrication of integrated circuits. A complete review of the process may be found elsewhere,[1]
and we will include only a cursory review which emphasizes those points important for our
discussion. The spin coating process is carried out by dispensing a sufficient amount of liquid
onto a wafer to flood it with the casting solution and then rapidly accelerating to the final spin
speed. Rotation proceeds for a specified amount of time during which the film thins by a
combination of fluid flow and evaporation. Finally the disk is decelerated to rest. This
procedure may then be followed by a baking step to remove residual solvent, or in the case of
polyimide films, to cure the polymeric precursor to its desired final form.

Because of the importance and complexity of the spin coating process a considerable amount
of effort has been directed to establishing empirical relations which describe the dependence of
spin coating on process parameters. The basic results of these investigations were that the final
film thickness was proportional to a function of the the initial solvent concentration and to some
power of the rotational speed:[2-7]

$$h_f \sim f(C_o) \omega^b .$$

(1)

Often b was found to be -1/2, although some authors concluded that it depended on the
evaporation conditions. Daughton and Givens[6,7] described an extensive experimental
investigation of spin coating parameters. They found the film thickness to be independent of the
amount of fluid dispensed, the disk rotational speed at which the fluid was dispensed, and the
acceleration rate to the final spin speed. The film thickness, however, was found to depend
strongly on the final rotational speed and spin time as well as the material properties of the fluid.

Many attempts have been made at modeling the spin coating process in varying degrees of
complexity. The essence of these models[8-15] is to derive an kinematic expression for the film
thickness,

$$\frac{\partial h}{\partial t} = -\frac{\rho \omega^2}{3 \eta} \frac{1}{r} \frac{\partial (r^2 h^3)}{\partial r} - e ,$$

(2)

where e stands for film thinning due to solvent evaporation. The pioneering effort by Emslie,
Bonner and Peck[8] described a non-volatile fluid where solvent evaporation did not occur. They
found that an uniform film would remain uniform and that initially nonuniform film profiles

would reduce to uniform films as spinning proceeded. The common result from models which included solvent evaporation is that the rate of film thinning is initially dominated by fluid convection and then by mass transfer,[13-17] prompting many to claim that the spin coating process can be effectively modeled by splitting it up into two processes. During the first stage only fluid flow is allowed to occur while in the second stage mass transfer occurs exclusively. We will show that significant errors can develop following this "split model" approach.

CONSERVATION AND CONSTITUTIVE EQUATIONS

The solutions to both the flat substrate and planarization problem are based on the continuity equation, Cauchy momentum balance, and convective diffusion equation. As boundary conditions, we stipulate no slip and no solvent flux at the solid surface and that mass, momentum and solvent mass be conserved at the free surface. The functional dependence of viscosity and diffusivity on concentration were chosen to exemplify polymeric solutions and were fitted to experimental data on two Du Pont polyimide precursor solutions, PI2525 and PI2545. The viscosity data were taken using a Cannon-Fenske No. 300 bulb viscometer resulting in the following expressions:

$$\eta_{PI2525} = 3.89 \times 10^4 (1 - w_s)^5 \qquad \eta_{PI2545} = 3.70 \times 10^5 (1 - w_s)^5 \quad , \quad (3)$$

where w_s is the weight fraction solvent in the solution and the viscosity is measured in poise. The diffusivity data were calculated using mass loss data taken on a Cahn Electrobalance and then fit to find the following equation for the diffusion coefficient in unit of cm^2/s:

$$D = 2.1 \times 10^{-10} \exp(8.3 \text{ wfs}) \quad . \tag{4}$$

Details about the operation of the electrobalance can be found elsewhere,[18-20] while a description of bulb viscometry can be found in Rodriguez.[21] Figures 2 and 3 show the data and correlations.

SPIN COATING MODEL

An ordering analysis[22] is performed on the above system of equations in cylindrical coordinates to determine the appropriate dimensionless quantities and to reduce the equations to a manageable form. Inertial, surface tension and certain viscous forces are found to be negligible. A complete description of this procedure will be described in a forthcoming paper. By further assuming that the film thickness is uniform across the wafer one can derive the following set of one-dimensional integropartial differential equations first suggested by Bornside et. al. :[15]

$$\frac{\partial v_z}{\partial z} + 2 \rho \omega^2 \int_0^z \frac{(h - z')}{\eta} dz' = 0 \tag{5}$$

$$\frac{\partial w_s}{\partial t} + v_z \frac{\partial w_s}{\partial z} = \frac{\partial}{\partial z}\left(D \frac{\partial w_s}{\partial z}\right) \tag{6}$$

$$\frac{\partial h}{\partial t} - v_z|_{FS} + k\left(w_s^{g*} - w_{s\infty}^g\right) = 0 \tag{7}$$

Equation (7) is the kinematic expression which includes both the fluid convection and solvent mass transfer terms. In addition to solving the complete model based on simultaneous simulation

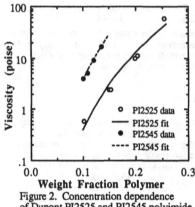

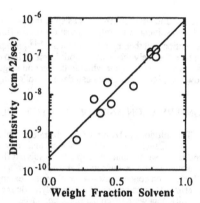

Figure 2. Concentration dependence
of Dupont PI2525 and PI2545 polyimide
precursor solutions fit to a power law.

Figure 3. Concentration dependence of
Dupont PI2525 polyimide precursor
solution fit to an exponential function.

of Equations (5-7), we will solve for a split model which consists of solving only Equation (5)
and the first two terms of Equation (7) with a constant viscosity. Then, when the second and
third terms of Equation (7) become of equal magnitude, we will remove the solvent to arrive at
the final dry film thickness. The wet film thickness at which this switch in mechanisms occurs
depends only on the external resistance to mass transfer and not on the diffusive part. We will
take this film thickness, H_c, as the characteristic length for mass transfer and will define a
Sherwood Number, Sh_c, which is a ratio of the external to internal resistances to mass transfer,
based on this characteristic length. These quantities can be shown to be:

$$H_c = \left(\frac{3\,k'\,\eta}{2\,\rho\,\omega^2}\right)^{1/3} \qquad\qquad Sh_c = \frac{k'\,H_c}{D} \tag{8}$$

RESULTS AND DISCUSSION

The above equations were solved using the finite element method for a variety of values of
H_c and Sh_c. Each value of H_c represents an order of magnitude change in the mass transfer
coefficient when all other parameters remain unchanged. Figure 4 compares the values calculated
from the complete model to those determined from the split model. In each case the initial film
thickness was taken to be large enough that fluid convection initially dominated the rate of film
thinning. The results are presented as a ratio of the film thicknesses for the complete model to
the split model. If the latter model was appropriate this ratio would be unity. Obviously this
model is not quantitative as the error can be as much as a factor of three. Some overlap in the
data appears when plotted in this manner, although this is not quite true in the sloped region.
Figure 5 shows a plot of the actual final film thickness as a function of the two parameters,
H_c and Sh_c. Notice that for a given H_c, at low Sh_c (high diffusivity) the final film thickness is
independent of the Sh_c (diffusion coefficient). In this region the concentration fields are flat and
the mass transfer is controlled exclusively by the external resistance. For high Sh_c (low
diffusivity), the final film thickness again becomes independent of the diffusion coefficient.
Here the concentration profiles are flat with an extremely sharp gradient at the free surface. The
mass transfer is now controlled completely by the internal resistance, however, the concentration
does not significantly decrease until long after the fluid flow has stopped. There is, however, a
range of Sh_c values the final film thickness depends on Sh_c. It is here that the diffusion is fast
enough to compete with external mass transfer and fluid convection and to affect the final film
profiles. The film thickness versus Sh_c follows a power law dependence. If we recall our
definition for the H_c and Sh_c, we can derive the previously mentioned power law dependence of
the final film thickness versus the spinner speed, i.e. Equation (1).

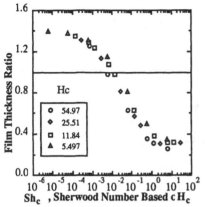

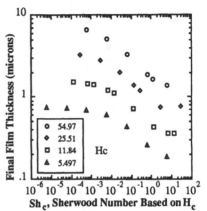

Figure 4. Comparison of the complete model with the split model. The ordinate is the ratio of the film thickness calculated by the two methods respectively.

Figure 5. The final film thickness prediction using the complete model as a function of H_c and Sh_c. These two parameters depend only on material properties and the spinning conditions.

PLANARIZATION MODEL

Finally, as a first attempt at solving the planarization problem, the steady state flow field over an axisymmetric step was solved. Solvent evaporation was neglected and the analytical flow field developed by Emslie, Bonner and Peck[8] was used for the inflow and outflow conditions. Figure 6 shows the results of this simulation for a variety of surface tension forces. Lengths have been scaled with the feature height and the surface tension value represents the ratio of the surface tension forces to the centrifugal driving force. Notice that as the importance of surface tension is increased, the free surface profile becomes increasingly asymmetric, although the feature profile below is symmetric. The profile tends to lean upstream against the flow field and body force. Obviously more work and comparison to results from other authors are required before any further conclusions can be drawn.

CONCLUSIONS

The spin coating process can be quantified in terms of two characteristic parameters, H_c and Sh_c, which depend only on material properties and spinning conditions. These parameters measure the relative importance of the external mass transfer and internal mass transfer to fluid convection. The split mechanism model is shown to give inaccurate predictions when compared to the complete solutions of the differential equations. The results of the flat substrate simulation can be used as the boundary conditions for the planarization problem. Finally, surface tension causes asymmetric film profiles over symmetric topographical features.

ACKNOWLEDGEMENTS

This work has been sponsored by the Office of Naval Research, Grant Number N00014-87-K-0211.

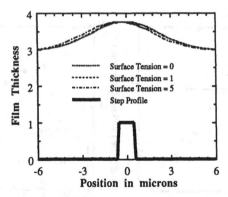

Figure 6. Film profile over an axisymmetric step with no solvent evaporation. The value given for the surface tension is ratio of the surface tension forces to the centrifugal driving forces. Lengths are scaled by the feature height.

REFERENCES

1. D.E. Bornside, C.W. Macosko and L.E. Scriven, J. Imagining Technol. 13 (4), 122 (1987).
2. G.F. Damon, Proc. Second Kodak Seminar on Microminaturation, Rochester NY, p. 36 (1967).
3. J.H. Lai, Poly. Eng. Sci. 19 ,1117 (1979).
4. B.T. Chen, Polm. Eng. Sci. 23 (7), 399 (1983).
5. P. O'Hagan and W.J. Daughton, Proc. Kodak Seminar on Microelectronics, Rochester NY, p. 95 (1977).
6. F.L. Givens and W.J. Daughton, J. Electrochem. Soc. 126 (2), 269 (1979).
7. W.J. Daughton and F.L. Givens, J. Electrochem. Soc. 129 (1), 173 (1982).
8. A.G. Emslie, F.T. Bonner and L.G. Peck, J. Appl. Phys. 29 (5), 858 (1958).
9. P.C. Sukanek, J. Imaging Technol. 11 (4), 184 (1985).
10. P.C. Sukanek, J. Electrochem. Soc. 136 (10), 3019 (1989).
11. B.D. Washo, IBM J. Res. Dev. 21, 190 (1977).
12. T. Ohara, Y. Matsumoto and H. Ohashi, Phys. Fluids A 1 (12), 1949 (1989).
13. D. Meyerhofer, J. Appl. Phys. 49 (7), 3993 (1978).
14. W.W. Flack, D.S. Soong, A.T. Bell and D.W. Hess, J. Appl. Phys. 56 (4), 1199 (1984).
15. D.E. Bornside, C.W. Macosko and L.E. Scriven, J. Appl Phys. 66 (11), 5185 (1989).
16. S.A. Jenekhe, Ind. Eng. Chem. Fundam. 23, 425 (1984).
17. S. Shimoji, Jpn. J. Appl. Phys. 26 (6), L905 (1987).
18. Cahn Instruments, Instrument Manual for D200 Digital Recording Balance, Cerritos, Ca,
19. A.W. Czanderna and S.P. Wolsky, in Elsevier Methods and Phemomena Series (Elsevier Scientific Publishing Company, London, 1980).
20. M.G. McMaster, MS Thesis, University of California, Berkeley, 1988
21. F. Rodriguez, Principles of Polymer Systems , (McGraw-Hill, San Francisco, 1982).
22. C.C. Lin and L.A. Segal, Mathematics Applied to Deterministic Problems in the Natural Sciences , (Macmillian Publishing Co., Inc., New York, 1974).

POLYIMIDE PLANARIZATION WITH POLYSTYRENE BY RIE ETCH-BACK

HUNG Y. NG, T. LII and M. JASO
IBM T. J. Watson Research Center, Yorktown Heights, NY 10598

ABSTRACT

This paper reports the results of a planarization process of polyimide which has been utilized as a passivation material in multilevel interconnection. The degree of planarization with spin-coated polyimide over metal topography is only 10-20% due to crosslinking and shrinking of the polymer during the curing process. Polystyrene used as sacrificial planarizing film can achieve <95% planarity over $100\mu m$ metal pad with $1\mu m$ height. A conformal polystyrene film can be spin-coated onto the polyimide surface without adhesion promoter. The topography of polyimide over the metal pattern can be planarized with thermal reflow of the polystyrene by baking the film below 250°C. The planarity of this sacrificial film, polystyrene, can be transferred to polyimide by etch-back without degrading the polyimide surface properties which has been examined by X-ray photoelectron spectroscopy.

INTRODUCTION

Polymers are increasingly used in microelectronics as dielectric materials to form multilayered structures for wiring interconnections and packaging. For these applications, the polyimide(PI), a class of high-temperature polymers, are well suited owing to their low dielectric constant and high temperature stability. PI films provide excellent step coverage and can be precisely etched both by wet process and dry process with an O_2 plasma. The fabrication of multilevel interconnections requires a planarized interlevel dielectric layer. However, spin-coated PI film only can reduce the original topography by 10 to 20% because the evaporation of spinning solvent causes film shrinkage and degradation of planarization. Many planarization processes and new materials have been examined by many investigators such as spun-on-glass, thermal reflow of BPSG, bias-sputtering of dep-etch oxide and even chemical-mechanical polishing[1] for multilevel interconnection, but no report was found on planarizing polyimide topography. A low molecular weight polystryrene(PS), previously reported for SiO_2 etch-back planarization[2,3], is utilized as a spin-coated resin to flatten the surface as sacrificial layer. Polyamic acid was spin-coated onto the substrates and thermally cured at various temperature in order to remove the solvent and causing the imidization of the polyamic acid into polyimide film. Outgassing diffusion of the solvent and H_2O, a by-product of imidization, tends to form voids in the narrow spacings of fine patterns during the thermal curing. This paper describes (1) the void formation of spin-coated PI between high aspect ratio metal line with cross-section scanning electron microscopy(SEM) inspection, (2) the investigation of PS film coating on PI as planarizing layer, (3) the characterization of the etch-back process by O_2 plasma and (4) the polyimide surface analysis by XPS after O_2 plasma treatment.

Mat. Res. Soc. Symp. Proc. Vol. 227. ©1991 Materials Research Society

EXPERIMENTAL

BPDA(biphenyl tetracarboxylic)-PDA(phenylene diamine) polyamic acid in NMP(N-methyl-pyrrolidone) solvent, Dupont Pyralin 5810D, was used in this investigation. Polyamic acid was spin-coated onto the test pattern substrates and thermally cured with final temperature at 400°C to form polyimide film.

Polystyrene was purchased from Pressure Chemical Company as special polystyrene polymer(Mw/Mn < 1.09) with wide variety of molecular weight(Mw). Mw1500 and Mw2500 were examined in this study. Polystyrene which had been dissolved in solvent, was spin-coated and baked for 60 minutes at 200°C. Viscosity of polystyrene was measured by means of Brookfield rotational viscometer while the surface tension of PS solution was measured with a CAHN model DCA-312.

An Applied Material Precision 5000 magnetically-enhanced single wafer etcher was used for this application. The electrode temperature was maintained at 15°C during the etching process by backside helium cooling. Oxygen was used as reactant gas. The parameter space consisted of power of 200-500W, magnetic fields of 0-70Gauss, pressure 20-80mTorr, and flow ratesof 20-80sccm. These ranges were chosen in order to avoid overheating of polymer films during etching.

The XPS-spectra have been taken on as Perkin-Elmer model with the excitation by X-ray Mg-radiation source. Incident angle used for obtaing spectra was at 45°. All samples were well outgassed before exposure to the X-ray. The survey spectra at a binding energy of 0-1000 eV and spectra of single-photoelectron lines C 1s, N 1s, O 1s were taken.

RESULTS AND DISCUSSION

POLYIMIDE GAP-FILL

Polyamic acid in solvent is commonly spin-coated onto the substrate and thermally cured into a polyimide film. This polyamic acid solution, which has low viscosity, is able to fill any narrow spacings between the metal lines. However, void formation between metal pattern with high aspect ratio, the ratio of width to height, was detected after the final curing at 400°C. Trapping of the outgassing solvent and H_2O, by-product of imidization, was believed to be the cause of PI-void formation during the curing process. A silicon trench pattern with difference aspect ratios ranging from 0.5 to 2.5 was utilized to examine the PI gap-fill characteristic. Preliminary results showed that the numbers of PI-void increased with increasing aspect ratio. The major occurrences of PI-void were detected below the temperature 250°C which suggested that solvent out-diffusing through the PI was the prime factor of PI-void formation mechanism. Inspection with cross-section SEM indicated that PI cured with hot plate under nitrogen purge could fill without void up to aspect ratio 1.2, whereas PI cured under vacuum could only achieve below an aspect ratio of 1 with the same heat ramping rate at 2°C/min. It is believed that rapid evaporation of solvent from PI surface under vacuum curing causes the formation of a dry skin near the surface which becomes a barrier for the remaining solvent out-diffusing.

As mention above, solvent out-diffusion through PI over topography with high aspect ratio can be the fmain determinant of PI-void formation. Multilayer coating technique was examined to circumvent this problem. The effectiveness of multilayer coating depends on the coating of the initial layer which is used to reduce the topography as well as the aspect ratio of the pattern. The first coating thickness shall be sufficient to reduce the topography of high aspect ratio feature without filling up the spacings in between the patterns. This thin conformal coating over the topography allows the solvent diffusing out in all directions without trapping in the gap between topography during the curing. Figure 1 demonstrates the dual-layer coating of polyimide over Al pattern with aspect ratio less than 2, cured on hot plate in nitrogen ambient.

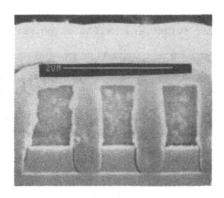

Fig. 1. Two layers of polyimide coated over metal pattern.

POLYSTYRENE ON POLYIMIDE

The effective planarization of PS over topography is directly correlated to molecular weight(Mw) of PS as illustrated in figure 2. A value 100% was assigned to fully planarized surfaces with values less than 100% were assigned when surface topography remained. The melt viscosity decreases with increasing temperature and with reduced PS molecular weight. Viscosity of 450cp was measured for PS Mw2500 at 200°C while 320cp was obtained for Mw1500 at the same temperature. Excellent planarized PS over metal lines is demonstrated as shown in figure 3. This suggests that low molecular weight PS with low melt viscosity flows easily to refill the recesses and planarized surface topography over a wide range of pattern widths range from 2μm to 700μm. Baking time has no effect on the planarization characteristic of PS. Temperature greater than 250°C caused reticulation of the PS film.

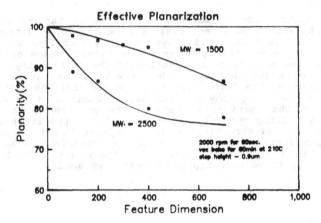

Fig. 2. Effective planarity after baking at 210°C as a function of pattern width for Mw1500 and Mw2500 of polystyrenes.

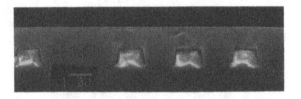

Fig. 3. Planarized polystyrene over metal lines.

Polystyrene in powder form was dissolved in solvent and spin-coated onto PI surface. Toluene was the solvent initially used for the study. Two major visible adhesion problems were observed of coating PS over PI. First, PS film pulled back from the edge of the wafer because the static force around the edge. O_2 RIE roughening the PI surface was introduced prior to PS coating in order to increase the interfacial contact area.[4] The second adhesion problem is the classic fish-eye formation with diameter range from 1-5 mm. Toluene is a good solvent for dissolving PS in room temperature. However, the high volatility, low viscosity and high surface tension of toluene makes it a poor liquid adhesive. PS mixed with toluene exhibits a maximum surface tension at 0.25 weight fraction of PS as shown in figure 4. The surface tension of the mixture decreases with increasing weight fraction of PS. This is also verified with the finding that the visual amount of fish-eye occurrence decreases with increasing amount of PS in the solution. Therefore, binary solvent system was examined to enhance the adhesion of PS over PI.[5] Anisole with higher molecular weight and polarity than toluene was introduced into the mixture in order to increase the wettability of the surface. Although the surface tension with anisole, 15 volume percent of solvent mixed with toluene, has changed very little compared with pure toluene at the same weight fraction of PS, no visible fish-eye adhesion problem was observed with the PS prepared with the binary solvent of toluene and anisole.

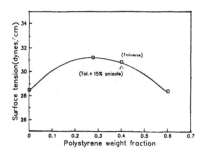

Fig. 4. Surface tension of solutions with different solvents as a function of weight fraction of polystyrene.

ETCH-BACK PLANARIZATION

Statistically designed experiments were performed which determined the etch rate, uniformity, and etch selectivity of polystyrene to polyimide as function of power, pressure, magnetic field, and gas flow. The average etch rate increases with RF power and pressure as expected. As power increased from 200W to 500W, etch rate increases from 0.4μm/min to 1.2μm/min. Since the glass transition temperatures for PS is about 100°C, temperature rise during etching can significantly degrade the polymer surface smoothness. Therefore, it was necessary to suppress the temperature rise in etch-back by lowering RF power. The etch rate increased monotonically with magnetic field from 0 to 80G, but the uniformity(2σ) degraded from 2.4% to 4.7% as the magnetic field increased. It is worth noting that the etch-back process requires removal of more than 1.5μm of PS and PI films, so etch rate uniformity is very important to the success of the planarization scheme. Between 20 and 80mTorr the etch rate increased with pressure, as did the non-uniformity.

Laser interferometry was employed to measure etch rate and detect endpoint. Because of the different transmittance of PS and PI films, clear RIE endpoint at the PS/PI interface is detected by a change in signal intensity as shown in figure 5. The etch rate selectivity between PS and PI primarily determines the planarity of PI after etch-back. Within the parameter space investigated, PS always exhibits a higher etch rate than PI. Reducing RF power and increasing gas pressure reduces etch rate selectivity between PS and PI, and results in better planarity. This may be rationalized in terms of a difference in chemical and mechanical properties for the two films, such as the different main-chain bond energies(73 kcal/mole for C-N and 83 kcal/mole for C-C) resulting in different chain scission yields of ion bombardment. The results show that low RF power and low gas pressure give good uniformity and planarity. Profilometer measurements show that better than 90% planarity of a PI layer over 200μm wide metal pads has been obtained after etch-back as shown in figure 6. Figure 7 demonstrates two levels of metal with planarized PI as the interlevel dielectric.

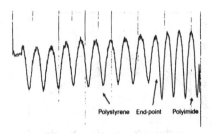

Polystyrene End-point Polyimide

Fig. 5. Laser interferometry end-point trace of etch-back process.

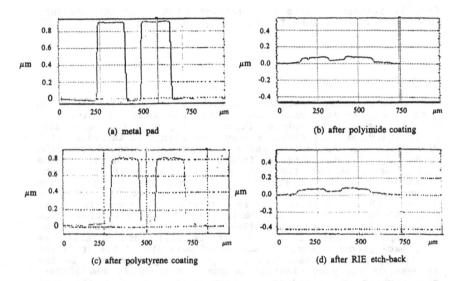

(a) metal pad

(b) after polyimide coating

(c) after polystyrene coating

(d) after RIE etch-back

Fig. 6. Profilometer measurements of polyimide topography during the entire planarization sequence (a) metal pad (b) after polyimide coating (c) after polystrene coating (d) after RIE etch-back.

Fig. 7. Two levels of metal interconnection with planarized polyimide as the interlevel dielectric.

POLYIMIDE SURFACE ANALYSIS AFTER O_2 RIE

The results of O_2 RIE polyimide are summarized in Table 1 which gives both the composition of the PI surface, and the C 1s components for data reconstruction. Comparing the atomic composition or the curve fit components %C, %N and %O for polyimide before and after RIE indicates that the two surfaces are essentially the same. The component II of C 1s signal is simply lower after RIE as shown in figure 8. The shift in relative intensities of component II suggests the loss of the carbonyl component in the C 1s spectrum.[6] A multiplicity in C-N bonding which is believed in determining the selectivity of etching between PS and PI films. The breaking-up of C-N bonds in the polyimide chain is possibly the initial mechanism of etching PI in O_2 RIE.

Atomic composition of BPDA-PDA

	C(1s)	N(1s)	O(1s)	
I	285.0	400.5	532.1	
II	285.8		533.7	
III	288.6			
atom %	78.4	7.1	14.5	none
	79.0	7.8	13.2	after O2 RIE

	C(1s)	I	II	III	
area %		50	35	15	none
		65	18	17	after O2 RIE

Table 1. Curve fit components for BPDA-PDA and atomic composition before and after O_2 plasma.

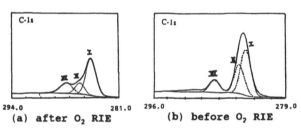

294.0 281.0	296.0 279.0
(a) after O_2 RIE	(b) before O_2 RIE

Fig. 8. BPDA-PDA polyimide: (a) the experimental C 1s XPS for polyimide after O_2 plasma. (b) the same spectral region before etching.

REFERENCE

(1) D.Moy et.al., IEEE VMIC Proc.(1989), p.26.
(2) S.Ngasawa,Y.Wada,H.Tsuge,M.Hidaka,I.Ishida and S.Tahara, IEEE Transaction on Magnetics, vol.25, no.2, March 1989, p.777.
(3) H.Gokan, M.Mukainaru, and N.Endo, J. Electrochem. Soc. vol.135, no.4, April 1988, p.1019.
(4) D.S.Dunn, J.L.Grant and D.J.McClure, J.Vac.Sci.Technol., A7(3), May 1989, p.1712.
(5) Fred W. Billmeyer, Jr., Textbook of Polymer Science, 2nd ed. (John Wiley & Sons, 1971, p.23)
(6) B.J. Bachman and M.J. Vasile, J.Vac.Sci.Technol. A7(4), July 1989, p.2709.

PROTONATION EFFECTS ON PPQ POLYMER DURING HF LIFT-OFF PROCESS USED IN MULTICHIP MODULE FABRICATION.

FRANÇOIS TEMPLIER*, J. TORRES*, C. CLARISSE** and J. PALLEAU*.
*France Telecom, CNET-CNS, BP 98, F38243 Meylan, France.
**France Telecom, CNET Lannion B, BP 40, F22301 Lannion, France.

ABSTRACT

The reactions between PPQ and several acidic solutions have been studied. Protonation reactions have been evidenced using U.V.- Visible spectroscopy. PPQ revealed three different states concerning these completely reversible reactions. Each state exhibits a specific colour: Pale yellow (as deposited PPQ), and gold yellow and purple red for, respectively, the first and the second protonation states. Using IR spectroscopy, it has been shown that the protons are fixed at the nitrogen sites of the polymer. A possible degradation of the PPQ related to the protonation reactions is also studied, and thus technological solutions are proposed.

INTRODUCTION

The increasing complexity of VLSI chips, characterized by a high clock frequency and a high number of input/output requires the development of multichip packages providing low signal delays and size reduction. Such systems involve the development of thin film multilayer interconnect modules built using integrated circuit technologies. Since the distance of the connections becomes much longer, new materials (metal and dielectric) are needed for these technologies. The fabrication of these modules undertaken in our laboratory involves copper as conducting material (very high conductivity) and a new polymer, PolyPhenylQuinoxaline (PPQ, from CEMOTA) as dielectric. This material offers low dielectric constant (epsilon < 3) and tan (delta) < 10^{-3}. In our process, a 10 μm thick layer of PPQ is first spun on the silicon substrate (100 mm wafer). Patterns are then defined on the polymer using techniques such as photolitography and etching processes. Conducting lines and vias are obtained by copper deposition and using a lift-off process. The next interconnection level is made by repeating the same technological steps with 5 μm thick layers. The lift layer used is polysiloxane (Spin On Glass, SOG), soluble in HF. During the process, HF was found to diffuse rapidly across the PPQ. HF solutions have been shown inducing at least two kinds of effects, involving both electrochemical and chemical reactions and occurring in the whole module. The electrochemical effect, involving formation of porous silicon at the backside of the wafer and a reduction of the polymer at the front side, is discussed elsewhere [1]. A chemical reaction between PPQ and HF is shown occurring with changes in the polymer coloration. In this paper we present the results in the study of the reaction between PPQ and acids (HF and also other acids by way of comparison).

EXPERIMENTAL

The polymer films were prepared using a spin-on technique. PPQ solutions are available in 35% o-xylene/65% m-cresol. The mixture is spun on silicon substrates at 3500 rpm. A 2.5 μm-thick pale yellow film is obtained with a 13 % solution. Solvent removal is achieved in a first low temperature annealing step. The polymer densification is performed at temperatures of 200 and 400 °C for 30 min in nitrogen ambient. U.V.-Visible and Infra-Red spectra were performed using, respectively, Cary 2300 and Perkin Elmer 1720 spectrometers.

RESULTS AND DISCUSSION

Visual observations:

When a PPQ film is dipped in the lift-off solution which consists of concentrated hydrofluoric acid (29M HF), it becomes orange after a few seconds. Other PPQ films were dipped in different acidic or non-acidic solutions. Observations concerning both colour changes and PPQ solubility are shown in Table I.

Solution	HF	HCl	H_2SO_4	CH_3SO_3H	NH_4F
Conc. (mol/l)	29	12	18	15	13
Colour	Or.	Or./Red	P. Red	P. Red	Pale Yellow
Solubility	SS	S	VS	VS	NS

Or. = Orange; P. Red = Purple Red; NS = not soluble;
SS = slightly soluble; S = soluble; VS = very soluble.

Table I : Behaviour of PPQ films in different media.

No modification is observed in NH_4F, so no effect of the fluorine anion is expected. As all colour modifications are observed in acidic media, they could be related to the presence of protons. The red coloration is obtained only in concentrated strong acids (completely dissociated), with a threshold effect at concentration of about 12M (case of HCl). This threshold effect is checked by using H_2SO_4 and CH_3SO_3H at increasing dilutions: the colour turns from purple red to orange. As this last coloration is obtained in concentrated weak acid (29M HF), it can be assumed that the coloration depends directly on the proton concentration (pH). Thus we believe that:

- The colour modifications of the PPQ dipped in acids are related to the protonation of the polymer (fixing of protons onto the PPQ). Such effects have been already observed on other materials such as Quinoxaline [2,3,4,5], Polyquinoxaline [6,7] and other nitrogen containing organic molecules [8].

- The colour depends on the proton concentration (that is, on the protonation level). Thus the purple red coloration represents the highest protonation level.

Table I also shows that the best solubility of PPQ is obtained when the polymer is highly protonated. The relation between protonation and solubility is well known for rigid chain and ladder polymers [9], and has been reported particularly in the case of ladder PolyQuinoxaline [6].

U.V.-Visible. spectroscopy:

U.V.-Visible absorption characterization was performed using dissolved polymer. A spectrum of the reference state (PPQ powder dissolved in dichloromethane) is shown in Fig.1. Two broad absorption bands are located respectively at 275 and 390 nm. The spectrum in fig.2 was obtained with PPQ dissolved in concentrated CH_3SO_3H. In that case, the polymer solution is purple red. Ethanol was then added to dilute the solution. The colour turned successively to orange and gold yellow. Spectrum 3 was performed on this last state. If further dilution is achieved, the PPQ solution recovers its initial pale yellow colour and exhibits the same U.V.-Vis. behaviour as in fig. 1. The 275 nm peak, present on the reference spectrum, is slightly shifted (15 nm) in the acidic media. This effect is attributed to a modification of aromaticity of the rings. The position of the shifted peak (290 nm) does not depend on the acid concentration. Concerning the other peaks, i.e. 390, 435 and 530 nm, the shifts are much more important and thus reveal three different states of the polymer. In this case the shift is attributed to a modification of the electronic states, involving the formation of charge transfert complexes.

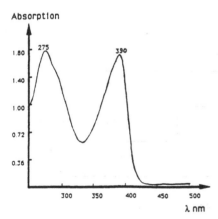

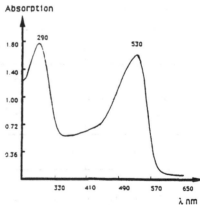

Fig.1: U.V.- Vis. spectrum of PPQ in Dichloromethane.

Fig.2: U.V.- Vis. spectrum of PPQ in conc. CH_3SO_3H.

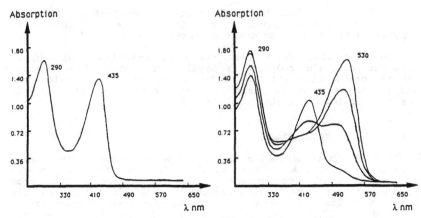

Fig.3: U.V.- Vis. spectrum of
PPQ in diluted CH₃SO₃H.

Fig.4: Intermediate spectra
between fig. 2 and 3.

Fig. 4 shows the intermediate states obtained with the solution containing concentrated CH₃SO₃H (having purple red colour) and the ones being partly diluted in ethanol up to the gold yellow state. The spectra show that increase at the 435 nm peak is correlated with decrease at the 530 nm peak. As this effect is completely reversible, it can be assumed that these two peaks reveal two different protonation states of the polymer.

U.V.-Visible spectroscopy was also performed using PPQ solutions in H_2SO_4 and $HClO_4$ media. The same behaviour was observed. Position of peaks obtained using concentrated acids are shown on Table II.

	Peak 1 (nm)	Peak 2 (nm)
CH_3SO_3H	290	530
H_2SO_4	290	552
$HClO_4$	290	547

Table II: Position of absorption peaks with concentrated
CH_3SO_3H, H_2SO_4 and $HClO_4$.

A small shift to higher wavelengths is observed compared to spectra obtained with CH_3SO_3H. It could be due to an effect of the anion.

In summary, we found two protonation states on the PPQ. This behaviour has been already reported on Quinoxaline [2]. With the PPQ, the two states exhibit specific colour: gold yellow for the first state, purple red for the second. The pure second state of protonation was obtained only when PPQ was dipped in concentrated H_2SO_4, $HClO_4$ or CH_3SO_3H. The orange coloration, observed in concentrated HF and HCl, is due to a mixture of the two states. Changes between the two protonation states and the unprotonated state are completely reversible.

Structure of the protonated phases:

IR transmission spectroscopy was performed on 50 μm-thick films. One sample was dipped in 12M $HClO_4$ and dried for 10 hours at 80 °C in air. The spectrum of this film exhibits a strong absorption band at 3450 cm^{-1}. This band, which is not present on a spectrum of a virgin PPQ, was assigned to N - H binding. A such behaviour was already observed on protonated quinoxalines [7]. Thus it can be assumed that the protons are located at the nitrogen sites of the polymer (see figure 5). The creation of the N-H binding could give rise to a modification of the electronic configurations of the ring, evidenced in the UV-Visible spectra. The first protonation level, obtained at the lower protons concentrations, could be related to the bonding of only one proton on each nitrogen containing ring, i.e. two per monomer. Thus the second state of protonation could be related to the fixing of two additional protons on the remaining nitrogen sites of the monomer, creating quinoid structures.

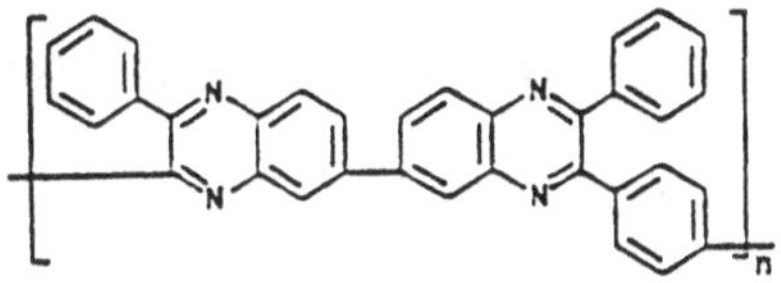

Fig.5: PolyPhenylQuinoxaline.

Consequences of protonation on interconnect modules:

We have to consider the possible polymer degradation related with such effects in our particular process, i.e. PPQ films in HF. We have been showing that the reaction of protonation itself is reversible. But solubilization effects, evidenced in some acids, may damage the dielectric layers. To evaluate the solubility of PPQ in HF, we dipped 5 μm thick films in 29M HF. After a period of 72 hours, the weight loss was less than 0.8 %. As the duration of lift-off performed in our process is less than one hour, the effect has been considered negligible. On another hand, it has been decided to use diluted HF solutions (5 %) to avoid protonation effects discussed above, which can possibly modify the dielectric behaviour of the polymer. These possible modifications will be evaluated: electrical tests are now in progress in our laboratory.

CONCLUSION

We have shown that the changes in coloration of PPQ are related to reversible protonation effects. In our study, two different states of protonation have been evidenced, depending on the acidity of the solutions. The consequences on the possible degradation of the PPQ during the lift off process have been discussed. In order to improve the process, technical solutions have been described.

REFERENCES

[1] F. Templier, J. Torrès, A. Halimaoui, J. Palleau and J.C. Oberlin, to be presented at the 179th Meeting of the Electrochemical Society, Washington DC, May 5-10, 1991.

[2] A. Grabowska, J. Herbich, E. Kirkor-Kaminska and B. Pakula, J. of Luminescence, 11, 403 (1976).

[3] S.M. Beck and L.E. Brus, J. Am. Chem. Soc., 103, 2495 (1981).

[4] J. Waluk, Chem. Phys. Letters, 63, 579 (1979).

[5] J. Herbich and A. Grabovska, Chem. Phys. Letters, 46, 372 (1977).

[6] S.A. Jenekhe and P.O. Johnson, Macromolecules, 23, 4419 (1990).

[7] S.A. Jenekhe, Macromolecules, 24, 1 (1991).

[8] Y. Kobayashi, T. Nomizu and Y. Ujihira, J. Am. Chem. Soc., 101, 537 (1979).

[9] C.P. Wong and G.C. Berry, in "Structure Solubility Relationships in Polymers", Harris F.W. and Seymour R.B. editors, Academic Press, New York 1969, p. 71-88.

PART IV

Thermotropic Liquid Crystalline Polymers

LIQUID CRYSTALLINE THERMOSETS AS MATERIALS FOR MICROELECTRONICS

C. K. OBER AND G. G. BARCLAY
Materials Science & Engineering, Bard Hall, Cornell University, Ithaca, NY 14853-1501

ABSTRACT

New liquid crystalline thermosets have been prepared from end-functional monomers and oligomers of varying molecular weight. Both triazine and epoxy networks were explored. Of primary interest was the exploitation of the mesophase properties of these networks for developing polymers with high thermal stability and low coefficients of thermal expansion (CTE). Curing was carried out either within the nematic mesophase or the isotropic phase of the prepolymers. Transition temperatures associated with the mesophase were observed to change after curing under these two sets of conditions. The networks with the highest crosslink density were found to exhibit the lowest CTE values. Crosslinking of these thermosets was also carried out in the presence of a 13.5 Tesla magnetic field to determine the effect of orienting fields on the curing of the LC network. Orientation parameters as measured by wide angle x-ray diffraction were as high as 0.6. Values of the coefficient of thermal expansion as low as 15 ppm were achieved in the aligned direction. Of the two resin types, those with the triazine crosslinks had the lowest thermal expansion coefficient. Other thermal properties of these networks will be discussed.

INTRODUCTION

Recently the theoretical and experimental development of both nematic as well as rigid-rod thermosets has taken place [1,2]. Such materials may provide polymers with high thermal stability, low coefficient of thermal expansion and suitable dielectric properties for various applications in electronic packaging, and as matrix materials for advanced composites.

While conceptually and structurally similar, rigid-rod and nematic thermosets have distinctly different physical characteristics. Thermosets based on nematic networks have sufficiently low levels of crosslinking that the mesophase properties of the prepolymer are retained in the network. Much of the preparation of nematic networks has been focused on formation of siloxane based low T_g elastomers [3]. Such networks are being investigated for their optical and piezoelectric properties. Possibilities also exist for the crosslinking of otherwise chemically identical networks in either the nematic or isotropic

Mat. Res. Soc. Symp. Proc. Vol. 227. ©1991 Materials Research Society

states with expected differences in their physical properties. Curing may also be undertaken in applied fields to freeze in orientation of the networks.

In contrast, the molecular components of rigid-rod networks need not be mesophase formers, but do possess the stiff units typical of mesogenic groups leading to high T_g materials. Theory suggests that while deformation of rigid-rod networks can be expected to occur, it takes place by rotation about crosslink sites. It is this rotation rather than chain extension which is responsible for the theoretical entropy contribution to elasticity in rigid-rod networks. Nematic networks instead deform due to classical chain extension which must be modified for their mesomorphic behavior. Divergence from theory can be expected in both sets of materials since crosslinking agents and crosslinking sites may be more flexible than the mesogenic groups and may also perturb the nematic state. It is potentially unusual behavior like this that makes such networks interesting.

The earliest mesomorphic polymers consisted of acrylate networks which formed a smectic-like network structure [4]. Rigid-rod networks have been more recently prepared using rigid aromatic segments with both ester [5] and amide links [6]. Epoxy thermosets have also been produced from glycidyl endcapped mesogens and were explored for their resistance to wear. Finally, the fractal nature of rigid-rod networks has also been explored [7].

In this paper, we describe our efforts to prepare both rigid-rod and nematic networks. Our intention was to produce thermosets with low coefficients of thermal expansion, a property common in many liquid crystalline materials. The networks were assembled from both epoxy and triazine terminated mesogenic structures. The crosslinking of these thermosets was also explored in the presence of magnetic fields for the production of networks that were highly anisotropic. Orientation was established using X-ray diffraction and their mechanical properties were investigated.

EXPERIMENTAL

Three families of thermotropic thermosets, based on prepolymers shown below, are described in this report. The prepolymers consisted of two types of epoxy resins produced from the mesogenic bisphenol, dihydroxy-α-methyl-stilbene (I and II), the preparation of which is given in Reference [8]. Epoxy prepolymer II was prepared with a number of molecular weights corresponding to x values ranging between 1 and 9. Polymethylene segments of spacer length, n, equal to 5 and 7 were combined in equimolar quantities to produce a highly soluble material.

Curing of prepolymers I and II was carried out by mixing the appropriate amount of methylene dianiline and heating into the mesophase.

(I)

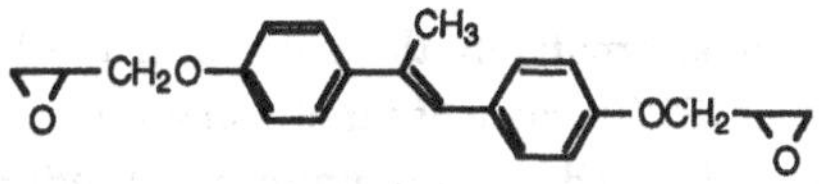

(II)

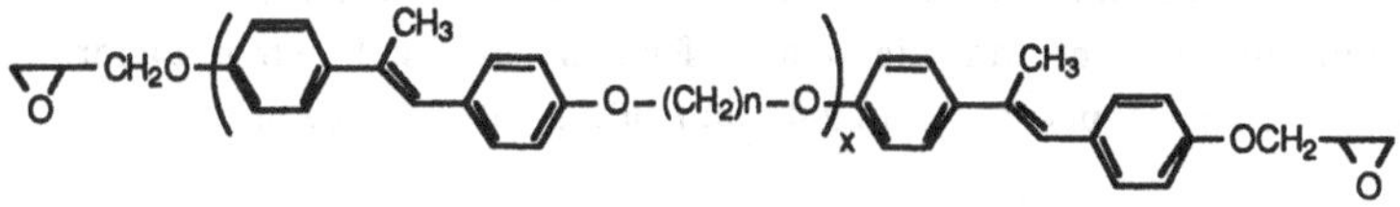

A rigid rod dicyanate with the general formula given in III was also investigated. As described in Reference [9] substituents ($R = CH_3$, Cl) were used in order to produce a processable material. Curing reactions were performed by heating to the desired temperature in the presence of a transition metal salt such as zinc stearate.

(III)

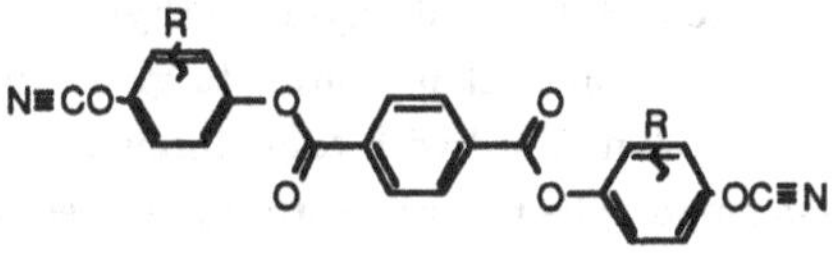

Magnetically induced orientation during the curing process was carried out at the Bitter Magnet Laboratory using an Intermagnetics General Corporation superconducting magnet with a field strength of 13.5 Tesla. The epoxy mixture was placed into thin-walled quartz tubes (10 mm in length and 3 mm in diameter). Once placed in a cylindrical graphite heater, the sample was cured within the mesophase temperature range while exposed to the magnetic field under nitrogen atmosphere. Wide angle X-ray diffraction (WAXD) exposures were taken of the aligned networks using a wavelength of 1.54 Å at the Cornell High Energy Synchrotron Source (CHESS). To estimate the degree of planar orientation of the liquid crystalline domains within these aligned networks, the orientation function f was calculated from data on the azimuthal distribution intensity [10].

Compositions of the oligomers were determined using a Varian XL-200 [1]H-nmr in d-chloroform. Infrared spectroscopy was performed using a Mattson FTIR 220. Thermal transition temperatures were obtained on using a Perkin Elmer DSC-2C at a heating and cooling rate of 20°C/min. Liquid crystal mesophases were examined with a Leitz polarizing optical microscope at 200x magnification equipped with a Mettler FP-52 hot stage and a Canon AE-1 35mm camera. Number and weight average molecular weights were determined by GPC in THF using RI and UV detectors with Ultrastyragel[TM] columns of 500, 10^3, 10^4, and 10^6 Å pore sizes calibrated with monodisperse polystyrene standards.

Samples for CTE measurements were cured in the capillary tubes described above for both unaligned and magnetically aligned samples. The thermosets were removed from the tubes and cut to give samples typically 1mm^2 x 5 mm for CTE measurements. DSC measurements were performed on sections of these larger samples or alternatively those cured in the DSC pan. Coefficient of thermal expansion (CTE) was measured using a Du Pont 943 Thermomechanical Analyzer calibrated with aluminum standards. Dynamic mechanical analysis was performed using a Polymer Laboratories DMTA at 1 Hz frequency, strain x 4 and heating rate of 5 °C/ min.

PROPERTIES OF ORDERED THERMOSETS

Thermosets from Nematic Prepolymers

Polyethers were chosen for the preparation of liquid crystalline epoxy prepolymers in order to avoid effects such as sequence randomization during the epoxidation process. The mesogen, 4,4'-dihydroxy-α-methylstilbene (DHMS), compounds I and II, was chosen since it is known to give soluble liquid crystalline (LC) polyethers when polymerized with α,ω-dibromoalkanes [13]. The most suitable oligomer was found to be the copolyether prepared from DHMS and a 1:1 mixture of dibromopentane and dibromoheptane. This composition was soluble in organic solvents such as tetrahydrofuran and chloroform, allowing accurate molecular weight evaluation and determination of structural compositions by ^{1}H nmr. Endcapping of the hydroxy terminated oligomers with epichlorohydrin yielded the required epoxy-terminated prepolymers.

The epoxy terminated LC prepolymers were thermally cured using methylenedianiline as the crosslinking agent [8]. An aromatic amine was chosen since it is less basic than aliphatic amines, thereby allowing the oligomers to melt completely to form the nematic mesophase before a substantial degree of crosslinking had occurred. Relatively slow reaction rates also permitted orientation in magnetic fields prior to complete curing. Determination of the stoichiometry of the curing reaction assumed that the LC prepolymers were difunctional and the amine was tetrafunctional.

Four moleculr weights of poepolymers were prepared: 3600, 2700, 1700 and 1000 g/mol. Using polarized light optical microscopy, it was observed that the resultant epoxy networks were extremely birefringent. The networks from the three highest molecular weight oligomers formed large domains, the boundaries of which gave a characteristic "thread-like " nematic texture (Fig. 1). However, the networks from the lowest molecular weight oligomer and the DHMS diglycidyl ether described later exhibited a schlieren texture which could not be readily associated with any mesophase (Fig. 2).

Wide angle x-ray diffraction (WAXD) of the networks exhibited a diffuse

Figure 1 Photomicrograph of Texture of Low Crosslink Density Epoxy
Networks (200 x)

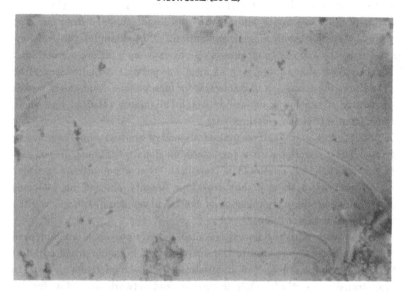

Figure 2 Photomicrograph of Texture of High Crosslink Density Epoxy
Networks (200 x)

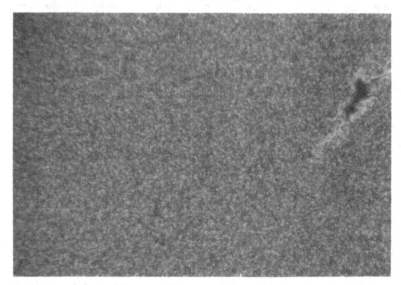

diffraction ring with a d-spacing of 4.5 Å characteristic of a nematic mesophase. This value was comparable to the nematic phase of the uncrosslinked oligomers. Interestingly, WAXD of the networks prepared from the lowest molecular weight oligomers showed an additional inner diffraction ring indicating a smectic molecular organization. For example, the network prepared from the 1000 M_n oligomer as well as showing a diffuse outer ring at 4.5 Å exhibited a sharp inner ring at 39 Å. A possible explanation for the observation for the formation of a smectic-like mesophase is that the narrower polydispersity of the lower molecular weight oligomers resulted in a more uniform distance between crosslink sites.

DSC studies of these crosslinked oligomers showed some interesting behavior. The networks with a low crosslink density (highest prepolymer molecular weight) still exhibited a transition from a nematic to isotropic phase. Networks with a higher crosslink density showed no isotropic transition, the molecular organization of the LC phase being "frozen" into the thermoset until decomposition.

The network with the lowest crosslink density possessed an increased clearing temperature compared to the uncrosslinked oligomer when cured at 140°C in the nematic state. The uncrosslinked prepolymer had a nematic to isotropic transition at 176°C while the crosslinked network showed a clearing transition at 192°C. Theoretically, the distribution of the crosslink sites may stabilize the LC order during the curing reaction, resulting in the increase in clearing temperature of the crosslinked network. To verify this, the same prepolymer was cured at 190°C in the isotropic state. DSC measurements demonstrated that both cured samples exhibited the same T_g; however the clearing transition was reduced to 157°C in the case of the network cured in the isotropic state. This observation, confirmed by optical microscopy, is in qualitative agreement with theory.

Rigid-Rod Thermosets

From studies of the effect of crosslink density on the coefficient of thermal expansion, it became clear that a molecular feature dominating physical properties was the crosslink density. The higher the crosslink density, the lower the measured CTE values. For this reason, it was decided to prepare rigid-rod networks based on the glycidyl endcapped DHMS mesogen. This thermoset, while showing only indirect evidence of LC behavior via microscopy and DSC, proved to have the lowest coefficient of thermal expansion of the epoxies (~60 ppm/°C). X-ray diffraction, however, showed that there was a smectic-like order in the network with two predominant spacings of 4.5 Å and 22 Å, the latter *d*-spacing corresponding to the distance between crosslinks. This surprising result occurs despite the fact that there is 30 mole-% non-LC crosslinking component. These result further suggests that improved CTE characteristics might be gained from the preparation of a rigid-rod structure with a stiffer crosslinking group.

The formation of networks via the cyclotrimerization of the cyanate group to the 1,3,5-triazine crosslink site was of interest, because of the selectivity of the reaction forming the triazine ring with very little side reactions [11]. Triazine networks were expected to be better model systems for the investigation of rigid-rod networks. The work reported here summarizes the preparation of new liquid crystalline 1,3,5-triazine networks from dicyanate compounds of ring substituted di(4-hydroxyphenyl)terephthalate [12].

Hotstage cross-polarized optical microscopy showed that the dicyanate monomers formed isotropic fluids just after melting. However, as the temperature increased, birefringent droplets appeared which coalesced to form a continuous mesophase with a schlieren texture. Initially the mesophase was mobile, stir-opalescent and could be sheared. As the thermal crosslinking proceeded, the mesophase was eventually "frozen" into the crosslinked network. Therefore some molecular changes were taking place during the curing reaction to form the liquid crystalline phase.

A possibile explanation is that the ether linkages connecting the mesogens to the 2,4,6-positions of the triazine crosslink site are flexible enough to accommodate the formation of colinear rigid-rod species during the initial stages of the curing process. These rigid-rod species would have to possess a sufficiently high length to diameter ratio to coalesce into a liquid crystalline phase. WAXD of these unaligned rigid-rod networks showed a diffuse ring at 5.2 Å characteristic of a nematic mesophase; however, a low intensity inner ring at 19.6 Å was also present.

Dynamic mechanical analysis (Fig. 3) of the unaligned LC triazine networks indicates that these networks have a high glass transition temperature (185°C, tan δ max) and good mechanical strength (Young's modulus ~2500 MPa). Thermogravimetric analysis indicated that these materials have good thermal stability, 10% decomposition occurring at approximately 440°C in both air and nitrogen. Triazine thermosets are known to exhibit good thermal stability due to the aromatic nature of the triazine ring, the thermal stability of which is close to that of benzenoid structures [13].

Magnetic Alignment

An attempt was made to prepare thermosets with oriented architectures, that is, polymeric networks with a distinct directional

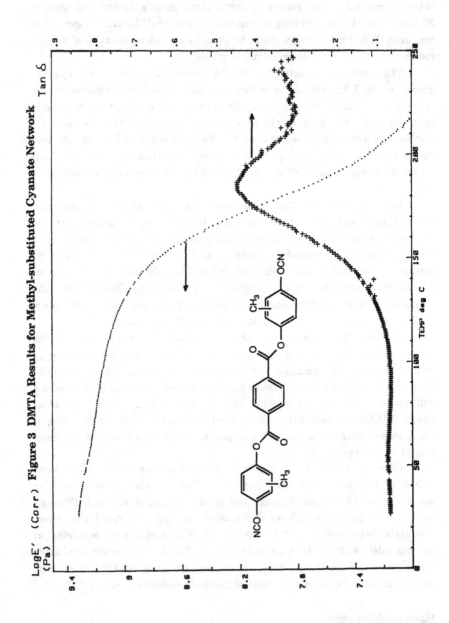

Figure 3 DMTA Results for Methyl-substituted Cyanate Network

orientation of the rigid-rod molecules within the crosslinked material. Highly crosslinked oriented networks are of interest, because of the possibility of creating high T_g, highly anisotropic structures. The inherent anisotropy of these networks would be expected to increase the modulus and lower the CTE behavior of the thermoset parallel to the direction of alignment. The process for orienting the molecules within these thermosets involved the crosslinking of these compounds within the liquid crystalline phase under the influence of a magnetic field (13.5 Tesla).

Orientation was achieved in both the epoxy and triazine networks, thus suggesting that neither the formation of the bulky epoxy or the triazine crosslink hindered the ability of the liquid crystalline phase to align during the curing process. Orientation was measured by wide angle x-ray diffraction (WAXD). For the purposes of discussion, only the results of the orientation of the triazine networks will be described.

WAXD of the aligned rigid-rod networks showed a layered molecular organization indicative of an aligned smectic network (Fig. 4a). As well as equatorial arcs at 5.2 Å, meridional arcs at 19.6 Å could be seen. The former arcs are the result of diffraction from chains parallel to the applied field, while the latter arcs are the result of diffraction from layers perpendicular to the applied field. Molecular models of the rigid-rod mesogen between triazine crosslink sites (Fig. 4b) were in close agreement with the x-ray data for the layered structure. The mesogen length calculated from molecular modelling is 19.8 Å which is consistent with with the d-spacing calculated for the meridional arcs (19.6 Å). Therefore it would seem that magnetic field alignment of the LC phase during the curing reaction has induced the formation of more highly organized networks than the unaligned case, which showed only a very weak inner diffraction ring at 19.6 Å. The flexibility of the ether linkages connecting the mesogens to the triazine crosslink site as previously suggested may accommodate this highly organized layered network structure.

It was found that the degree of orientation of these triazine networks could be influenced to some extent depending upon the type of catalyst used. For example, when Zn(II) stearate, a highly active catalyst, is used an orientation parameter of 0.42 is achieved. Using the less potent catalysts, Co(III) acetylacetonate and Cu(II) acetate, higher orientation parameters of 0.46 and 0.54 respectively were obtained. Variable temperature WAXD showed that the orientation "frozen in" during the curing process was stable to at least 280°C for the aligned network prepared from the dicyanate of di(4-hydroxy-methylphenyl)terephthalate (III, R = Me) using Cu (II) acetate as catalyst.

The inherent anisotropy of the magnetically aligned rigid-rod networks prepared from the III (R = Me) is readily apparent from CTE measurements

Figure 4 Molecular Structures of the Triazine Network

a) Schematic of network structure

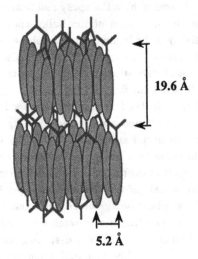

19.6 Å

5.2 Å

b) Molecular dimensions between crosslinks

19.8 Å

parallel and perpendicular to the direction of the applied field. For the unoriented sample, the CTE for the X/Y plane and the Z plane are comparable, ~60 ppm/°C. The oriented samples showed a considerable difference in CTE measurements between the X-Y and Z planes. In the oriented sample the CTE value parallel to the direction of the applied field (Z-plane) was remarkably low (17 ppm/°C), whereas the CTE value perpendicular (X/Y) to the applied field was much greater (94 ppm/°C).

Acknowledgements

NSF grant DMR-8717815 is acknowledged for partial support of this work. We would also like to acknowledge the use of facilities of the Materials Science Center at Cornell University. Financial support on the part of IBM Corporation Systems Technology Division, Endicott, NY is gratefully appreciated as well as much productive and enjoyable interaction with Drs. D. Wang and K. Papathomas. Acknowledgement is made to the Donors of the Petroleum Research Fund, administered by the American Chemical Society, for partial support of this research. We would also like to acknowledge the use of the Francis Bitter National Magnet Laboratory and the very kind help of Dr. L. Rubin.

REFERENCES

1. T. A. Vilgis, F. Boué and S.F. Edwards, *Molecular Basis of Polymer Networks*, Springer Proceedings in Physics, 42 (Springer Verlag, Berlin, 1989), p. 170.
2. M. Warner, K. P. Gelling, and T. A. Vilgis, *J. Chem. Phys.*, 88, 4008 (1988).
3. H. Finkelmann, Advances in Polymer Science, 60/61 (Springer Verlag, Berlin, 1984), pg. 99.
4. S.B. Clough, A. Blumstein, and E.C. Hsu, *Macromolecules*, 9(1), 123, (1976).
5. A. E. Hoyt and B. C. Benecewicz, *J. Polym. Sci: Part A: Polym. Chem.*, 28, 3403 (1990).
6. A. E. Hoyt and B. C. Benecewicz, *J. Polym. Sci: Part A: Polym. Chem.*, 28, 3417 (1990).
7. S.M. Aharoni, N.S. Murthy, K. Zero and S.F. Edwards, *Macromol.*, 23, 2533 (1990).
8. C. K. Ober. G. G. Barclay, K.I. Papathomas and D. Wang, in Electronics Packaging Materials V, (Mat. Res. Soc. Proc. xx , Pittsburgh, PA 1991) in press.
9. G. G. Barclay and C.K. Ober, "Liquid Crystalline Thermosets 2. Triazine Networks", *J. Polym. Sci.: Polym. Chem. Ed.*, in preparation.

10. L. E. Alexander, "X-ray Diffraction Methods in Polymer Science", Krieger, Huntington, NY (1979).
11. V. V. Korshak, V. A. Pankratov, A. A. Ladovskaya and S. V. Vinogradova, *J.Polym. Sci., Polym. Chem. Ed.*, 16, 1697 (1978).
12. G.G.Barclay, C.K.Ober, K.I. Papathomas and D.W. Wang, U.S. Patent Application.
13. V. A. Pankratov, S. V. Vinogradova and V. V. Korshak, *Russian Chem. Rev. (Engl.)*, 46 (3), 278 (1977).

THERMOTROPIC POLYESTERS CONTAINING CARBONYL OR ETHER LINKAGES

ROBERT S. IRWIN, Pioneering Research Laboratory, Du Pont Fibers and Central Research and Development, E. I. du Pont de Nemours and Co., Inc., Experimental Station, Wilmington, DE 19880-0302

ABSTRACT

In an assembly of aligned, thermotropic polyester macromolecules, such as a fiber, the flexibility and deviations from rectilinearity occasioned by in-chain ArCOAr or AroAr units are reflected sensitively by overall stiffness (initial modulus) and elasticity (elongation at break). The influence of frequency and placement of these units, as well as the effect of ring substituents, is discussed.

In a few cases where carbonyl or other units are regularly placed at suitable intervals along the chain, the latter apparently is extended as a low amplitude, regular helix which, in terms of physical properties, approximates to a straight, rod-like conformation. In situations where the angulation created by a carbonyl unit is immediately cancelled by an adjacent m-phenylene ring, the extended macromolecule is virtually rectilinear and, though marginally liquid crystalline, may provide some of the strongest organic fibers known.

INTRODUCTION

The basic requirements for polyester melts to exhibit liquid crystallinity are the capability of the macromolecules to adopt a rod-shaped conformation, and a significant degree of stiffness. There should be significant energy barriers to rotation serving to resist rearrangement to a less extended conformation. Under the proper circumstances chains adopt a parallel packing arrangement (liquid crystallinity) in the melt, within local regions or domains as long as there are no obstructions to close-packing, such as very large substituent groups, or frequent chain branches. An additional condition for liquid crystallinity (or anisotropy) is that the melting temperature of the polymer should be appreciably below the temperature of thermal decomposition, whether it be pyrolytic fragmentation, cross-linking, or other kind of chemical change. Most homopolyesters of interest are aromatic, for reasons of high chain stiffness and high thermal stability. Aromatic units favor high melting points so that a degree of controlled disorder is usually necessary to reduce melting points for facility in processing. This usually invokes a degree of random copolymer character, together with certain compositional features such as asymmetric ring substitution, flexible backbone units such as 1,2-ethylenedioxy, or even minor proportions of flexible, non-linear units.

Liquid crystalline polyester melts typically have lower viscosities than isotropic materials of comparable molecular weight. Mild stresses, notably shear in flowing across a stationary surface, as in an extrusion process, or elongational attenuation as in a fiber-spinning threadline, cause a very high degree of macromolecular alignment in the melt as a whole along the direction of stress. On removal of the stress this macro-alignment may be eventually erased as groups of aligned macromolecules, or domains, relax to adopt a random distribu-

Mat. Res. Soc. Symp. Proc. Vol. 227. ©1991 Materials Research Society

tion of orientations with respect to one another, with no net directionality. Relaxation is, however, sufficiently slow that it is possible to capture the oriented condition virtually intact by rapidly cooling the polymer below its freezing point to produce a so-called frozen nematic state. This is an essential feature of the melt-spinning of thermotropic polyesters, wherein a very high degree of macromolecular alignment in the axial direction is the key to very high fiber tensile strength (T) and modulus (Mi). Tensile loads are thus shared much more evenly by practically all the molecules in contrast to poorly aligned, conventional fibers where fewer chains bear the applied load.

Unlike Mi, the T in a fiber depends critically on the molecular weight, i.e., length of the polymer chain, although at high levels of chain length, T approaches asymptotically an upper limit. For commercial fibers such as 66 nylon or Kevlar® aramid [poly(p-phenyleneterephthlamide)], the number of monomeric repeat units is about 100 which represents balance between a fairly close approach to the tenacity ceiling and high levels of viscosity beyond which processing becomes excessively difficult. With thermotropic polyesters the viscosity limit is reached well ahead of the tenacity limit, so that fibers as-extruded, despite good chain orientation, have comparatively low strength. However, it is possible to take advantage of their excellent dimensional stability at elevated temperatures (a function of macromolecular alignment and orientation) for a heat-treatment process close to, but below the softening temperatures, which causes a large growth in molecular weight by continued polymerization in the solid state. In this way molecular orientation is preserved while tensile strength is advanced, typically, from 2-10 gpd to 20-25 gpd. Modulus is not usually increased significantly for macromolecules which are inherently rod-like (it may happen occasionally that crystallization or related processes may improve the relative alignment of chains to produce a rather large increase in modulus). Note that grams per denier units (gpd), which are common fiber usage, may be converted into the possibly more familiar pounds per square inch (psi) units by factoring by 12,800 x density.

Experience has shown that variations in the perfection of macromolecular alignment in the melt, inherent in the chemical composition, are captured sensitively in measurements of fiber T and Mi. In this paper we are concerned with variations in macromolecular alignments caused by the introduction of the carbonyl or ether components into the polymer backbone. Both units have a bond angle of about 120°, i.e., as chain components they introduce locally a deviation of 60° from linearity. In aromatic ethers there is considerably freer rotation about the oxygen bonds than about carbonyl groups because the latter tend to conjugate with adjacent aromatic rings. In both cases the angularity of carbonyl or ether units occasion intervening rod-like segments to show an angular distribution about the macromolecular axis. This has the same effect as misalignment of rod-like macromolecules as a whole, reducing T and Mi. A concomitant effect is increased strain as indicated by elongation-at-break (E). A thermotropic polyester typically has T/E/Mi value of about 20-25 gpd/4%/400 gpd. These are broadly similar to Kevlar® aramid fibers (from lyotropic solutions) (23 gpd/3%/600-900 gpd). In both cases stress-strain curves are essentially linear.

The effect on tensile properties of progressively introducing a non-linear unit, meta-phenylene, into an otherwise rod-like, thermotropic copolyester fiber, consisting initially of 60% (mole) 4-hydroxybenzoic acid, 20% terephthalic acid, and 20% 2,6-naphthalenediol, is shown in Table I. The progressive replacement of the latter diol by resorcinol is reflected initially by a drop in modulus and a parallel increase in elongation, and later by a drop in tenacity; ultimately T and Mi fall to very mediocre levels and liquid crystallinity disappears. (The initial increase in T is an unrelated effect whereby the intial small introduction of non-linear monomers seems to facilitate heat-strengthening of an otherwise rather rigid matrix.)

Table I. Effect on tensile properties of 4-hydroxybenzoic (60%)/terephthalic acid(20%)/2,6-naphthalenediol(20%) copolymer when latter is progressively replaced by resorcinol.

Resorcinol (%)	T (gpd)	E (%)	Mi (gpd)	Reference
0	(14)	3.5	420	[1]
10	18	4.7	390	[2]
15	18	6.3	290	[2]
20	8	20.0	40	[2]

The stress-strain curves of the Table I compositions are represented schematically in Figure 1.

Stress-strain behavior is a much more sensitive measure of macromolecular alignment than wide angle X-ray orientation angle, which pertains basically to crystalline regions. Conversely, liquid crystallinity, as may be measured in terms of birefringence in the melt between crossed polarizers, indicates appreciable, but not necessary, the highest degree of alignment, and is no guarantee that very high strength fibers therefrom will result.

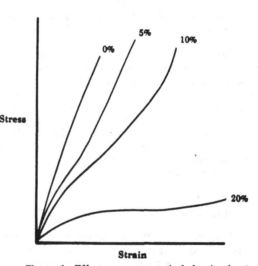

Figure 1. Effect on stress-strain behavior for 4-hydroxybenzoic acid/terephthalic acid/2,6-naphthalenediol when latter is progressively replaced by resorcinol (schematic).

POLYESTERS FROM MONOETHER OR CARBONYL-CONTAINING MONOMERS

In general Mi is a more sensitive measure of chain alignment than T because the latter is subject to some variability due to differences in heat-strengthenability between various compositions. The progressive introduction of 4,4'-carbonyldiphenol or 4,4'-oxydiphenol into a rod-like thermotropic polyester composition such as poly(chlorohydroquinone terephthalate) produces the same general alteration in tensile properties as described for resorcinol (Table I). A comonomer such as 2,6-naphthalenedicarboxylic acid, which does not materially reduce molecular linearity, combined with poly(chlorohydroquinone terephthalate) (as is necessary to reduce melting temperatures to a processible level) gives a high Mi of about 500 gpd. By contrast, a non-linear comonomer, 4,4'-oxydiphenol at the same replacement level (12-15 mole %), causes a substantial modulus reduction to 245 gpd (Table II) [3]without significantly reducing tenacity. The isomer 3,4'-oxydiphenol, by contrast, permits modulus to be retained at a high level. As an asymmetrical monomer, 3,4'-oxydiphenol is particularly effective in lowering polymer melting point compared with the symmetrical isomer, 4,4'-oxydiphenol. Although the 3,4'-oxydiphenol unit can exist in a variety of conformations, in the present context it prefers the extended conformation which is consistent with a rod-like macromolecule, even though flexibility is higher. In comparison, Mi is not diminished below levels usual for linear thermotropic polyesters (Table II). The higher stiffness of the carbonyl group in a comparison of 3,4'-oxydiphenol versus 3,4'-carbonyldiphenol is reflected in a significantly higher Mi (590 versus 425 gpd). Compared with 3,4'-oxydiphenol, 3,4'-oxydibenzoic acid or 3-carboxy-4'-hydroxydiphenyl ether as comonomers with chlorohydroquinone and terephthalic acid provide higher levels of Mi indicating higher stiffness. It is not clear why this is so.

Table III shows copolymerizations involving a basically AB polymer system, analogous to those made with a basically AA-BB system in Table II. Thus, 4,4'-oxydiphenol, at a rather higher level (20 mole % versus 12-15% in Table II), combined with terephthalic acid (20%) and 4-hydroxybenzoic acid (60%), actually precludes liquid crystallinity and causes a drastic loss of T and Mi and increase in E. In the comparison involving 3,4'-isomers, the carbonyl unit actually gives a lower Mi (with high T) than the ether, despite its greater stiffness - the opposite of the AA-BB case. This can be explained in terms of the greater ease with which the more flexible ether monomer can adapt to provide helical rods which can align well.

Homopolyesters from Monocarbonyl-Containing Monomers

The stiffness of the benzophenone unit is such that the melting point of poly(4,4'-carbonyldiphenyl terephthalate) is comparable with decomposition temperatures. The isomeric polyterephthalate (I) from 3,4'-carbonyldiphenol melts much lower, at about 280°C. Although rotations about the carbonyl-to-phenyl bonds can provide a large number of conformations, the liquid crystalline nature of the melt indicates that the extended linear form is preferred [6]. The asymmetry of the 3,4'-carbonyldiphenol allows it to enter the polymer chain in either a head-to-head or head-to-tail manner, depending on its directionality (Figure 2) [7].

Table II. Effects of carbonyl or ether units in a rod-like polyester: AA-BB type.

12-15 mole % R—⟨O⟩—X—⟨O⟩—R   in $\left[\!\!-O—⟨O⟩—O_2C—⟨O⟩—O—\right]$ Where R=OH, CO$_2$H
X=-CO-, -O-

	TENSILE STRENGTH (gpd)	INITIAL MODULUS (gpd)	COMMENTS
(1) Isomer Effect			
	20	245	
	20	425	3,4'-Isomer in extended conformation
(2) Ether vs. Carbonyl			
	20	425	
	22	590	Added stiffness of carbonyl
(3) Diol vs. Hydroxyacid vs. Diacid			
	20	425	Diol is most flexible
	18	550	
	26	525	

<u>Table III</u>. Effects of carbonyl or ether units on a rod-like polyester: AB type.

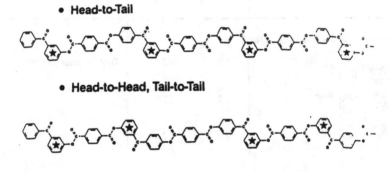

	Tensile Strength (gpd)	Elongation (%)	Initial Modulus (gpd)	Comment
(1) ISOMER EFFECT				
	20 (Ref. 4)	3.2	460	
	7.5	26	17	Isotropic melt – catastrophic change
(2) ETHER VS. CARBONYL				
	20	3.2	460	
	25 (Ref. 5)	5.1	275	Reverse effect as in AA-BB case

The existence of two directional possibilities is tantamount to two different comononers, with respect to melting point depressions. The linearity of the extended conformation of this polymer is evident in Figure 2.

- **Head-to-Tail**

- **Head-to-Head, Tail-to-Tail**

<u>Figure 2</u>. Directionality of 3,4'-carbonyldiphenyl unit in polyterephthalate.

When the polymer is converted into a fiber and heat-strengthened, it develops a remarkably high degree of crystallinity; the 3,4'-carbonyldiphenoxy units are evidently isomorphous even when directionality is reversed:

I is characterized above all by outstanding tensile strength as shown in Table IV, even though the polymer is not quite as rectilinear as many other thermotropics. Despite the small lateral displacement in the 3,4'-carbonyldiphenol unit, it nevertheless is able to provide unusually high macromolecular alignments over long distances. It is speculated that the limited flexibility inherent to an aromatic ketone group permits a higher degree of polymerization in heat-strengthening while ability to crystallize enhances macromolecular orientation (orientation angle measured by X-ray is a remarkably low 5° or less [7]). This effect has been observed for other crystallizable fibers, especially from lyotropic polymers, but is not particularly evident in many poorly crystallizable, thermotropic copolymers. The minor irregularity in chain linearity in I, i.e., lateral displacement, restricts Mi to lower levels (Table IV) than is possible with a more rectilinear chain, such as poly(phenylhydroquinone terephthalate) (Mi >600 gpd).

Table IV. Tensile properties of poly(3,4'-carbonyldiphenylterephthalate).

	Average	Highest	cf. Kevlar® 29
Tenacity, gpd	36	46	23
Elongation, %	9	9	4
Modulus, gpd	420	580	490
Orientation (2Ø °)	4.6		

A comparison of the T and Mi of compositions related to I shows that there is a very delicate margin between ability to form a truly excellent fiber and very mediocre performance, i.e., quite small changes in composition can have quite dramatic influences on ability to form a well-aligned assembly of rod-like polymer chains in the melt.

The criticality of the carbonyl unit in I for providing chain stiffness adequate for melt anisotropy becomes clear in the comparison with its more flexible ether analog, poly(3,4'-oxydiphenylterephthalate) (II) (Table V) which is incapable of melt anisotropy. When 3,4'-oxydiphenol is copolymerized with a larger proportion of stiff monomers such as 4-hydroxybenzoic acid or terephthalic acid (Tables II and III), the propensity of the stiff chain segments therefrom to form a liquid crystalline phase is such that the 3,4'-oxydiphenol units are induced to conform by adopting their most extended conformation. Melt anisotropy requires the average stiffness of the chain to exceed a certain threshold.

In poly(hydroquinone-3,4'-oxydibenzoate) (III), an isomer of II, the 3,4'-oxydibenzoyl units are apparently stiffer than the 3,4'-oxydiphenoxy units of II (frequently noted with other analogous diol-diacid pairs, but not clearly understood). The overall stiffness of III is conducive to melt anisotropy but fiber T and Mi (8 gpd and 200 gpd; Table V) are quite mediocre, indicating that chain flexibility still prevents really good alignment. Further chain stiffening by replacing 20% of the 3,4'-oxydibenzoyl units in III by terephthaloyl units considerably increases T and Mi (to 16 gpd/500 gpd); in Table II, where 85% of the 3,4'-oxydibenzoyl acid units were replaced thus, T and Mi were pushed still higher (to 26 gpd/525 gpd). However, the improvement in modulus was very minor because short-range alignments of perhaps 10 or more repeat units (as distinct from long-range alignments) had already been maximized by the 20% substitution by terephthaloyl units.

Comparison of the melt anisotropic I with the isomeric poly(hydroquinone-3,4'-carbonyldibenzoate) (Table V) shows a jump in melting point to a level where extrusion is not possible. This is again attributed to increased overall stiffness of the polymer but the reason for this is not clear.

Ring substitution can profoundly affect the properties of I in terms of melting (or flow) temperature and molecular alignment (fiber Mi). As Table VI shows, a single methyl substituent in the 2 position considerably reduces the temperature range in which the melt exhibits anisotropy (274-320°C versus 280-355°C, as might be expected with poorer molecular alignment. Modulus is reduced considerably (360 to 250 gpd) in the fiber as extruded and parallels a deterioration of orientation angle (21° to 28°) as measured by wide-angle X-ray (on heat-treatment of I, the development of high crystallinity is paralleled by a spectacular increase in orientation, as manifested by a very low orientation angle of 5° and high tensile strength of over 36 gpd [7], but is not well reflected in modulus change (360 to 420 gpd).

The closeness of the poly(3,4'-carbonyldiphenol terephthalate) (I) structure to the threshold of melt isotropy, despite its remarkable mechanical properties, is illustrated by the disappearance of liquid crystallinity on heating to 355°C, only 75° beyond its melting point. Most thermotropic polyesters of practical value for fiber formation, albeit more rigid, retain anisotropy to well beyond decomposition temperatures.

Apparently minor ring substitutions in I have profound effects on macromolecular alignment and the very existence of liquid crystalline structure. As illustrated in Table VI, this is undoubtedly the result of sterically imposed deviations from the rod-like helical structure of I [7] imposed. [Note that modulus data in Table VI pertains to as-extruded fibers, because their melting points are often below temperatures (ca. 270°C) where molecular growth and heat-strengthening by the usual acetolysis mechanism can occur at effective rates. For such low M.W. materials, tenacity differences and levels, except for very low values, are not significant; very low T levels are associated with comparatively high E, low Mi and negligible anisotropy.]

Table V. Isomer comparisons.

KETONE — Anisotropic, T/Mi=35/420

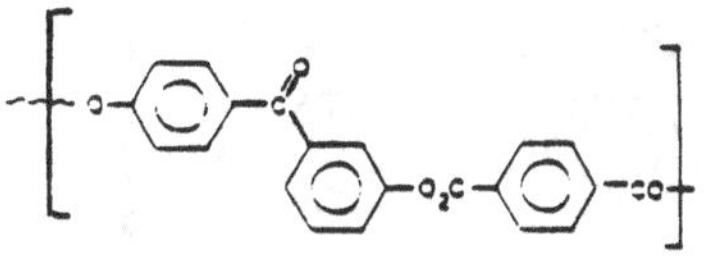

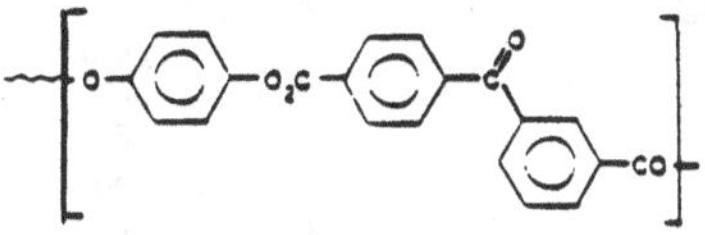

Very high melting

ETHER — Isotropic

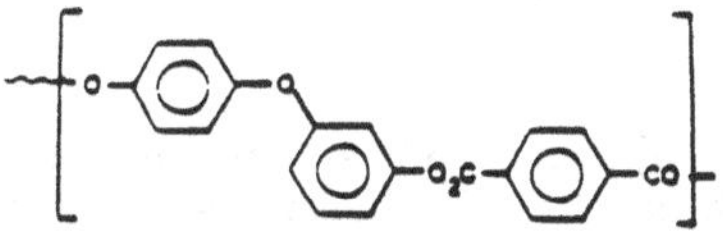

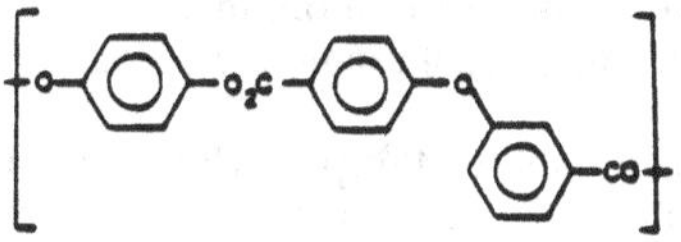

Anisotropic
T/Mi=8/200
Rather flexible (8)

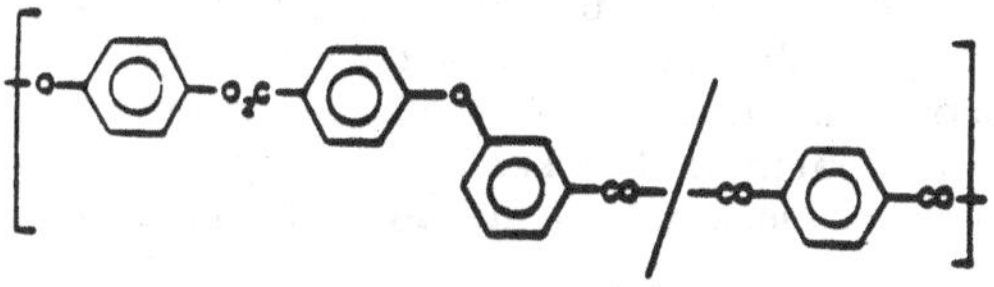

Stiffer T/Mi=16/500

(80/20)

3'-Methyl substitution has little effect on melting point but diminishes chain alignment as evidenced by a lessening of the range in which the melt is anisotropic, a significant increase in orientation angle and decrease in modulus. 3'-Chloro-substitution has almost precisely the same result except that static anisotropy is not present at all in the melt. However, this composition must be only marginally beyond the anisotropy threshold as indicated by considerable molecular alignment, indicative of high anisotropy under conditions of flow. Since chloro- and methyl-substituents are almost identical in size, and chain conformations are influenced most of all by steric effects, it is not at all clear why the chloro group does not exhibit static anisotropy. The larger ethyl group in the 3'-position completely inhibits melt anisotropy, even under flow conditions;

Table VI. Effects of ring-substitution in poly(3,4'-carbonyldiphenyl terephthalate) (I)

Substituents	M.P. (°C)	Anisotropic Range (°C)	Tenacity (gpd)	E (%)	Mi (gpd)	Orientation Angle (°)
Hydrogen	280	280-355	7	4	360	21
3'-Methyl	274	274-320	5	3.2	250	28
3'-Chloro	-	Nil	5	2.0	250	26
3'-Ethyl	260	Nil	1	29	32	Large
3',5'-Dimethyl	246	Nil	1	26	27	Large
Chloroterephthaloyl	234	+	-	-	-	-

mechanical properties - very low T, high E, and very low Mi - are characteristic of a completely isotropic, flexible polymer. A similar result is achieved by 3',5'-dimethyl substitution. When a chlorine substituent is placed on the terephthaloyl unit in I, it substantially reduces the melting point, as might be expected from a combination of two asymmetric monomers, but melt anisotropy, although appearing in a lower temperature range, is not lost.

With larger, stiff units, notably 4,4'-biphenylene, substituted into the I composition (Table VII), while there is no particular reduction in modulus, tenacity after heat-strengthening is reduced. Possibly with stiffer units long-range order is much less easily obtained than short-range alignment, but experience suggests that the more likely cause of limited tenacity is poorer response to heat-strengthening. Table VII shows that whereas 4,4'-bibenzoyl units in place of terephthaloyl units in I causes an increase in melting point, where the biphenyl is incorporated in the ketodiol unit as shown, thus exaggerating the asymmetry of the monomer, the melting point is reduced (from 280°C to 250°C).

Whereas the isomer of I, poly(hydroquinone-3,4'-carbonyldibenzoate) (III) is excessively high melting and cannot be satisfactorily extruded, fibers from a third isomer, the random copolymer (IV) of equal amounts of 4-hydroxybenzoic acid and 4-hydroxy-3'-carboxybenzophenone, gave an anisotropic melt extrudable to well-oriented fibers giving T and Mi of 18 gpd and 400 gpd, respectively [9]. It failed to provide very high T levels. A key to obtaining a satisfactory fiber in this case was to ensure quite random distribution of comonomers because any blockiness would result in an excessively high melting point. The homopolymers from the constituent monomers are very high-melting.

Thermotropic Polyesters Based on Oxy- or Carbonyldi(phenyl-4) Units

Random copolymers (V) of 4-hydroxybenzoic acid with the angular 4-hydroxy-4'-carboxybenzophenone also gave an anisotropic melt, extrudable as a

Table VII. Poly(3,4'-carbonyldiphenylterephthalates) containing larger rigid ring units.

Structure	Flow (°C)	Liquid Crystal	Heat-Strengthened			
			Ten. (gpd)	E (%)	Mi (gpd)	O.A.
(structure 1)	250	+	11	-	400	27->23 Modest crystallinity
(structure 2)	325	+	9.4	-	340	Poor H.T. response
(structure 3)	298	+	17+	4.7	350+	V. good H.T. response

well-oriented fiber [9]. Figure 3 shows a two-dimensional representation of IV and V in their most extended conformations (assuming ideal alternating distribution o the comonomers). V is clearly not a linear, rod-like monomer as is IV but can achieve a liquid crystalline melt structure because it can conform to rod-like helix, and provide fiber properties comparable with IV. A second example of this comparatively rare type of thermotropic polyester, also shown in Figure 3, is poly(chlorohydroquinone 4,4'-oxydibenzoate [VI]) [3]. An earlier study [10] showed that as long as the helix was regular and the straight segment between bent units (oxy or carbonyl) close to a certain average length (about 16 in chain atoms), and of course give a suitable melting point, an anisotropic melt, extrudable to a high T/Mi fiber results. The carbonyl- or oxy-containing units had to be somewhat stiff. Thus poly(4,4'-oxydiphenylchloroterephthalate), the low melting isomer of VI, and the higher-melting poly(4,4'-oxydiphenyl-terephthalate) are isotropic in the melt because of their greater flexibility. Thermotropic polyesters such as VI are extremely sensitive to introduction of comonomers which make the macromolecular helices less regular and therefore less rod-like. Thus the introduction of isophthalic acid or 4,4'-biphenol into the VI structure (Figure 3) caused a rapid drop in fiber T/Mi, although anisotropy persisted to rather higher levels of modification.

As shown in Table VIII [11], poly(4,4'-carbonyldiphenylterephthalate) is a very high melting polymer so that any melt anisotropy propensities are masked. However, the modestly substituted analogue, poly(3-methyl-4,4'-carbonyldiphenylchloroterephthalate) melts much lower (304°C) but displays no anisotropy

(1) <u>4-Hydroxybenzoic acid/3-Hydroxy-4'-carboxybenzophenone</u>
T/Mi/m.p. = 18/450/284°C

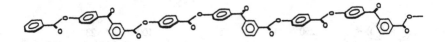

(2) <u>Isomer of above</u>   T/Mi/m.p. = 18/490/325°C

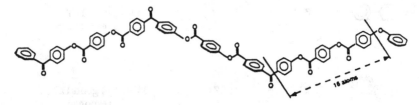

(3) <u>Chlorohydroquinone/4,4'-oxydibenzoic acid</u>
T/Mi/m.p. = 24/430/315°C

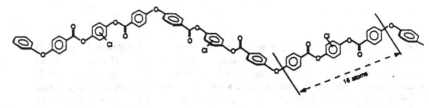

Figure 3. Extended chain conformations.

in the melt [11], although its moderate fiber properties indicate that some anisotropy is induced in the spinning process. Most surprisingly, poly(3,5-dichloro-4,4'-carbonyldiphenylterephthalate) showed anisotropy above its melting point (250°C) and provided fiber properties characteristic of well-oriented macromolecules. The dimethyl analog, melting at the same temperature, was melt isotropic but gave fiber properties indicative of flow birefringence during processing.

Table VIII. Substituent effects in the polyterephthalates in various carbonyldiphenols.

Diacid	R1	R2	R3	R4	M.p.(° C)	Aniso-tropy	T	Mi	O.A.	Ref.
T	H	H	H	H	>400					
T	Me	H	H	H	>400					
CTT	Me	H	H	H	304	-	9	138		
T	Cl	Cl	H	H	250(?)	+	19	620	28°	(10)
T	Me	Me	H	H	250	-	10	160	31°	(10)
T	Cl	Cl	Cl	Cl		-				(10)

It is hypothesized that in both polymers steric hindrance by the ortho-situated dichloro or dimethyl groups forces the oxygen of the adjacent ester group carbonyl into a position between them so that the phenylene rings are forced well out of coplanarity. With the dichloro groups electronegative repulsion would position the oxygen exactly between the chloro groups, so the rings would be positioned exactly at right angles. This stiffening at some of the ester groups would be adequate to produce melt anisotropy. With the far less polar methyl groups, the oxygen can oscillate to an extent between two extreme positions as shown, so that there is a much lower degree of stiffening at the ester bonds in question, than with dichloro groups.

SUMMARY

Progressive introduction of non-linear comonomer units into a rod-like thermotropic polyester structure causes sequential loss of modulus, tensile strength, and thermotropicity. The polyterephthalate of 3,4'-dihydroxybenzophenone is an essentially linear chain wherein adjacent carbonyl and m-phenylene groups offset each other. Although close to the lower limit of stiffness needed for thermotropicity, it provides outstanding tensile strength. Certain compositions having a non-linear unit in each repeat can adopt a helical, rod-like conformation to provide high tensile properties providing (a) it has stiff segments of sufficient length, (b) helical symmetry is not depressed by non-conforming comonomers or excessively broad helical amplitude distribution (in copolymers), and (c) certain flexible angled units, e.g., resorcinol, are excluded. Ortho substituents in diol can provide additional stiffening. Thermotropic polyesters containing a modest amount of isophthaloyldiphenol provide the unique combination of high macromolecular orientation and high elongation. This may be explained in terms of axially folded chains in the nematic melt.

REFERENCES

1. G. Calundann, U.S. Patent No. 4,184,996 (22 January 1980).
2. R. S. Irwin, U.S. Patent No. 4,188,476 (12 February 1980).
3. T. C. Pletcher and J. Kleinschuster, U.S. Patent No. 4,066,620 (3 January 1978).
4. R. S. Irwin, U.S. Patent No. 4,262,965 (26 May 1981).
5. R. S. Irwin, U.S. Patent No. 4,232,143 (4 November 1980).
6. R. S. Irwin, U.S. Patent No. 4,245,082 (13 January 1981).
7. R. S. Irwin, W. Sweeny, K. H. Gardner, C. R. Gochanour and M. Weinberg, Macromol. 22, 1065 (1989).
8. R. S. Irwin, U.S. Patent No. 4,487,916 (11 December 1984).
9. A. H. Frazer, U.S. Patent No. 4,398,015 (9 August 1983).
10. R. S. Irwin, "Structure-Property Relationships in Mesophase Polyester Fibers", Symposium on "Carbon and Other High Performance Fibers", 189th A.C.S. National Meeting, April 1985, Miami Beach. Abstract Papers, No. 64.
11. A. H. Frazer, U.S. Patent No. 4,399,270 (16 August 1982).

UNSYMMETRICAL SPACERS: A NEW CONCEPT FOR SYNTHESIS OF MAIN-CHAIN LIQUID CRYSTALLINE POLYMERS

R. Kosfeld, G. Poersch, M. Hess
University of Duisburg, Dept. of Physical Chemistry,
Lotharstrasse 1, D-4100 Duisburg, FRG

ABSTRACT

12 polyesters with branched spacers and 7 polyesters with linear spacers were synthesized and characterized.

INDRODUCTION

In recent years there have been a lot of investigations on the structure-properties relationship of liquid-crystalline polymers [1-3]. One way to modify the properties of LC-main-chain polymers is the use of alkyl-groups, either as flexible moieties in the polymer backbone or as pendant substituents in the mesogenic unit. Even a combination of both, as flexible spacers in the main-chain and as additional substituents in the mesogenic moiety was investigated [4], but little is known about the effects of alkyl-substitutents in the flexible spacer [5]. Therefore this work deals with the synthesis and characterization of LC-main-chain polyesters with branched alkyl-spacers.

SYNTHESIS

The polyesters T m,n (Figure 1) were synthesized by solution polycondensation in chloroform using 4,4"-dichloroformyl-p-terphenyl and mono- and disubstituted 1,3-propandiols.

4,4"-dichloroformyl-p-terphenyl was prepared from p-terphenyl by a Friedel-Crafts-acylation with oxalylchloride and $AlCl_3$ [6] and subsequent treatment with thionylchloride. The substituted 1,3-propandiols required were obtained by Malonestersynthesis and reduction with $LiAlH_4$.

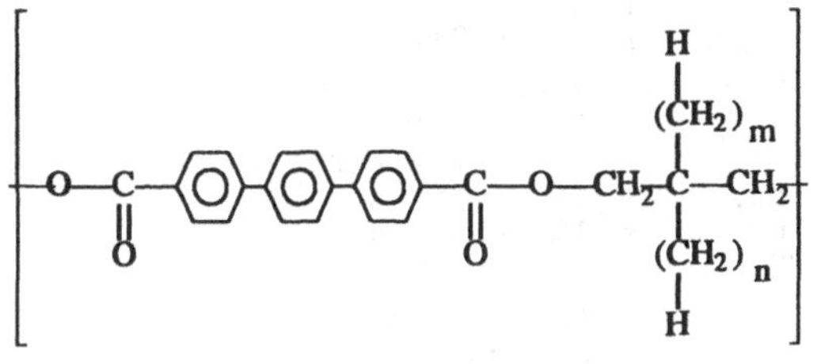

$$m = 0,1 \; ; \; n = 1\text{-}5,10$$
Figure 1: Terphenylester T m,n

The linear polyesters T x (Figure 2) were prepared like their branched analogs by solution polycondensation of 4,4"-dichloroformyl-p-terphenyl and the corresponding α,ω-diols.

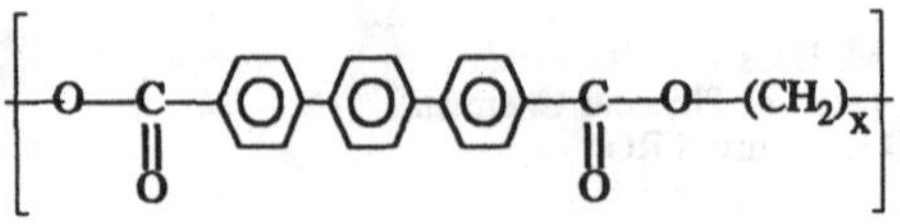

x = 3,4,6-9

Figure 2: Terphenylester T x

RESULTS AND DISCUSSION

The thermal properties of the polyesters which had been synthesized, were investigated by Differential Scanning Calorimetry and polarizing microscopy (Table I, II and Figure 3 to 5). All polyesters show birefringent liquid phases.

The branched polyesters T m,n (Table I) exhibit at least one smectic phase. Except the polyesters T 0,5, T 0,10 and T 1,10 there was another mesophase observed, most probably a nematic one.

As shown in Table I the monosubstituted polyesters T 0,n show higher transition temperatures than their disubstituted analogs T 1,n. In both series it is seen that an unsymmetrical substitution causes a reasonable decrease in melting temperature. With increasing length of the alkyl-substituent the influence of an additional CH_2-group on the melting temperature decreases. Hence for n≥4 the melting temperatures of both series remain almost constant. The melting points remain almost constant while the clearing temperatures decrease with increasing number of CH_2-groups, especially if the side-chain is short. Consequently, the interval of the mesophase is reasonably broad at first and decreases to 3/4 of that of n=1 for n>5.

Table I: Transition temperatures of
Polyesters T m,n

Polymer	Transition temperatures[a][°C]						
T 0,1	cr	209	s	291	n	367	i
T 0,2	cr	219	s	306	n	341	i
T 0,3	cr	206	s	306	n	334	i
T 0,4	cr	220	s	278	n	316	i
T 0,5	cr	222	s	313	i		
T 0,10	cr	221	s	296	i		
T 1,1	cr	275	s	301	n	363	i
T 1,2	cr	224	s	287	n	309	i
T 1,3	cr	197	s	264	n	287	i
T 1,4	cr	148	s	233	n	262	i
T 1,5	cr	155	s	239	n	262	i
T 1,10	cr	152	s	244	i		

a) cr = crystalline; s = smectic[+];
n = nematic; i = isotrop

+ = The type of the smectic phase is still under investigation at present.

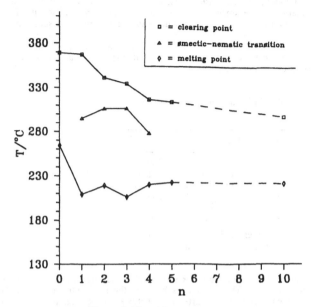

Figure 3: Transition temperatures of
Polyesters T 0,n

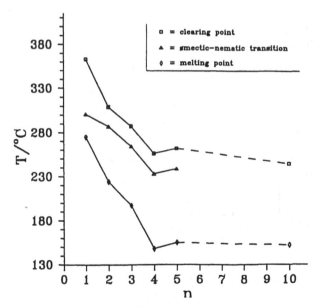

Figure 4: Transition temperatures of
Polyesters T 1,n

The clearing temperatures decrease almost in the same magnitude as the melting temperatures. The interval of the smectic phase increases with increasing length of the side–chain while the interval of the nematic phase decreases and vanishes at some chain–length in between $5<n<10$.

The linear polyesters T x (Table II and Figure 5) exhibit at least one mesophase, probably a smectic one. Polymers T 3, T 4 and T 6 show also an additional mesophase at higher temperatures. Its nature is still unidentified.

Table II: Transition temperatures of
Polyesters T x

Polymer	Transition temperatures[a][°C]				
T 3	cr	264	lc	369	i
T 4	cr	356	lc	>380i	
T 6	cr	245	lc	371	i
T 7	cr	225	lc	341	i
T 8	cr	270	lc	343	i
T 9	cr	223	lc	307	i

a) cr = crystalline; lc = unidentified
mesophase; i = isotrop

In comparison the linear polymers T x show higher melting and clearing temperatures than their branched analogs. The differences between the monosubstituted and the linear polymers are much smaller than the differences between the disubstituted and the linear polyesters.

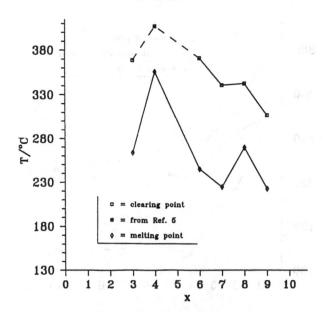

Figure 5: Transition temperatures of
Polyesters T x

Corresponding to the decrease of thermal transition temperatures there is an increase in solubility with increasing amount of CH_2-groups in the spacer side-chain. It was observed that the solubility of the monosubstituted polymers lies between that of the linear and the disubstituted species. For example: all of the linear polyesters are almost insoluble, even in dichloroacetic acid, where all of the substituted polymers are soluble. Furthermore the polymer T 1,5 is considerably soluble in chloroform and tetrahydrofuran, where polyester T 0,5 is by far less soluble.

As shown by Ober et al. [9] the transition temperatures depent on the molecular weight (MW), but levels off at fairly low MW, so that the transition temperatures of the polyesters under investigation should not be effected by MW.

The properties mentioned above can be explained as follows:
a) substituting a hydrogen-atom from the spacer by an alkyl–chain will disturb inter-molecular interactions and packing of the polymer, especially in the solid state;
b) with increasing length of the substituent the alkyl-chain will be able to arrange parallel to the mesogenic unit [7], so that there are only minor changes for longer substituents;
c) geminal substituents, even if the second is as small as a methyl-group, will disturb the intermolecular interactions much more than only one substituent does;
d) with increasing length of the lateral alkyl-chain it will be more and more able to act as "bounded solvent" [8];
e) "odd–even" effects are in substituted spacers suppressed in contrast to spacers of the linear type.

REFERENCES

1. H. Finkelmann, Angew. Chem. 99, 840 (1987)
2. W. Brostow, Kunststoffe 78 (5), 441 (1988)
3. R. W. Lenz, Faraday Discuss. Chem. Soc. 79, 21 (1985)
4. Q.-F. Zhou and R. W. Lenz,
   J. Polym. Sci, Chem. Ed. 21, 3313 (1983)
5. P. Meurisse, C. Noel, L. Monnerie and B. Fayolle,
   Brit. Polym. J. 13, 55 (1981)
6. T. W. Campbell, J. Am. Chem. Soc. 82, 3126 (1960)
7. B. Reck and H. Ringsdorf,
   Makromol. Chem., Rapid Commun. 7, 389 (1986)
8. M. Ballauff, Angew. Chem. 101, 261 (1989)
9. Ch. K. Ober, J.-I. Jin and R. W. Lenz,
   Adv. Polym. Sci. 59, 103 (1984)

ACKNOWLEDGEMENT

We gatefully thank the Deutsche Forschungsgemeinschaft, Bonn, for their financial support.

PHASE TRANSITIONS IN BLENDS OF
LIQUID CRYSTALLINE POLYMER/POLYETHER IMIDE

Joonhyun NAM, Tomohiro FUKAI and Thein KYU
Institute of Polymer Engineering, The University of Akron, Akron, OH
44325-0301

ABSTRACT

A thermotropic liquid crystalline copolymer consisting of bisphenol
E diacetate, isophthalic acid and 2,6-naphthalene dicarboxylic acid was
blended with polyether imide by dissolving in a mixed solvent of
phenol/1,1,2,2-tetrachloroethane in a ratio of 60/40 w/w and
co-precipitating the ternary solution in non-solvent (methanol).
Wide-angle x-ray diffraction and differential scanning calorimetry
studies revealed that the blends were completely amorphous with a single
glass transition temperature. The single phase was probably entrapped
during solvent removal, but these mixes were unstable and phase
separated upon heating. Mesophase structure developed in the LCP rich
region with continued annealing. The evolution of crystalline texture
was monitored by time-resolved wide-angle x-ray diffraction following a
temperature jump from ambient to 265 °C. The recrystallization process
of LCP was found to slow down in the blend state relative to that of the
neat LCP.

INTRODUCTION

In recent years, thermotropic liquid crystalline polymer (LCP)
blends, commonly known as situ-LCP composites or self-reinforced LCP
blends, have received considerable attention for their improved
mechanical performance [1-5]. These LCP blends were melt-blended in an
extruder or an internal mixer and subsequently injection molded to
fabricate desired shapes. When LCP chains were subject to flow during
processing, microfibrills developed in the continuum of flexible chains.
It has been speculated that the oriented microfibrills serve as
reinforcing entities, thereby improving modulus, tensile strength or
fracture toughness in some compositions. The above studies were
primarily concerned with the flow characteristics, morphological
characterization and its relationship to mechanical properties. It has
been realized that miscibility or partial miscibility between LCP and
flexible polymers could have appreciable effects on the fiber/matrix
adhesion and their ultimate properties. Consequently, there are
considerable reports addressing the issue of polymer miscibility in such
LCP blends and their thermally induced phase separation behavior [5-10].
We have been involved in the studies of phase transitions in some
neat LCPs and phase segregation in their blends with thermoplastics
[5,10,11]. In a previous paper [11], it was demonstrated that the
premelting transition in a thermotropic liquid crystalline copolymer
(LCP) containing bisphenol E diacetate, isophthalic acid and 2,6
naphthalene dicarboxylic acid was a consequence of a melting -
recrystallization process. At high temperature annealing near the
melting transition, the solid crystals lost their positional ordering.
Then the rigid macromolecules reorganize to form a new crystal and
finally terminal melting to a mesophase occurs. This reorganization
process was attributed to be responsible for the dual differential

scanning calorimetric (DSC) endotherms.

In this paper, DSC, wide-angle x-ray diffraction (WAXD), optical microscopy and light scattering have been employed to elucidate the miscibility of this LCP with polyether imide blends. The phenomena of phase separation in the amorphous blend state has been investigated. Subsequently, phase transition within the phase separated LCP rich regions was explored in comparison with that in the pure LCP state.

EXPERIMENTAL

Materials

The thermotropic LCP with inherent viscosity $\eta = 0.59$ dl/g was supplied by Monsanto Co. This LCP was produced by copolymerizing bisphenol E diacetate, isophthalic acid and 2,6 naphthalene dicarboxylic acid in a co-monomer ratio of 50:40:10 [12]. Polyether imide (PEI) is a commercial grade Ultem 1,000 from Generic Electric Co. The inherent viscosity was determined to be 0.52 dl/g. The two polymers were dissolved in a mixed 60/40 phenol/tetrachloroethane solvent at a polymer concentration of 1 wt %. These tenary solutions were co-precipitated in methanol and subsequently vacuum dried at 70 °C for one week. Some blend films were solvent casted in Petri dishes which turned out to be transparent.

Methods

DSC scans were acquired under nitrogen circulation on a Du Pont thermal analyzer (Model 9900) interlinked with a heating module (model 910). The heating rate was 20 °C/min unless indicated otherwise. Indium standards were used for temperature calibration. Wide angle x-ray diffraction experiments were carried out on a 12 kW Rigaku rotating anode with a pin hole collimator. The generator was operated at 150 mA and 40 kV using a Ni filtered CuKα line.

The wavelength of x-rays was 0.15406 nm. The 2θ scans was performed from 10 to 40 degree at a scan rate of 10 °/min in the early part of time resolved studies and 0.5 °/min for the longer time and static measurements.

Optical micrographs were obtained on a Leitz polarizing microscope (Model Pol 12) using a heating stage (THM-600, LINKAM Scientific Co.) controlled at an arbitrary heating and cooling rate of 10 °C/min, unless indicated otherwise.

Cloud point measurements were undertaken using laser light scattering with a Vidicon camera attached to a Optical Multichannel Analyzer (OMA III, EG&G Princeton Applied Research, Inc.). A He-Ne laser with a wavelength of 632.8 nm was utilized as a light source.

Inherent viscosities were determined using the mixed solvent 60/40 phenol/tetrachloroethane. Measurements were made at 0.5 g/dl using a Ubbelohde viscometer (Size 100) in a water bath controlled at 30 °C.

RESULTS AND DISCUSSION

Figure 1 shows the DSC traces of pure LCP, pure PEI and their blends as obtained by co-precipitation. A glass transition (Tg) was evident at about 220 for the pure PEI whereas a Tg (110 °C) and dual endotherms (275 and 295 °C) were observed for the LCP in the 2nd run. The intermediate blends exhibited a single Tg, shifting systematically with composition. These blends are totally amorphous since there is no indication of melting transition. Wide-angle x-ray studies, as will be

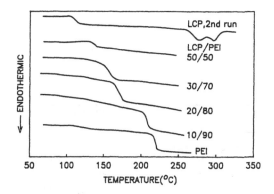

Figure 1. DSC heating traces of LCP, PEI and their blends.

shown later, showed the lack of crystallinity in the blends. This
single phase blend is probably frozen-in during solvent removal, thus it
may be thermodynamically unstable [10,13]. As can be seen in Figure 2,
the 50/50 blend showed two separate Tgs with a small endotherm during
the subsequent DSC heating cycle after cooling from 270 °C. With
continued heating cycles, the two Tgs progressively separated further.
Concurrently, the melting endotherms intensified and moved to a higher
temperature with progressive heating, suggesting crystal perfection or
reorganization.

Figure 3 shows an optical micrograph and a light scattering picture
of a solvent-casted 50/50 blend after heating to 300 °C. The
interconnected domain structure and a diffuse scattering halo is
suggestive of thermally induced phase separation by spinodal
decomposition. Cloud point measurements were undertaken for various
compositions. As typical for entrapped single phase blends [7], phase
separation occurs slightly above their Tgs (Figures 4a and b). The
scattered intensity increases abruptly again at a higher temperature
around 270 °C where reorganization of rigid LCP phase takes place. It
may be speculated that phase separation occurs first, then it is
followed by reorganization of LCP intracrystalline chains in the LCP
rich region.

To confirm this hypothesis, time-resolved WAXD experiments were
conducted on the 50/50 blend in comparison with the neat LCP films.
Figures 5a and b show the evolution of WAXD 2θ scans of the neat LCP
during annealing isothermally at 250 and 270 °C. At 270 °C, the
original crystalline peaks gradually disappear with annealing time, and
reappear at a later time, then gradually become stronger. This process
has been attributed to melting-recrystallization previously [11].
However, the low temperature annealing at 250 °C shows no indication of
melting. Instead, the WAXD become more pronounced probably due to
enhanced crystal perfection. In the case of the 50/50 blend (Figure
5c), the starting blend is completely amorphous as exemplified by a
single broad amorphous WAXD scattering peak. It required about 2 to 4
hours to see the development of LCP crystalline phase. This delayed
organization of LCP phase may be a consequence of entrapped miscibility
of LCP/PEI blends.

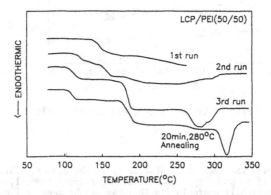

Figure 2. DSC traces of 50/50 LCP/PEI blend with progressive heat treatments.

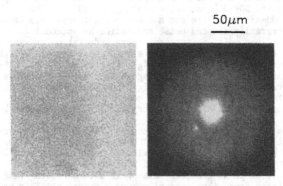

Figure 3. Optical micrograph and SALS pattern of 50/50 blend after phase separation.

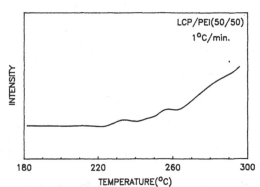

Figure 4a A typical cloud-point curve of 50/50 blend obtained by light
scattering. The heating rate was 1 °C/min.

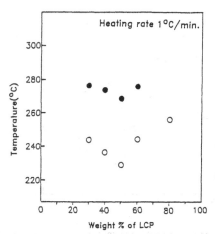

Figure 4b A cloud point phase diagram for LCP/PEI blends. Open circle
represents phase separation temperatures while closed circle
indicates temperatures at which phase growth resumed.

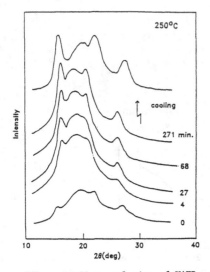

Figure 5a Time-evolution of WAXD
scans of LCP at 250 °C.

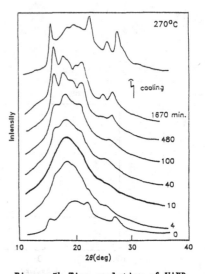

Figure 5b Time-evolution of WAXD
scans of LCP at 270 °C.

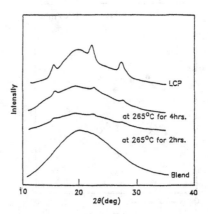

Figure 5c WAXD profiles of 50/50 LCP/PEI blend subject to thermally
induced phase separation.

CONCLUSIONS

Blends of thermotropic LCP/PEI were found to be completely amorphous with a single glass transition. This single phase may have been entrapped during solvent removal, thus it is not thermodynamically stable. Upon heating, the blends phase separated with a phase diagram very close to their Tgs. Some crystallinity developed during prolonged annealing. This recrystallization process appeared to be retarded relative to that in the neat LCP.

REFERENCES

1. Weiss, R.A., Huh, W. and Nicolais, L. Polym. Eng. Sci., 27, 684, (1987).
2. Isayev, A.I. and Modic, M.J. Polym. Composite, 8, 158 (1987)
3. Blizard, K.G. and Baird, D.G., Polym. Eng. Sci., 27, 653 (1987)
4. Brostow, W., Dziemianowicz, T.S., Romanski, J. and Werber, W., Polym. Eng. Sci., 28, 785 (1988)
5. Zhuang, P., Kyu, T. and White, J.L., Polym. Eng. Sci., 28, 1095 (1988).
6. Pracella, M., Dainelli, D., Galli, G. and Chiellini, E., Makromol. Chem., 187, 2387 (1986).
7. Friedrich, K., Hess, M. and Kosfeld, Makromol. Chem. Macromol. Symp. 16, 251 (1988).
8. Paci, M., Barone, C. and Magagnini, J. Polym. Sci. Polym. Phys. 25, 1595 (1987).
9. Nakai, A., Shiwaku, T., Hasegawa, H. and Hashimoto, T., Macromolecules, 19, 3010 (1986).
10. Zheng. J.Q. and Kyu, T. in Liquid Crystalline Polymers, Ed. R.A. Weiss & Ober, C.K., ACS Symp. Ser. # 435, p. 458 (1990).
11. Nam. J., Fukai, T. and Kyu, T., Macromolecules, submitted.
12. Deex, O.D. and Weiss, V.W., U.S. Patent No. 4,102,864 (1978).
13. Schultz, A.R. and Young, A.L., Macromolecules, 13, 663 (1980).

PART V

Surfaces, Interfaces and Adhesions

PREPARATION OF MOISTURE RESISTANT POLYIMIDE FILMS BY PLASMA TREATMENT

MARK A. PETRICH AND HSUEH YI LU
Department of Chemical Engineering, Northwestern University, Evanston, Illinois 60208

ABSTRACT

Polyimides are an important class of polymeric materials used in microelectronics fabrication. These polymers could be used even more extensively if it were possible to improve their moisture resistance. We are using plasma processing techniques to modify the moisture resistance of polyimide films. Films are exposed to nitrogen trifluoride plasmas to introduce fluorine into the surface of the polyimide. Fluorination is monitored with x-ray photoelectron spectroscopy and Fourier transform infrared absorption spectroscopy. Water contact angle measurements are used to assess the hydrophobicity of the treated surfaces. Thus far, we have demonstrated that this plasma treatment is a good way of introducing fluorine into the polyimide surface, and that these treatments do enhance the hydrophobic nature of polyimide.

INTRODUCTION

Polyimide films are used as insulating materials in microelectronic device fabrication. The low dielectric constant, ease of processing, and good thermal stability make polyimide an attractive candidate for interconnection and packaging applications [1]. However, the widespread use of polyimide is limited by concerns about its resistance to moisture [2]. Polyimides used as passivation layers are expected to protect underlying circuit elements from moisture. These components may corrode and prematurely fail in the presence of moisture. Another reason for concern about moisture resistance is that the dielectric properties of the polyimide change upon moisture absorption [2].

In this project, we use plasma processing techniques to modify the moisture resistance of polyimide thin films. By exposing the films to reactive fluorinated compounds such as NF_3 in the plasma, we can introduce fluorine into the film and modify the surface properties of the polyimide.

The interaction of polyimides with plasmas has been studied previously [3-7]. However, the emphasis of most of this research has been on etching of polyimide, or on determining the resistance of polyimide to chemical processing used to etch other integrated circuit materials. It has been found that the surface of polyimides are fluorinated by plasmas of CF_4 and oxygen. It was also found that the fluorine does not remain at the surface of the polymer, but diffuses into the bulk of the polymer. These researchers were not trying to surface modify polyimide, and did not measure any properties of the fluorinated films.

There have been several reports of surface fluorination of polymers using plasmas, most notable for biomedical materials applications [8,9]. NF_3 has been used as the fluorinating species in a previous study involving polymers other than polyimide [10].

EXPERIMENTAL

To prepare the polyimide films, we first spin-coat polyamic acid onto two-inch silicon wafers with a 4000 rpm spinning rate. The films in this report are DuPont PI-2611. Films are cured using a recommended procedure of heating in vacuum for 30 minutes. Cured film thickness is between 3.5 and 4.0 micrometers. The structure of this polyimide is shown in Figure 1.

The polyimide films are exposed to NF_3 plasmas in a homebuilt plasma processing system. The system has 250 cm^2 parallel-plate electrodes with a 3.5 cm gap. The chamber walls are electrically grounded. Wafers to be treated are placed on the lower, grounded electrode. Reactor pressure is sustained by a rotary vacuum pump and is controlled electronically. Power is applied at 13.56 MHz through a matching network to the powered electrode of the reactor. Table 1 shows the plasma conditions used in this study.

The etching effect of the NF$_3$ plasmas is determined by masking a portion of the polyimide films with a glass cover slip during treatment. The difference in height of the masked and exposed film is measured with a Tencor Alpha Step mechanical stylus.

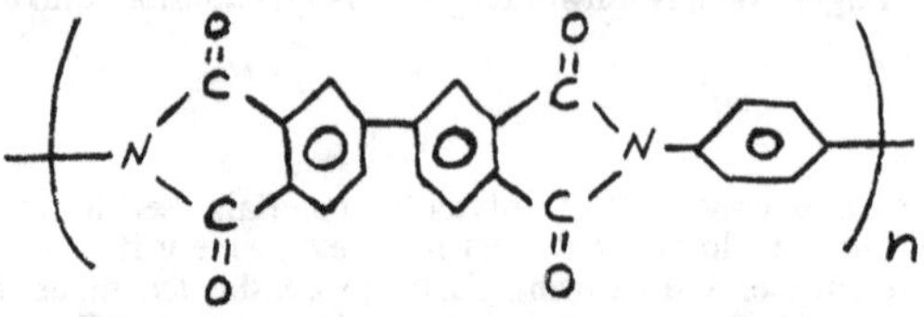

Figure 1 — Structure of polyimide used in this study.

Treated and untreated films are studied with Fourier transform infrared spectroscopy (FTIR) and x-ray photoelectron spectroscopy to observe chemical changes caused by the plasma. Moisture resistance of the films is determined by sessile water drop contact angle measurements [11]. We photograph small water drops on the film surfaces, and measure the contact angles in the photographs.

Table 1. Plasma Treatment Conditions

Flow rate NF$_3$	10 sccm
Substrate temperature	25 °C
Reactor pressure	200 mTorr
rf frequency	13.56 MHz
rf Power	15 W
Exposure Time	0, 0.5, 1, 2, 5, 10 minutes

Infrared absorption spectra are obtained with a Mattson Alpha-Centauri Fourier-Transform Infrared Absorption Spectrometer. Plots of absorption coefficient vs wavenumber are calculated by normalizing the absorption curves by film thickness, which is measured with a mechanical stylus. Attenuated total reflectance (ATR) FTIR spectra are acquired with a Nicolet spectrometer with special ATR attachment.

X-ray photoelectron spectra are acquired with a VG Instruments ESCA/SIMS system. Spectra are acquired for carbon (C(1s)), nitrogen (N(1s)), and fluorine (F(1s)) using a magnesium Kα x-ray source.

RESULTS AND DISCUSSION

We are able to fluorinate the surface of polyimide films using the treatment conditions shown in Table 1. Figure 2 presents C(1s) XPS spectra of a film exposed to NF$_3$ for ten minutes without a plasma and a film exposed to an NF$_3$ plasma for five minutes. Figure 3 presents the C(1s) XPS spectrum for a film that was exposed for one minute. Note that Figures 2 and 3 have different x-axes. It is apparent from these spectra that the NF$_3$ plasma fluorinates polyimide even after very short times. The peaks seen in the untreated film at 287.3 eV and 291 eV correspond to ring carbons and carbonyl-group carbons. After one minute, the intensity of the 287.3 eV peak is reduced and a new peak appears at 293 eV. The peak at 291 eV is unchanged. After five minutes, all three peaks shift to higher binding energies of 289, 292 and 295 eV. These spectra suggest that fluorination of the polyimide film begins with

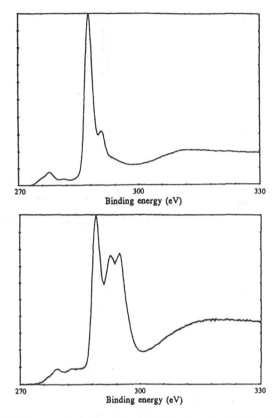

Figure 2 — C(1s) XPS spectra for unexposed polyimide (top) and polyimide after 5 minutes of plasma exposure (bottom).

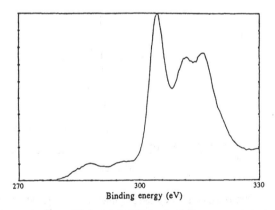

Figure 3 — C(1s) XPS spectrum for polyimide after one minute of plasma exposure.

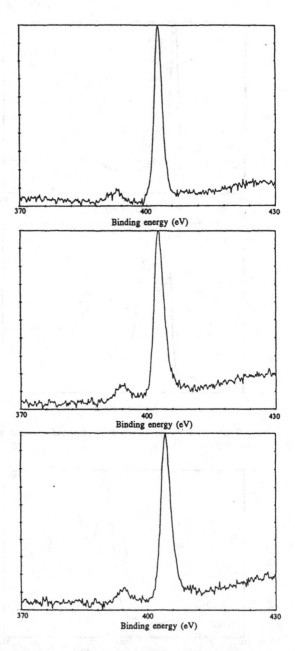

Figure 4 — N(1s) XPS spectra for unexposed polyimide (top), polyimide after one minute of plasma exposure (middle), and polyimide after five minutes of plasma exposure (bottom).

fluorination of the ring carbons, followed at longer times by fluorine attack of the carbonyl groups.

Support for this picture of the fluorination process is provided by nitrogen XPS spectra shown in Figure 4. After one minute of plasma exposure, the N(1s) peak broadens slightly, but there is no shift in binding energy. This indicates the fluorinated carbons observed in the carbon spectra are distant from the nitrogen atoms. The ring carbons fit this description. After five minutes of plasma exposure, the N(1s) peak shifts by approximately +1 eV. The presumed attack of carbonyl groups by fluorine after five minutes of exposure leads to a large enough change in the nitrogen environment to cause this binding energy shift. In both the carbon and nitrogen spectra, there is almost no change between five and ten minutes of plasma exposure. The surface appears "fluorine saturated" after five minutes.

Fourier transform infrared absorption (FTIR) measurements taken in transmission mode do not reveal any differences between the treated and untreated films. We interpret this as evidence that plasma fluorination is a surface phenomenon at the conditions used in this study. We are currently using attenuated total reflectance FTIR [12] to study the chemical structure of the fluorinated surface.

The physical effects of the surface treatments are shown in Table 2. These treatments involve relatively low power plasmas, but for exposure times longer than one minute, an appreciable amount of etching occurs. Approximately 700 Å of the polyimide is removed in only two minutes time. However, it is possible to fluorinate the surface without severe etching as demonstrated by the extremely small amount of etching in the 0.5 and 1.0 minute exposed films.

Table 2. The effects of plasma treatment on polyimide film surfaces

Exposure time (minutes)	Etched Depth (μm)	Water Contact Angle
0	—	84°
0.5	—	81°
1	—	84°
2	0.07	97°
5	0.18	102°
10	0.50	92°

The water contact angle measurements shown in Table 2 confirm that surface fluorination can improve the moisture resistance of polyimide films. The contact angle for water on untreated polyimide (84°) improves to 102° after five minutes of plasma exposure which is comparable to the 109° reported for polytetrafluoroethylene [13]. Notice that there is an optimum plasma treatment time of approximately 5 minutes for the processing conditions used here. After this time, we presume that surface roughening leads to the decrease in water contact angles.

CONCLUSIONS

We have demonstrated the plasma fluorination is a promising means of improving the moisture barrier properties of polyimide dielectric thin films. It is possible to fluorinate polyimide surfaces without introducing excessive damage from etching by using low power NF_3 plasmas. The water contact angle of the most hydrophobic sample prepared in this work is 102°, which indicates that this treated polyimide is considerably more hydrophobic than polyethylene (94°) but is not quite as hydrophobic as polytetrafluoroethylene (109°). More careful variations in processing conditions, especially plasma exposure time, should help us to improve on this result.

In our future work, we will quantify the effects of film preparation conditions on ease of plasma fluorination, and will quantify the relationship between fluorination and moisture resistance.

328

ACKNOWLEDGMENTS

This work was supported by a research initiation grant from the Engineering Foundation. We thank Dean Fjelstul for his assistance in spin-coating the polyimide films. We are grateful to Dr. Joe Machado (Air Products and Chemicals, Inc.) for providing the NF$_3$ used in this study and for the donation of a gas regulator. We also thank Dr. Luc Verdet (IFP Enterprises, Inc.) for providing polyphenylquinoxaline (PPQ) resin which is being used in our continuing work. This work made use of Central Facilities supported by the National Science Foundation through Northwestern University Materials Research Center, Grant No. DMR 8821571.

REFERENCES

1. Rao R. Tummala and Eugene J. Rymaszewski, eds., *Microelectronics Packaging Handbook*, Van Nostrand Reinhold, New York, 1989.

2. David S. Soane and Zoya Martynenko, *Polymers in Microelectronics: Fundamentals and Applications*, Elsevier, New York, 1989.

3. V. Vukanovic, G.A. Takacs, E.A. Matuszak, F.D. Egitto, F. Emmi, and R.S. Horwath, J. Vac. Sci. Technol. **B6**, 66 (1988).

4. L.J. Matienzo, F. Emmi, F.D. Egitto, D.C. VanHart, V. Vukanovic, and G.A. Takacs, J. Vac. Sci. Technol **A6**, 950 (1988).

5. P.M. Scott, L.J. Matienzo, and S.V. Babu, J.Vac.Sci.Technol. **A8**, 2382 (1990).

6. F.D. Egitto, F. Emmi, R.S. Horwath, and V. Vukanovic, J.Vac.Sci.Technol.**B3**, 893 (1985).

7. G. Turban and M. Rapeaux, J. Electrochem. Soc. **130**, 2231 (1983).

8. H. Yasuda, Plasma Polymerization, Academic Press, Orlando, 1985.

9. A.S. Chawla, in Polymeric Biomaterials, E. Piskin and A.S. Hoffman, eds., Martinus Nijhoff, Boston, 1986, p.231.

10. Toshiharu Yagi and Attila E. Pavlath, J. Appl. Polymer Sci., Appl. Polym. Symp. **38**, 215, 1984.

11. Paul C. Hiemenz, Principles of Colloid and Surface Chemistry, Marcel Dekker, New York, 1977.

12. H.A. Willis and V.J. Vichy, in Polymer Surfaces, D.T. Clark and W. J. Feast, eds., Wiley, New York, 1978.

13. Arthur W. Adamson, Physical Chemistry of Surfaces, John Wiley and Sons, New York, 1982.

SURFACE PROPERTIES OF PHOTOABLATED THERMOSTABLE POLYMERS

Hiroyuki Hiraoka*, Sylvain Lazare**, Alain Cros***, R. Gustiniani***
* Hong Kong University of Science and Technology
Chemistry Department, Kowloon, Hong Kong
** Laboratoire de Photophysique et Photochimie Moleculaire
Universite de Bordeaux 1, F-33405 Talence, France
*** Universite d'Aix Marseille II, URA 783 du CNRS, 13288 Marseille, France

Abstract

Surface properties of photoablated thermostable polymers are considerably different from those of original, unexposed ones, and of mercury UV lamp exposed surfaces. They also depend on the laser exposure conditions. This paper reports such non-linear phenomena observed on polymer surfaces, caused by excimer laser exposures, like increased hydrophilicity of polymer surfaces, imagewise wetting and metallizations, and new metal compound formations on laser exposed polymer films. Because these features are limited to polymer films with high glass transition temperature, like polyimides and poly(phenylquinoxaline), photo-mechanical effects may be important. In some cases, new interfacial reactions appear between exposed polymer surfaces and deposited metal films. Swelling of polyimide in pre-photoablation irradiation is discussed.

INTRODUCTION

When polymer films are exposed to intense short pulses of photons, many photochemical, and photothermal/photophysical events appear on the polymer surfaces, depending on pulse duration, fluence, light absorption by films, ambient gases present, and others. These changes caused by pulsed laser exposures are not the same as the ones observed by CW-photoexposure, even though the exposures were carried out at the same wavelength with the similar UV-light amount absorbed under the same ambient conditions.

Failures of the reciprocity was reported with photosensitivity of inorganic resists when exposed to an excimer laser irradiation,[1] although organic photoresists generally did not show the anomaly.[2] Some new photochemical reactions are also reported under excimer laser irradiation.[3] This kind of non-linear behaviors is often observed when rigid polymer surfaces are exposed to intense excimer laser, which results in photoablation, increased wetting and others. We report here mainly these rigid polymer surface-related, non-linear behaviors.

EXPERIMENTAL

Polymer films
Polyimide Kapton and poly(ethylene terephthalate) PET films were used as supplied from Du Pont. Other polymer films were spin-casted onto glass or quartz substrates, and

then baked on a hot plate at 100°C.   For polyamic acid P(MDA-ODA) **1** and PPQ **2**, poly(phenylquinoxaline), the baking temperature was about 300°C to insure imidization and curing.

| | |
| Polyimide P(MDA-ODA) **1** | Poly(phenylquinoxalene), PPQ **2** |

Equipments
    An excimer laser (Lamda Physics, EMG 210) was run at 248 nm (KrF) and 193 nm (ArF) and used in connection with a power meter.   Metal deposition was carried out in a vacuum chamber with a turbo pump for aluminum and in an UHV system for gold and copper.

RESULTS

Increased hydrophilicity
    When the laser pulses are above a threshold fluence, polymers will undergo photoablation. The extend of the photoablation depends on fluence, absorption coefficient, polymer structure, and others. With laser fluence far below a threshold value, chemical reactions will take place on the polymer surfaces, providing many specific characteristics to the polymer surfaces.   One of these changes is increased wettability and increased hydrophilicity after the laser exposure in air. The surface characteristics are demonstrated in contact angle measurements in advancing aand receding modes, as shown in Figure 1 for Kapton films. The increasing contact angle in advancing mode was caused by increasing redeposited materials on the polymer surfaces.   Spray-washing with water removes most of the redeposited materials from the surfaces, resulting in smaller increments in the advancing contact angle.[4]   The changing chemical nature of the surfaces is demonstrated in the contact angle measurements in receding mode. The rapid decrease in the receding contact angles as a function of the laser fluence is the demonstration of increased hydrophilicity. The water contact angles did not change much when the laser exposure was carried out in vacuum.
    As shown in Fig. 2, mercury lamp irradiation of the polymer surfaces in air requires about 100 times more UV-dose than those required for the excimer laser exposure to bring about the similar magnitude of the hydrophilicity.   This increased hydrophilicity was not obtained from the films exposed in vacuum. Fig. 3 shows the imagewise wetting of PPQ films after the excimer laser exposure in air through a mask. The liquid used for Fig. 3 was glycerol, but other hydroxy group containing liquid like water or ethylene glycol can be used.   Such a liquid won't stay on the hydrophobic polymer surfaces, like cured polyimide and others.   However, once exposed to the excimer laser in air, this hydroxy group containing liquid stays only where exposed. The resolution is limited by the mask

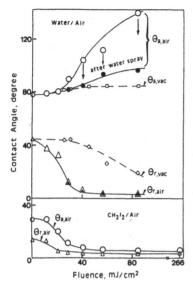

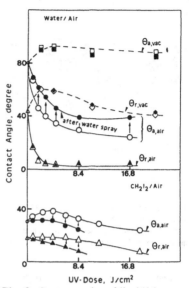

Fig. 1. Contact angles of liquid/air on Kapton film exposed on one pulse of ArF laser as a function of the laser fluence: (upper) water contact angle; the solid marks were obtained after water spraying and drying in air stream, (lower) di-iodomethane contact angle. $\Theta_{ad,air}$ is the advancing angle on the film exposed in air, $\Theta_{ad,vac}$ the one exposed in vacuum, $\Theta_{rd,air}$ the receding angle exposed in air, and $\Theta_{rc,vac}$ the one exposed in vacuum.

Fig. 2. Contact angles of liquid/air on Kapton film after UV (254/185 nm) exposure as a function of UV dose; (upper) water contact angles, (lower) di-iodomethane contact angle.

and viscosity of the liquid used. The imagewise wetting is obtained only with polyimide and PPQ films under the excimer laser exposures. The mercury lamp irradiation required very large UV doses, more than 10 J/cm^2, to obtain a similar effect. Polymers other than polyimide and PPQ did not show such imagewise wetting, although they demonstrated reduction of similar receding contact angles with increasing laser fluence. As discussed in the next section, gas permeability through the laser exposed films increased. These results indicated certain photo-mechanical effects, like uneven surfaces in the exposed areas. Shock waves propagation in the gas phase is reported in excimer laser exposures.[5]

Metal film depositions

After ArF excimer laser exposures in air or in vacuum, metal depositions were carried out in vacuum: aluminum in a vacuum chamber with a turbo pump, and gold or copper in a UHV system. The metal films on the laser exposed areas looked dark, dull-colored, while metal films on unexposed areas are shiny, and reflective. We studied their surface conductivity, gas permeability through polymer films with a quartz microbalance, and XPS measurements of PPQ surfaces with and without 50 Å thick gold depositions.

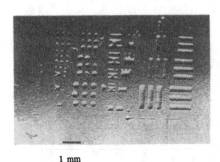

1 mm

Fig. 3　Imagewise wetting of PPQ film after imagewise laser exposure in air at a fluenece of 95 mJ/cm². Glycerol was used for this picture. The smallest image here was 100 μm.

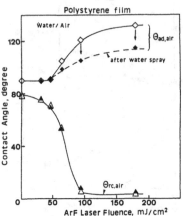

Fig. 4　Water contact angles on polystyrene surfaces exposed to ArF excimer laser.

Surface conductivity of deposited metal films

　　Surface conductivity of aluminium films of 200 to 500 Å thickness deposited on polymer films exposed to ArF excimer laser in air is shown in Table 1 for several polymer films　as a function of the laser fluence.　Polymer films with a high glass transition temperature cause anomaly on the surface conductivity of metal deposited.　A part of the increased resistivity of the metal film on the laser exposed polymer surfaces is the surface roughness.　With the spray water washed areas have less resistivity than those unwashed. However, the surface roughness by redeposited materials alone cannot explain the dependency of the resistivity on polymers and others.　As shown in Fig. 4, polystyrene showed a similar　increase of advancing contact angles as a function of the laser fluence, as demonstrated with polyimide and PPQ films.

Table 1　Surface resistivity of 300 Å Aluminum films on ArF laser exposed areas.

Polymer	Fluence	Resistance
Polyimide (Kapton)	0　mJ/cm²	40 Ohm/cm
	28	48 kOhm/cm
	41	∞
	60	∞
	93	∞
PPQ	0	20 Ohm/cm
	100	∞
PPES¹	0	38 Ohm/cm
	71	1300
	93	∞
	200	∞
PS² (PET³, PMMA, PC⁴)	0	13 Ohm/cm
	60	13
	86	18
	95	18
	127	25
	200	50

1: polyphenylethersulfone, 2: polystyrene, 3: poly(ethylene terephthalate), 4: polycarbonate.

Air absorption and its rates through metal films

A quartz crystal microbalance (QCM) was used to follow the weight change of the metal deposited PPQ films when the freshly prepared metal deposited films were exposed to air. The results are shown with aluminum on PPQ in Fig. 5. Fig. 5(a) shows a QCM response of unexposed 2 μm thick PPQ film; first, 23 nm thick aluminum film was deposited, then air was introduced into a chamber; after attaining nearly equilibrium, again the chamber was pumped down to about 10^{-6} Torr, and then air was reintroduced. The permanent weight increase due to the air introduction is due to the oxidation of the aluminum surface, amounting to about 8% oxidation of all of the aluminum. Fig. 5(b) was the similar QCM response of the ArF excimer laser exposed PPQ film. Except the laser exposure in air, the polymer film was treated in an exactly in the same way as in Fig 5(a). Comparison of Fig. 5(a) and Fig. 5(b) reveals the following differences: (1) the air absorption rate is much faster with the laser exposed polymer films; about 4 times in the first order rate constant ratio, (2) the oxidation amount of the metal deposited is much larger with the laser exposed films; about 15% vs 8% oxidation. The result demonstrates the increased capacity of gas absorption of the laser exposed films probably due to to increased porosity.

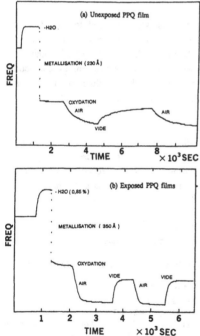

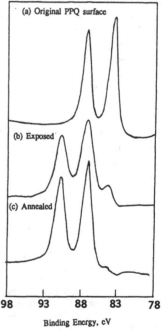

Fig. 5 Quartz crystal microbalance responses of aluminum depositions on PPQ films in vacuum; (a) 230 Å thick Al on unexposed PPQ film, followed by repeated cycles of air introduction and pumping. (b) 350 Å thick Al on laser exposed PPQ film in air, followed by the same repeated cycles as in (a).

Fig. 6 The XPS signals of 50 Å Au $4f_{1/2, 3/2}$ deposited on polymer films: (a) original PPQ, (b) ArF laser exposed PPQ films, (c) annealed in UHV at 200°C.

XPS study

A long mercury UV lamp exposure of PPQ films in air resulted in photo-oxidation, as revealed by a new carbon signal at 288 eV in binding energy, an indicative of a carbon with a carbonyl group. Similar appearance of a new carbon signal in photo-oxidation is reported in novolak resins and others.[5] The excimer laser exposed films in vacuum did not show any change from the original PPQ film. The laser exposed films in air showed a slight broadening of the carbon signal, but no distinct new peaks appearing, different from mercury UV lamp exposures in air.

When gold was deposited to a 50 Å thickness on the PPQ film exposed to the KrF excimer laser in air with a fluence of about 150 mJ/cm^2, a new doublet XPS signal of Au $4f_{1/2, 3/2}$ appeared at 87 and 91 eV binding energy, as shown in Fig. 6, whereas the small original Au 4F doublet signals at 83 and 87 eV remain. After annealing at 200°C in UHV, the small original gold peaks disappeared. The similar change was also observed with the carbon signal. The photo-electron spectra of the Au $4f_{1/2, 3/2}$ doublet signals of Γ ray irradiated and unirradiated Au/PTFE sample.[6] A gold compound/complex may have been formed in the interface between gold and exposed PPQ. A further study on this problem is in progress.

Swelling in preablation exposures

The first observation of swelling of polymer films when exposed to pulsed laser beams was made with doped polyimide films exposed to the 2nd harmonic of Nd-YAG laser.[7] This experiment was done in multiple pulses of the laser, and concluded that the peak pulse fluence was more important for ablation rather than the total energy absorbed in the laser exposure. More recent studies[8] indicate that the swelling in preablation "incubation" period are more general rather than anomaly in the laser photoablation.

CONCLUSIONS

Intense laser pulse exposures yield quite different results from mercury lamp irradiation. Surface related phenomena discussed here like imagewise wetting and metallization were partly caused by simultaneous photo-mechanical effects of short, intense pulsed photons, as indicated by fast gas absorption through metal films and others.

References

1   K.J. Polasko, D.J. Ehrich, J.Y. Tsao, R.F.W. Pease, E.E. Varinero, *IEEE Trans. Electron Device Lett.*, EDL-5 (1984) 24.
2   K. Jain, S. Rice, B.J. Lin, *J. Polym. Eng. Sci.*, 23 (1983) 1019; S. Rice, K. Jain, *IEEE Trans. Electron Devices*, ED-31 (1984) 1.
3   N.J. Turro, M. Aikawa, I.R. Gould, *J. Am. Chem. Soc.*, 104 (1982) 5127.
4   H. Hiraoka, S. Lazare, *Applied Surface Science*, 46 (1990) 264-271.
5   R. Srinivasan, B. Braren, K. G. Casey, *J. Appl. Phys.*, 68 (1990) 1842-1847.
6   C.J. Sofield, C.J. Woods, C. Wild, J.C. Riviere, L.S. Welch, *Mat. Res. Soc. Symp. Proc.*, 25 (1984) 197.
7   H. Hiraoka, in *Photochemistry on Solid Surfaces* ed. by M. Anpo and T. Matsuura (Elsevier, Amsterdam 1989) pp. 448-478.
8   H. Fukumura, N. Mibuka, S. Eura, H. Masuhara, to appear in *J. Appl. Phys.*; see also ref. 5.
10 S. Kuper, M. Stuke, *Microelectron. Eng.*, 9 (1989) 475.

A CALORIMETRIC EVALUATION OF THE PEEL ADHESION TEST

J. L. Goldfarb, R. J. Farris, Z. Chai and F. E. Karasz
Polymer Science & Engineering Department
University of Massachusetts
Amherst, MA 01003

ABSTRACT

Simultaneous mechanical and calorimetric measurements of the work and the heat of peeling have been made for polyester films bonded to rigid metal substrates with a pressure sensitive adhesive. Most of the work expended in peeling is consumed by plastic deformation and viscous dissipation. 70% to 85% of the peel energy is dissipated as heat. The peeled polyester films were tightly curled. Energy stored in the peeled polyester films was measured using solution calorimetry and a significant fraction of the peel energy not dissipated as heat was found to be stored in the peeled films.

INTRODUCTION

The peel test is a standard mechanical test which is used extensively to measure adhesion. In a peel test, a flexible film is pulled away from a substrate to which it is bonded with the angle between the detached film and substrate $90°$ or $180°$. In the absence of energy dissipation due to plasticity or viscoelasticity, the energy required to separate the film from its substrate is a direct measure of the adhesion. However, for thin films which strongly adhere to rigid substrates or films which are bonded with soft rubbery adhesives, dissipative mechanisms dominate the peel behavior. For thin films, strongly adhering to rigid substrates, the peel force is sufficient to cause inelastic deformation near the point of detachment where the material is subjected to severe curvature.[1] When films bonded with soft rubbery adhesives are peeled, cohesive failure in the adhesive layer is caused by liquid-like flow.[2] Under these conditions, the peel energy will significantly exceed the true adhesion. Thus, the peel test is extremely sensitive to energy dissipative mechanisms. Detailed theoretical analysis of the effects of plastic deformation and viscous dissipation during peeling have been presented by Kim [3,4] and Gent.[5,6]

Interestingly, not all of the energy consumed by dissipative processes is dissipated as heat. In this paper, the results of simultaneous mechanical and calorimetric measurements of the work and heat of peeling are presented. Calorimetric measurements of the heats of solution of the peeled materials are also included.

Mat. Res. Soc. Symp. Proc. Vol. 227. ©1991 Materials Research Society

EXPERIMENTAL

The peel test samples used in this study were prepared by spin coating a thin, approximately 10μm, film of pressure sensitive adhesive (Scotch-Grip 4910-NF), consisting of a synthetic elastomer and a low molecular weight polymerized terpene tack agent in a solvent, onto .004" aluminum substrates. The substrates were etched in chromic acid and solvent wiped prior to coating. After coating, the substrates were placed in a 70°C vacuum oven, and dried for 2 hours. Poly(ethylene terephalate), PET, films of differing thicknesses were then bonded to the adhesive coated side of the aluminum. The assembly was placed in a press and subjected to a pressure of 15,000 psi at room temperature. Peel test samples were prepared by cutting strips from the polymer/metal sheets and rigid steel wires were bound to the back of the aluminum strip, prohibiting the aluminum from bending while the polyester film was peeled from the aluminum.

The work done on the peel test sample by the applied force per unit area peeled, ΔW, can be equated to the internal energy change per unit area peeled, ΔU, of the sample and the heat flowing from the sample per unit area peeled, ΔQ.

$$\Delta W = \Delta U - \Delta Q \qquad (1)$$

The heat and work of peeling were measured using a deformation calorimeter.[7] The instrument operates by measuring pressure changes in a gas surrounding the sample, which is contained in a sealed chamber, relative to a sealed reference chamber. The entire apparatus is contained in a constant temperature bath. The sample and reference chambers are connected to a mechanical testing device using tungsten pull-wires which pass through gas tight mercury seals. The reference chamber is identical to the sample chamber so that no relative pressure change results from volume changes due to motion of the wires. The sample volume is small compared to the total volume of the chamber so that volume changes due to Poisson's effects do not significantly affect the gas pressure. Any change in gas pressure is due to the emission or absorption of heat by the sample. The mechanical tester is equipped with a load cell and displacement transducer. All of the electronic transducers are connected to a computer which collects and analyzes the signals. The work is calculated from the force-displacement data and the heat is calculated from the pressure-time data. The instrument is calibrated using electric resistive heating elements. A minimum heat flow of 84 microwatts is required to produce a pressure deflection equal to twice the signal-to-noise ratio. The minimum detectable heat is about 0.42 millijoules and the precision is +/-3%.[8]

RESULTS

The work and heat of peeling PET films ranging in thickness from 34μm to 180μm was measured in the deformation calorimeter. All of the films were peeled at a constant rate of 2.71 (cm/min). After peeling through a distance of 2 to 4 (cm), the peeled film was completely unloaded releasing any elastic energy in the film and negating the heat effect due to thermal expansion. A detailed description of peel tests in the calorimeter can be found elsewhere.[8] The work, heat and the internal energy change, ΔU, of peeling are shown versus film thickness in Figure 1. 70 to 85% of the work expended in peeling is dissipated as heat. The remainder is the internal energy change of peeling. If mechanisms of energy dissipation other than heat flow are assumed to be negligible, for example, acoustic and light emission, conservation of energy requires that the difference between the work and heat increased the internal energy of the peeled specimen.

The peeled polymer is tightly curled. Thermodynamically, changes in the structure of the polymer due to peeling induce a non-equilibrium high energy state. At high temperatures, or over time periods which are long compared to the duration of experimental observation, these changes may relax out of the material due to molecular rearrangement. When the peeled polymer is placed in an oven at 200°C for 30 minutes, relaxation appears to occur and it uncurls to a flat state, indicating it has released its stored energy.

Direct quantitative measurements of the energy stored in the peeled films can be made using solution calorimetry. In solution, a polymer is able to move freely and quickly comes to an equilibrium state which does not depend on its prior history. Thus, the enthalpy of the polymer in solution should be the same regardless of whether or not the polymer was deformed prior to dissolution. Therefore, variations in the heat of solution between polymers of the same chemical structure and molecular weight distribution are equivalent to enthalpy differences of the solid polymers. $34\mu m$, $45\mu m$ and $120\mu m$ PET films, which were peeled from aluminum in the deformation calorimeter were placed in one cell of a Setaram C.80 double cell Calvet type solution calorimeter, manufactured by Setaram of Lyon, France. Undeformed PET films were placed in the other cell of the calorimeter. Before placing the samples in the calorimeter, both the peeled and undeformed samples were placed in methylene chloride to remove residual traces of the pressure sensitive adhesive from the peeled polymer. The films were then dried under vacuum at room temperature for several days before being dissolved in the calorimeter using a solution of phenol and tetrachloroethane. For all of the samples, the enthalpy of the peeled films was greater than that of the undeformed films. The enthalpy difference between the peeled and undeformed polymer is equal to the energy stored in the peeled polymer. In Figure 2, the work, heat and internal energy change of peeling measured in the deformation calorimeter are replotted in energy density units. The stored energies measured by solution calorimetry are also displayed on this figure. The thermodynamic data from

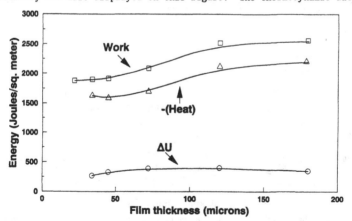

Figure 1. Work, heat and internal energy change of peeling, ΔU, for several different thicknesses of PET film, peeled from adhesive coated aluminum substrates in the deformation calorimeter .

the deformation calorimeter and the solution calorimeter are tabulated in Table I. The peel energy, heat and internal energy change were measured in the deformation calorimeter. The stored energy in the peeled polymer was measured by solution calorimetry.

The work and heat of peeling a 45µm PET film were measured in the deformation calorimeter over a limited range of peel rates, 0.271 (cm/min) to 13.3 (cm/min). The results of these measurements are shown in Figure 3.

DISCUSSION

In all of the peeling experiments, the aluminum substrate is assumed to be rigid and does not inelastically deform. Thus, very little energy dissipation occurs in the substrate. Both the pressure sensitive adhesive layer and the PET film are deformed and plastic deformation and viscous dissipation occurs in these layers. Separation of the layers occurs cohesively in the pressure sensitive adhesive layer. The pressure sensitive adhesive is a soft rubbery polymer and is subject to liquid-like flow during peeling. All of the energy consumed in deforming the pressure sensitive adhesive should be dissipated as heat. If the elastomer does not irreversibly strain crystallize or exhibit mechanical hysteresis, it is incapable of

Table I.
Work, heat and internal energy change of peeling vs. film thickness

Film Thickness (µm)	34	45	120
Peel energy (J/g)	47.3	30.6	15.1
Heat (J/g)	40.5	25.3	12.7
Internal energy change (J/g)	6.8	5.3	2.4
Stored energy in the peeled polymer (solution calorimetry) (J/g)	6.6	3.3	0.5

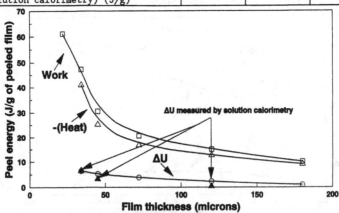

Figure 2. Work, heat and internal energy change of peeling, ΔU, for several different thicknesses of PET film, peeled from adhesive coated aluminum substrates in the deformation calorimeter and the stored energy in the peeled PET films measured by solution calorimetry.

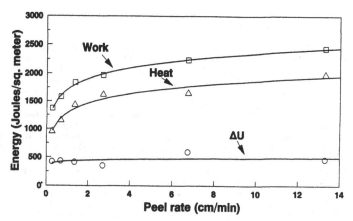

Figure 3. Work, heat and internal energy change of peeling, ΔU, for 45μm thick PET film, peeled from adhesive coated aluminum substrates in the deformation calorimeter at various peel rates.

internally storing any of the energy of deformation after the stresses imparted by peeling are removed.

The peeled PET film is bent through 180° during peeling and exceeds its elastic limit curvature near the point of detachment. It is the bending strains that are predominantly responsible for plastic deformation during peeling.[3] As a consequence of inelastic bending, the peeled polymer is tightly curled. The bending radius of the 34μm film approaches the film thickness. In this case, the maximum tensile strain in the bent film is one half the thickness of the film divided by the radius of the film or 0.5 at the outside edge and the maximum compressive strain in the bent film is -0.5 at the inside edge. The yield strain of PET films in tension is approximately .02. Thus, extensive plastic deformation occurs in the bulk of the film during peeling. The greater rigidity of the thicker films leads to a larger bending radius during peeling, but, the peel force still increases because a larger volume of film is undergoing plastic deformation. If the PET film exhibited ideal plastic deformation, such as the deformation of a Newtonian fluid, all of the work expended in plastic deformation would be dissipated as heat. When most materials are deformed, they undergo physical changes, storing some of the energy consumed by plastic deformation in the deformed material.[9] When glassy polymers like PET are drawn to high extensions, more than half of the energy under the stress strain curve is stored in the deformed material.[10] Peeling probably causes molecular rearrangement in the high strain regions of the polymer film similar to that which occurs when the polymer is subjected to homogeneous tensile or compressive deformation. Thus, it should be expected that the peeled PET film contains stored energy. The differences in the heats of solution of the peeled and undeformed films show that energy is stored in the peeled PET films. The reproducibility of the solution calorimetry measurements is approximately +/- 0.7 (J/g). From Table I, 97% of the internal energy change of peeling the 34μm film and 62% of the internal energy change of peeling the 45μm film was stored in the peeled film. The stored energy in the 120μm film is small on a joules/gram basis and is difficult to measure accurately because of instrument sensitivity limits. Thus, the heat measured with the deformation calorimeter and the stored energy measured with the solution calorimeter combine to account for 93 to 99% of the energy expended in peeling.

Previous calorimetric measurements of the work and heat of peeling polyimide films from aluminum substrates have shown that approximately one half of the work expended in plastically deforming the film is stored in the peeled polyimide film.[9] In this case, no adhesive was used to bond the polyimide to the aluminum. The adhesion of the film was a consequence of solidification and curing on the aluminum substrate. Polyimide and PET are both glassy polymers. Calorimetric measurements of the work and heat of uniaxially drawing polyimide and PET to strains well beyond their yield points, at room temperature, have also shown that approximately one half of the work expended in plastically deforming these materials is stored in the deformed materials.[10] It is thus reasonable to consider the stored energy in the peeled films as a measure of the amount of energy expended in plastically deforming the PET films during peeling. Using this argument, approximately 3/4 of the peel energy is expended in deforming the pressure sensitive adhesive and approximately 1/4 of the peel energy is expended in deforming the PET film. The peel energy increases with peel rate over the limited range of rates tested. Interestingly, the difference between the work and heat, the internal energy change, is approximately constant throughout the range. This suggests that the increase in peel energy with peel rate is a consequence of increased dissipation in the peeled layers.

CONCLUSIONS

The work of peeling greatly exceeds the thermodynamic work of adhesion. Most of the work of peeling is consumed by dissipative processes which accompany deformation of the peeled materials. Glassy polymeric adherands are capable of storing a large portion of the peel energy in the peeled material. The peel test measures deformation of the test specimen. It does not measure the true interfacial strength. However, a strong relationship does exist between the nature of the interface and the peel energy because the peeled material can only be subjected to stress (and thus to energy losses) if the interface is strong. Thus, the peel test can provide a qualitative measure of adhesion, providing that comparisons are not made between systems which would exhibit dramatically differing amounts of energy dissipation.

The authors gratefully acknowledge the support of the Materials Research Laboratory at The university of Massachusetts.

REFERENCES

1 K.S. Kim, J. Adhesion Sci. Technol. 3, 175 (1989).
2 A.N. Gent and R.P. Petrich, Proc. Roy. Soc. A. 310, 433 (1969).
3 K.S. Kim, J. Eng. Mat. & Tech. 110, 266 (1988).
4 K.S. Kim, Mat. Sci. & Eng. A107, 159 (1989).
5 A.N. Gent and G.R. Hamed, Polym. Eng. Sci. 17, 462 (1977).
6 A.N. Gent and G.R. Hamed, J. Appl. Polym. Sci. 21, 2817 (1977).
7 R.E. Lyon and R.J. Farris, Rev. Ssi. Instrum. 57, 8 (1986).
8 J.L. Goldfarb and R.J. Farris, to be published (1991).
9 G.W. Adams, The Thermodynamics of Deformation for Thermoplastic Polymers, Ph.D. Thesis (1987).
10 J.L. Goldfarb, (unpublished).

A NOVEL PHOTOCHEMICAL ROUTE FOR MODIFYING SURFACES: PROSPECTS AND APPLICATIONS.

MARK A. HARMER
Du Pont, Central Research and Development, Experimental Station, P.O. Box 80328, Wilmington, DE 19880-0328

ABSTRACT

Potential applications of a very general route to photochemically modifying a range of surfaces will be described. Surfaces which have been modified vary considerably and include polyimide, polyethylene, fluoropolymers, polyester, glass, tin oxide and aluminum. Photoactivated azides, for example 3-azidopyridine, have been found to be very effective at modifying a range of surfaces. Using this method organic type functional groups, in this case pyridine type groups, can be photografted on to the surface. As a result of photomodification the surface properties change. In the case of polyethylene for example the surface becomes more hyrophilic and shows improved adhesion to metals. These modified surfaces also display acid-base character. In the protonated form anions can be exchanged into the film. This method has been used to modify a metal oxide electrode surface to develop chemically modified electrodes. The process may also be used in lithography.

INTRODUCTION

Understanding and controlling the surface properties of materials has become an intense area of scientific study. In the area of polymer science for example there is often a need to tailor the surface properties while still maintaining the properties of the bulk. Simply altering the hydrophilic nature of a polymer or improving metal adhesion to the surface may require very sophisticated surface science. Quite often multiple reaction steps are required and reproducibility can be a problem. The chemistry often has to be developed by modifying specific functional groups on the surface. As new surfaces are developed new methodology has to be devised [1]. One major goal within this area may be to develop more general modification procedures that may be applicable to a wide range of surfaces.

We have recently found a photochemical technique which has been successfully used to modify a range of surfaces [2]. The applications of this new technique will be described in this paper. The general idea is to generate highly reactive intermediates at or near a surface, which due to their high reactivity, react with functional groups on the surface and hence modify the surface. The photomodification reaction discovered appears to be non-specific and successful modification of a number of different types of surfaces has been observed [2]. The reactive intermediates generated are nitrenes, generated photochemically from the corresponding heterocyclic azide.

A few reports have appeared describing the radiation induced modification of polymer surfaces [3-5]. Methacrylic acid has for example been photografted on to polyethylene in the presence of a sensitizer, for example aromatic ketones and quinones [3]. In these reactions however multicomponent systems are required as compared to the system described herein where both the surface modification and subsequent polymerization occurs from the same monomer. This kind of approach, using highly reactive intermediates, appears to be an attractive route for modifying surfaces.

EXPERIMENTAL

The heterocyclic azides used for the photomodification work, 3-azidopyridine [6], 4-azidopyridine [7] and 3-azidothiophene [8] were prepared using literature procedures. Pentafluorophenyl azide [9] was prepared as previously described. The substrates used in this study are all available commercially. These include polyimide film, polyethylene, polytetraflouroethylene, Teflon FEP, Tefzel, glass and aluminum. Conductive electrodes of tin oxide were supplied by PPG industries.

In a typical photolysis, a film was placed inside a quartz vessel which contained 0.1 ml of the liquid azide. This was cooled to 77 K and the system outgassed to a vacuum of about 10^{-5} Torr. The reaction vessel was then sealed and allowed to warm to room temperature. This generates a high enough vapor pressure of the volatile azides to ensure efficient surface modification. The surface and surface modifier was then irradiated using a Rayonet photochemical reactor using 254 nm light for periods of up to five minutes.

Contact angle measurements were made using a Rame-Hart model 100 contact angle goniometer.

Chemically modified electrodes were formed by applying 10 microliters of 3-azidopyridine to the surface of an tin oxide electrode, surface area of 0.5 cm^2, followed by photolysis in the Rayonet for 15 minutes. The modified electrode was washed with acetone and then treated for 0.5 hours in 0.1 M HCl and then washed in water. The acid treated surface was then placed in a 10 mM solution of ferricyanide ion for 30 minutes, removed and washed with water. Surface modification of the electrode was assessed using cyclic voltammetry. Dc cyclic voltammograms were obtained using a PAR potentiostat. The cells used were all glass construction incorporating a conventional three electrode system with a platinum auxiliary electrode and a silver/silver chloride reference electrode.

RESULTS and DISCUSSION

We have recently described the photomodification of surfaces using heterocyclic azides [2] and only a brief summary of the salient points will be presented here. We have found that heterocyclic azides, when photoactivated, are very effective at modifying a range of surfaces. Upon photoactivation of the azide, an highly reactive nitrene is generated. This highly reactive intermediate then reacts with surface functional groups, for example -CH or -OH groups, resulting in surface modification. Polymer formation then occurs as more material arrives at the surface. This type of chemistry including polymer formation is very characteristic of known reactions of nitrenes [10].

Surfaces which have been modified include, polyethylene, polyimide, glass, tin oxide, aluminum, tetrafluoroethylene and a number of fibers for example Dacron polyester and polyaramid Kevlar. All of these modified surfaces appear to be stable to extended washing in solvents ranging from water, acetone, acetonitrile, THF and hexane (except for fluoropolymer modified surfaces were most of the film is washed off). It seems likely that the surface modifier is covalently bonded. Surface coverage can range from submonolayer to microns depending upon the photolysis time. Figure 1(a) and Figure 1(b) show the surface modification of both a polyethlyene film and a polyester fiber using 3-azidothiophene. Uniform coverages (light brown in color) are obtained. In addition to azidothiophene both 3-azidopyridine and 4-azidopyridine can be used to photomodify surfaces. Figure 1(c) shows a thick film formed an aluminum surface masked in one area to show the surface morphology (photolysis time ca. 45 min). A number of different functional groups (from pyridines to thiophenes) have been photografted on to a number of different surfaces. This highlights the potential use of volatile heterocyclic azides as general purpose surface modifiers, for different surfaces.

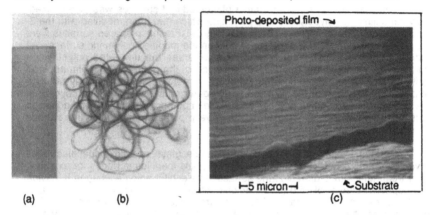

Figure 1. Photomodiffication of surfaces, by the gas phase photolysis of 3-azidothiophene showing modification of (a) a polyethylene film (2 cm wide), (b) a polyester fiber and (c) an SEM image to show a thick film formed on aluminum.

The resultant modified surface shows different properties from the untreated surface, consistent with photoattachment of the heterocycle. This leads to a number of different potential applications which are described in detail below.

Upon photomodification of polymeric surfaces the hydrophilic nature of the surfaces are altered. In the case of the more hydrophobic surfaces, for example polyethylene these become more hydrophilic after photomodification. Polyethylene modified surfaces show a reduction in the water contact angle from about 90 degrees for the untreated surface to 60 degrees for the photomodified surface. Similar results were also found on photolysis of 3-azidothiophene onto surfaces. This highlights the potential for using this simple technique for making surfaces more hydrophilic. Measurements of the water contact angle for a number of polymeric surfaces are shown in Table 1. In all cases of photomodification the water contact angle is reduced consistent with a more hydrophilic surface. It is also interesting to note that all of the fluoropolymers studied show an increase in hyrophilic character, although as mentioned above these modified surfaces appear to be generally less stable. The data is consistent with the photoattachment of pyridine type functionalities which would be expected to be more hydrophillic than the original surface.

POLYMER	CONTACT ANGLE UNTREATED	CONTACT ANGLE PHOTOMODIFIED
Polyethylene	90 deg.	65 deg.
Polyimide	75	64
PTFE	92	69
Teflon -FEP	90	55

Table 1. Photomodified polymer films, using 3-azidopyridine, to show increase in hydrophilic character.

The adhesion of metals to these photomodified surfaces was also studied. Polymeric films of polyethylene and polyimide which had been modified with the 3-azidopyridine showed improved adhesion to gold. Freshly prepared surfaces were coated by sputtering a 100 nm layer of gold on the modified polymeric surface. As judged by using simple Scotch tape tests, whereas the untreated polyethylene surface did not show any appreciable adhesion of the gold, the gold film adhered quite strongly to the photomodified surfaces. The improved binding may reflect a stronger coordination of the gold to the nitrogen group within the heterocycle. In the case of the flouropolymer modified surfaces no improvement in metal adhesion was observed which is consistent with a very poor interaction of the surface modifier with the more inert -C-F surface.

Surface treatment of oxides, for example glass and tin oxide electrodes revealed a number of interesting effects. On glass modified surfaces for example, after treatment with dilute acids the surface films exhibit anion exchange behavior. Bromophenol blue at pH 5.0 (in the anionic form) readily adsorbs on to these acid treated surfaces. The surface takes on a strong blue coloration of the dye. The presence of the bromophenol blue was confirmed by measuring the visible spectrum of a piece of quartz glass, showing the characteristic peak of the die at 580 nm. Without the acid step no anion adsorption occurs. This is consistent with protonation of the pyridine type functionalities, forming a quaternary ammonium salt, where the modified film becomes cationic and demonstrating typical anion exchange behavior. The films therefore exhibit acid-base behavior.

The anion exchange properties of these films has also been extended to develop chemically modified electrodes [11]. An electrically conducting indium-doped tin oxide electrode (with surface OH groups) can also be photomodified. Redox active anions, for example $Fe(CN)_6^{3-}$ can then be exchanged into the protonated film and the resultant electrochemistry is consistent with a surface bound redox couple, showing a distinctive electrochemical oxidation-reduction at a potential of 0.31V vs. Ag/AgCl. This is consistent with the expected redox potential of the one electron Fe^{2+}/Fe^{3+} redox couple. The films appear to behave in a somewhat similar fashion to protonated polyvinylpyridine, PVP. Electrodes modified with protonated PVP readily pick up redox centers bound as charge compensating anions [12].

Figure 2, shows the cyclic voltammogram of ferricyanide which has been anion exchanged on to the protonated modified electrode surface. The background supporting electrolyte contains 0.1M NaCl, with the pH of 4.0 to ensure the pyridine remains in the protonated form. The signal observed is due to material bound to the electrode and not free in solution. The magnitude of the

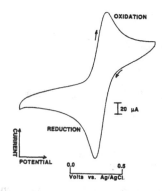

Figure 2. Dc cyclic voltammogram of $Fe(CN)_6{}^{3-}$, which has been exchanged onto a tin oxide modified electrode. Scan rate, 100 MVs^{-1}, pH 4.0, 0.1M NaCl.

electrochemical signal (ie. the current flow upon oxidation or reduction) is directly related to the amount of material on the surface. The current observed is consistent with multilayer coverage (ca.10^8 mole cm^{-2}) rather than a single monolayer. On repeated cycling the signal decays (two hours of cycling for complete loss of signal). This represents gradual loss of the redox active couple from the film. Without the protonation step exchange occurs. A similar decay has been observed on anion exchanged redox couples on PVP electrodes [12].

Photomodification of surfaces using heterocyclic azides can also be performed imagewise indicating that this method may be used for lithographic applications [2]. The liquid azide can be applied to a surface and the surface photolysed through a photomask. Upon removing the mask and washing with solvent an image is left behind. The behavior is very similar to conventional photolithography. Using this technique, fine lines and patterns can be developed on surfaces, for example well defined lines 10 microns in width have been formed on substrates although this has not as yet been optimized.

We have also found that photomodifying a glass surface (with 3-azidopyridine) can alter the electroless deposition of metals on these surfaces. Rather unexpectedly we have found that the untreated surface can be plated with copper (using standard electroless coating techniques) whereas the treated surface is slower to coat. Thus by patterning an image on the surface of the glass, as described above, followed by electroless deposition of copper for short periods allows one to build up a pattern of conducting copper lines, Figure 3. The copper deposits preferentially on the untreated area first.

Upon continued plating the whole area eventually coats. The resolution however of these patterned surfaces is quite poor. Further work is in progress to try to understand this unexpected result. We have recently found that glass surfaces treated with pentafluorophenyl azide, $C_6F_5N_3$, results in a surface which does not electrolessly coat even after extended times. It seems likely that the lithographic potential of this method coupled with the modification of deposition rates may be of use in developing conducting lines.

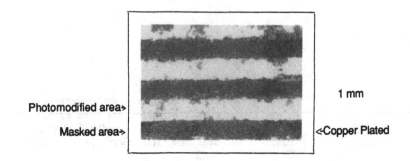

Photomodified area>

Masked area>

1 mm

<Copper Plated

Figure 3. Preferential copper plating on a glass modified surface, where a line
pattern was previously deposited using the heterocyclic azide.

In summary, photoactivated heterocyclic azides have been found to be very
effective at modifying a wide range of surfaces. The modified surface displays
properties consistent with that of the bound heterocycle. In the case of polymer
surfaces which have been modified these surfaces show a change in hydrophilic
character. Polyethylene and polyimide treated surfaces show improved metal
adhesion. We have also demonstrated that a glass modified surface displays well
defined acid-base character and this effect has been exploited to develop an
chemically modified electrode. The process can also be used in lithography. The
results found highlight the potential use of using highly reactive intermediates as
efficient surface modifiers.

REFERENCES.

1.   C. E. Carraher and M. Tsuda, Modification of Polymers. (ACS Symposium
     Series 121, Honolulu, Hawaii, 1979) 217-242.

2.   M. A.Harmer, submitted to Langmuir.

3.   H. Kubota, K. Kobayashi and Y. Ogiwara, Poly. Photoch., 7, 379 (1986).

4.   H. Mingbo and H. Xingzhou, Poly. Deg. and Stab., 18, 321 (1987).

5.   A. Hult, S.A. MacDonald and C. G. Wilson, Macromolecules, 18, 1804 (1985)

6.   H. Sawanishi and T. Tsuchiya, Chem. Pharm. Bull. 35, 4104 (1987).

7.   L. K. Dyalla and W. M. Wah, Aust. J. Chem., 38, 1043 (1983).

8.   P. Spagnolo and P. zanirato, J. org. Chem., 43, 3539 (1978).

9.   R. E.Banks, N. D.Venayak and T. A. Hamor, J. chem. Soc. Chem. comm.,
     900 (1980).

10. E. F. V. Scriven, Azides and nitrenes. (Acedemic press, New York 1984).

11. R. W. Murray, Acc. Chem. Res., 13, 135 (1980).

12. N. Oyama and F. C. Anson, Anal. Chem. 52, 1192 (1980).

Control of Polyimide Surface Property By Changing
The Surface Morphology

Kang-Wook Lee IBM, T. J. Watson Research Center, P.O. Box 218, Yorktown Heights, NY 10598

Abstract

The surface properties of polyimide can be changed without altering the bulk properties. The surface layer (around 200 Å) of fully cured PMDA-ODA polyimide is chemically modified to polyamic acid which is subsequently imidized at 230 °C for 30 min to give an unordered polyimide surface. This unordered layer seems amorphous since it swells well in NMP. The modified surfaces are analyzed by surface sensitive techniques such as contact angle measurement, XPS, ISS, and external reflectance infrared (ERIR) spectroscopy. Adhesion of polyimides and chromium on the amorphous polyimides is greatly enhanced. Interdiffusion and subsequent mechanical interlocking are the major contributions to the polyimide-polyimide adhesion.

Introduction

Polyimides are used as dielectric layers in a variety of microelectronic applications, as they have good processability, low dielectric constant, high thermal stability, low moisture absorption, and good mechanical properties.[1] Metal patterns are disposed on polyimide surfaces to form electrically conducting lines on chips and chip packages. However, direct adhesion of metals to polyimide is weak. Failure to activate a polymeric surface will normally cause subsequent coatings to be poorly adhered and easily cracked, blistered, or otherwise removed. Surface treatments and adhesion layers have been used to enhance the adhesion of metal layers to polyimides and the polyimide-polyimide adhesion. These methods can introduce foreign materials and (or) modified layers to the interface, resulting in possible reliability failures. Introducing new functional groups may also cause reliability problems as well. It is challenging to improve the adhesion without introducing foreign materials, foreign functional groups or adhesion promoters.

Polymers generally can be either semicrystalline or amorphous.[2] These categories are used to describe the degree of ordering of the polymer molecules. Amorphous polymers consist of randomly ordered tangled chains, i.e., amorphous polymers are highly disordered and interwined with other molecules. Semicrystalline polymers consist of a mixture of amorphous regions and crystalline regions. The crystalline regions are more ordered and segments of the chains actually pack in crystalline lattices. Some crystalline regions may be more ordered than others. If crystalline regions are heated above the melting temperature of the polymer, the molecules become less ordered or more random. If cooled rapidly, this less ordered feature is frozen in place and the resulting polymer is amorphous. If cooled slowly, these molecules can repack to form crystalline regions, and the polymer is crystalline.

One possible way to enhance the adhesion without introducing foreign materials to the interfaces is to change the morphology of the polyimide surface, coat the next layer (metal or PI) and then convert the morphology back to the original state.

Mat. Res. Soc. Symp. Proc. Vol. 227. ©1991 Materials Research Society

Crystalline polyimide consists of some amorphous polyimide and amorphous polyimide contains a small amount of crystalline polyimide. However, the properties of the two materials are very different. The adhesion property of polymer is related to the surface properties. The morphology of polyimide surface can be altered without changing the bulk morphology, thus the mechanical properties of the polymer remain unchanged. Here we report modification of polyimide surface morphology from a relatively more crystalline state to a relatively more amorphous state which provides an improved chromium-polyimide and polyimide-polyimide adhesion.

Experimental

Materials and Methods. Polyimide precursor, PMDA-ODA polyamic acid, was obtained from Du Pont. KOH, HCl, acetic acid, 1-methyl-2-pyrrolidinone (NMP) and isopropanol were obtained from Aldrich. Polyamic acid in NMP was spin-coated onto a Si wafer coated with chromium, and then cured to polyimide by slowly ramping the temperature to 400 °C. The purpose of the 200-500-Å-thick layer of chromium is to enhance wettability of polyamic acid solution onto the Si wafer. The 100-1000-Å-thick layers of polyimides were employed to obtain the ER IR spectra. The surface-modified samples for contact angle measurement and ER IR were dried under vacuum at ambient temperature for 12-24 h. Contact angle measurements were obtained with a Rame-Hart telescopic goniometer and a Gilmont syringe with a 24 gauge flat-tipped needle. Distilled water was used as the probe fluid. Dynamic advancing and receding angles were determined by measuring the tangent of the drop at the intersection of the air/drop/surface while adding (advancing) and withdrawing (receding) water to and from the drop. External reflectance infrared (ER IR) spectra were obtained under nitrogen using a Nicolet-510P FTIR spectrometer with a Harrick reflectance attachment. Thicknesses of polyimide were measured with a Dek-Tak for thicknesses greater than 1000 Å and with a Waferscan ellipsometer equipped with a HeNe laser ($\lambda = 6328$ Å) for the thickness of 100-1000 Å. The refractive index employed for BPDA-PDA is 1.848 at 6328 Å.[3]

Adhesion Measurement. For the chromium-polyimide adhesion, chromium (200 Å) and then copper (2 μm) were sputter-coated to form 3-mm peel strips followed by electro-plating a thick layer of copper (13 μm). The sample was treated at 400 °C for 60 min to convert the less ordered polyimide (or polyamic acid) layer to the more ordered polyimide. For the polyimide-polyimide adhesion, the solution of polyamic acid was spin-coated onto a chromium-coated Si wafer and cured at 400 °C for 60 min. Thickness of the polyimide layer is approximately 6 μm. A thin layer (200 Å) of gold was sputter-coated onto one side (20% of the total area) of the polyimide sample to initiate peel (polyimide has poor adhesion to gold). The surface of the polyimide in the exposed area was modified as described in the text. PMDA-ODA polyamic acid was spin-coated to the surface-modified polyimide film, and subsequently cured at 400 °C under nitrogen. Thickness of the adherend layer (peel layer) after curing is approximately 20 μm and the width of the peel layer is 5 mm. The peel strengths were measured by 90° peel using an MTS with 25 μm/sec peel rate. The reported values are the average of at least three measurements.

Results and Discussion

Formation of Amorphous Polyimide Surface. Poly(pyromellitic dianhydride-oxydianiline) (PMDA-ODA) polyamic acid was cured at 400 °C for 60 min to a semicrystalline state polyimide (PI).[4] The PI surface layer (100-200 Å) was converted to polyamic acid by a wet process in which polyimide reacted with aqueous KOH solution to yield potassium polyamate that was subsequently acidified in dilute acid.[5,6] Measurement of the modification depth has also been reported.[6] The polyamic acid surface can be imidized at 230 °C for 30 min to an unordered polyimide. To obtain vibrational spectra a thin layer (around 500 Å) of PMDA-ODA polyamic acid was coated onto a Cr-coated Si wafer and then cured to polyimide. The whole layer was modified to polyamic acid by the wet process. The external reflectance infrared (ERIR) spectrum displays the peaks 1727 (s), 1668 (m), 1540 (m), 1512 (m, sh), 1502 (s) and 1414 (m) cm^{-1}. The sample was cured at 230 °C for 30 min under nitrogen. The external reflectance infrared (ERIR) spectrum shows the same bands as those of fully cured polyimide at 1778 (w), 1740 (vs), 1726 (vw,sh), 1598 (vw), 1512 (m,sh), 1502 (s), 1381 (m,br), and 1247 (vs) cm^{-1}, indicating that the polyamic acid surface is fully imidized.

It has been known that PMDA-ODA polyimide cured at 230 °C has an unordered structure and that upon treating at 300 °C or greater it becomes more ordered to a relatively more crystalline state.[4] We carried out an experiment to test if this unordered polyimide is amorphous. Thin layer (around 500 Å) of PMDA-ODA was prepared on a silicon wafer and the whole layer was modified to polyamic acid followed by curing at 230 °C for 30 min to polyimide. As this sample was dipped into NMP solvent, NMP diffused into the modified layer and finally dissolved the whole layer. Another indication for the amorphous surface is that impurities such as ions can be rinsed out with a solvent. These results indicate that the unordered polyimide is relatively amorphous.

Adhesion of Cr-to-PI and PI-to-PI. Chromium (200-500 Å) was coated to the amorphous polyimide surface prepared as described above, followed by coating with 2 μm copper. Additional copper, 10-13 μm, was electro-plated to measure the peel strengths. The 90° peel strength was measured by peeling the metal Cr/Cu layer from the polyimide substrate. The peel strength was increased from 20-30 Joule/m^2 (control) to 500 J/m^2.

For the PI-PI adhesion, PMDA-ODA polyamic acid was spin-coated to the amorphous polyimide surface at 1700 rpm for 30 sec. The sample was baked at 85 °C for 30 min and then at 150 °C for 30 min. The second layer of polyamic acid was spin-coated at 1700 rpm for 30 sec and then cured at 400 °C for 60 min. This double coating gives around 20-μm thick PMDA-ODA peel layer. The 90° peel strength was measured by peeling the 20-μm thick polyimide coating layer. The peel strength was enhanced from 30 J/m^2 (control) to 1000 J/m^2.

The mechanism for polyimide-polyimide adhesion seems quite different from the dry process such as plasma treatment and ashing in which the chemical reaction is the most important factor. However, a chemical reaction for adhesion to the amorphous polyimide is excluded since the polyamic acid is not expected to react with fully imidized polyimide. Wettability and interdiffusion are the other possible contributions. Water contact angles of the amorphous (less ordered) polyimide surface are similar to those of the semicrystalline (more ordered) while the peel strength corresponding to the amorphous polyimide adhesion is 30 times greater than that corresponding to the semicrystalline polyimide adhesion. This result indicates that wetting of polyamic acid solution is good enough to the polyimide surface, but significant mechanical interlocking between the phases of polyamic acid solution and polyimide is involved only to the amorphous polyimide surface. Thus the polyimide-to-semicrystalline (more ordered) polyimide adhesion is weak and the polyimide-to-amorphous (less ordered) polyimide is strong. In the case of polyimide-to-amorphous polyimide adhesion, the incoming polyamic acid solution diffuses into the amorphous layer as NMP does. Upon curing, two polyimide phases are interlocking and the interlocking force will be greater for the more deeply modified surface. This turned out to be the case. The peel strengths of PMDA-ODA/PMDA-ODA adhesion are 40, 400, 850, and 1250 J/m^2 for 0, 10, 100 and 200 Å modified surface, respectively.

In summary, the morphology of the polyimide surface is changed from a more crystalline state to a more amorphous state. Chromium and polyimide adhesion to the amorphous polyimide surface is greatly enhanced. The surface property can be controlled by tailoring the surface morphology.

Acknowledgment I'd like to acknowledge D. Hunt and E. Adamopoulos at IBM for technical help.

References

1. M.I. Bessonov, M.M. Koton, V.V. Kudryavtsev, L.A. Laius, Polyimides: Thermally Stable Polymers, (Consultants Bureau: NY, 1987).

2. L. Mandelkern in Physical Properties of Polyimides, edited by J.E. Mark, A. Eisenberg, W.W. Grasessley, L. Mandelkern and J.L. Koenig, (ACS: Washington, DC, 1984).

3. W.A. Pliskin, J.D. Chapple-Sokol and K.-W. Lee, unpublished work.

4. N. Takahashi, D.Y. Yoon and W. Parrish, Macromolecules, 17, 2583 (1984).

5. K.-W. Lee, S.P. Kowalczyk, J.M. Shaw, Macromolecules, 23, 2097 (1990).

6. K.-W. Lee and S.P. Kowalczyk, in Metallization of Polymers, edited by E. Sacher, J-J. Pireaux, and S.P. Kowalczyk (ACS Sym. Series, vol 440; ACS: Washington, DC, 1990) chapter 13.

7. M. Ree and K.-W. Lee, further work to provide a direct evidence on the ordering of the modified polyimide surface is in progress.

THE ROLE OF SURFACE MODIFICATION ON ADHESION AT THE METAL/POLYMER INTERFACE

M. J. Berry, I. Turlik, P. L. Smith, and G. M. Adema

MCNC, Center for Microelectronics
PO Box 12889, 3021 Cornwallis Rd, RTP, NC 27709-2889

ABSTRACT

This paper presents studies of surface modifications of benzocyclobutene (BCB) using reactive ion etching (RIE), ion implantation, and combinations of those in conjunction with liftoff processing steps. It was found that O_2/N_2 RIE treatments of the BCB surface improve the adhesion of subsequently evaporated chromium and copper. Implant treatments with C^+, O^+, Si^+, As^+, and Sb^+, were also investigated. The implant treated samples exhibited improved adhesion prior to subsequent heat treatments in the Cr/BCB structures and some improvement subsequent to heat treating in the Cu/BCB structures. Combinations of the surface modification treatments were seen to have the greatest benefit, even after heat treatments.

INTRODUCTION

When choosing a material for use as a dielectric in a multilevel structure, many criteria must be considered. Electrical performance issues and reliability concerns both need to be addressed. In the case of multichip modules, reliability will be dependent on their structural integrity which is directly related to interlevel adhesion.

Benzocyclobutene (BCB) is a polymeric dielectric material which has received much attention for use in multichip modules. This material is very chemically stable, has a low dielectric constant, low dissipation factor, good planarizing ability, and low moisture uptake-all qualities which appear on the ideal dielectric wish list. The monomer for the polymer used in our experiments is shown in Fig. 1. It consists of a di-vinyl, siloxy linkage between the benzocyclobutenes. The monomer polymerizes via a Diels-Alder addition reaction where the cyclobutenes attach to one another. No volatile by-products are produced through this reaction. Combining low impingement energy deposition such as e-beam evaporation with this chemically stable polymer surface results in poor adhesion. It has been shown that

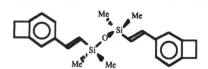

Fig. 1. BCB Monomer Structure

modification of a polymer surface by plasma ions [1,2] and ion beams [3,4] changes the surface chemistry and morphology of polymers. Therefore, these techniques can be applied to enhance the adhesion of subsequently deposited layers.

Mat. Res. Soc. Symp. Proc. Vol. 227. ©1991 Materials Research Society

The basic mechanisms which contribute to adhesion arise from mechanical interlocking, chemical bonding and adsorption from forces such as Van der Waals, London dispersion, H-bonds, and dipole moments. Most polyimides have the ability to achieve adhesion through these mechanisms. One of the assets of BCB, the material's chemical stability, contributes to a difficulty in adhesion of metal to the polymer. It was therefore worthwhile to study the effect of surface modification on adhesion.

The desired conductor for multichip modules is copper, because of its high conductivity, low cost, and solderability. Since copper has a low affinity for polymers, in general, it may diffuse into the polymer and agglomerate causing degradation of the dielectric characteristics of the polymer.[5] However, chromium generally binds more strongly to a polymer and therefore diffuses less into the polymer.[6] Chromium layers are generally used between the copper and the polymer. These tie-layers serve the dual purpose of providing a diffusion barrier and enhancing adhesion between the copper and the polymer. The interfacial characteristics of both chromium and copper with BCB is therefore important to investigate.

This study was undertaken to determine the effects of surface modification of benzocyclobutene on the adhesion of subsequently deposited metal. Surface modification techniques examined include reactive ion etching (RIE) and ion implantation. Various analytical techniques such as cross-sectional transmission electron microscopy (XTEM), secondary ion mass spectroscopy (SIMS), and Auger electron spectroscopy depth profiling (AES) were utilized to examine the interface between the deposited metal and the polymer.

EXPERIMENTAL DETAILS

Evaporation and sputtering are the most common metal deposition techniques employed in the fabrication of thin film structures. There is an order of magnitude energy difference between the atoms from an evaporant stream and a sputter cloud. Due to the nature of the sputter deposition process, the metal atoms impinge on the substrate with significant kinetic energy while the energy transfer to the substrate in an evaporation process is much lower since it is almost entirely from the heat of condensation of the evaporant. Therefore metal atoms deposited through sputtering are capable of intermixing with the polymer surface to a higher degree than evaporated atoms in the 'as-deposited' condition. This can be seen in Fig. 2 by comparison of the Auger analyses of the interfaces between evaporated and sputtered Cr on untreated BCB. The effect of this intermixing is readily evident when "Scotch Tape Tests" are performed. The evaporated samples fail the test, but the sputtered films are adherent.

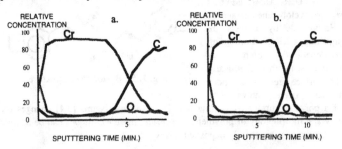

Fig. 2. Auger Depth Profiles a.) Evaporated Cr, b.) Sputtered Cr.

During the fabrication of the multichip module, the substrate is exposed to both chemicals and heat throughout processing. One of the processes investigated during this study was the use of a liftoff stencil as an additive patterning technique in conjunction with metal evaporation. The surface of the BCB may be exposed to photoresist, amine vapor, developer, a forming gas plasma descum, and an ultrasonic solvent bath during the liftoff process. The substrate is exposed to temperatures of 100°C for about 3 hours total as a result of these process steps. A full description of this process is given by Jones et. al. [7]. Adhesion of evaporated metals deposited through these liftoff stencils yielded poor performance, .3 to .5 ksi.

In order to improve the adhesion of evaporated films to BCB, various surface treatments were performed. These treatments included reactive ion etching and ion implantation. A matrix of samples were prepared so that the metal/polymer interface could be examined after a specific surface treatment or a combination of different surface treatments. The use of the liftoff process sequence was also included in the experimental matrix so its effect on interfaces in combination with other treatments could be examined.

Reactive ion etching in a split cathode magnetron system was performed using various etch chemistries. Different gas mixtures ($O_2/5N_2$, $9O_2/CF_4$, and SF_6) were introduced at a constant flow of 60 sccm. The treatment consisted of two minutes at 1.25kW incident power. In Fig. 3, an Auger spectra showing the interfacial region after O_2/N_2 treatment of the BCB with subsequently evaporated Cr is shown. Comparing this to the previously shown control spectra (Fig. 2a) indicates an increased interfacial width as well as a slight increase in the relative concentration of oxygen on the chromium side of the cross-over point.

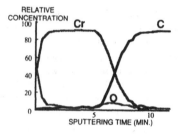

Fig. 3. Auger Depth Profile of Evaporated Cr on O_2/N_2 RIE Treated Surface.

A heat cycle consisting of a second full cure cycle was performed on these samples to assess the reliability of the surface treatment. The maximum temperature during this cycle is 250°C with a dwell period of 1 hour in a N_2 ambient. All RIE treated samples passed the tape test prior to heat treatment. Table I shows the pull test results for these samples both before and after the heat treatment. In almost all cases, the heat treated samples showed reduced pull strength.

Table I. Pull Test Adhesion Results for RIE Treatments

(ksi)	CONTROL		O_2/N_2		SF_6		O_2/CF_4	
		HEAT		HEAT		HEAT		HEAT
EVAP Cr	2.6	1.6	5.6	5.6	4.5	1.8	3.7	2.2
EVAP Cu	.5	.3	5.2	3.6	.9	<.1	5.5	.9

Cross-sectional transmission electron microscopy was performed on many of the samples generated for this study. In Fig. 4 an XTEM micrograph of a 100nm evaporated copper film deposited onto an O_2/N_2 treated BCB sample can be seen.

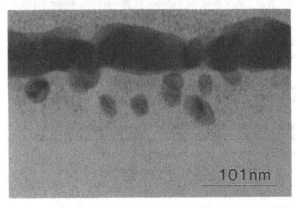

Fig. 4. XTEM of Evaporated Copper deposited on O_2/N_2 RIE Treated Surface.

There are large crystalline Cu agglomerates within the BCB. Interestingly, these nodules are present after having subjected the sample only to the temperature (<140°C) required for XTEM sample preparation [8].

Along with the reactive ion etch studies, ion implant studies were conducted. The ionic species used for these studies were C^+, O^+, Si^+, As^+, and Sb^+. The implant energies employed were 10, 40, and 70 keV. The ion dose was 1E15 ions/cm^2 for all experiments. A portion of each sample was subjected to the same additional cure cycle previously described.

Analysis of the interface through XTEM showed an increase in the density of the material in the implanted region. The depth of the implant was determined using XTEM and verified by SIMS. A summary of the tape test results for the implant treated samples before and after heat treatments is given in Table II. The samples exhibit improved adhesion prior to heat treatment in the Cr/BCB structures and some improvement subsequent to heat treatment in the Cu/BCB structures.

Table II. Tape Test Results for Implant Treatments

IONS OF ➡ ENERGY (KeV)	CARBON		OXYGEN			SILICON		ARSENIC	ANTIMONY
	10	40	10	40	70	40	70	40	40
Cr	P	P	P	P	P	P	P	—	—
Cr + HEAT	F	F	F	F	F	F	F	—	—
Cu	F	F	F	F	—	F	—	F	F
Cu + HEAT	F	F	F	P	—	F	—	F	P

• HEAT = 250°C IN N_2 FOR 1 HOUR AFTER METAL DEPOSITION

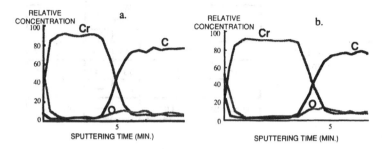

Fig. 5. Auger Depth Profiles a.) As implanted b.) After heat treatment.

An increase in the oxygen peak on the BCB side of the interface has been induced through the implant treatments. This can be seen from the Auger spectra of the 40 keV Si^+ treatment, in Fig. 5a. Fig. 5b show a shift in this peak towards the Cr side of the interface after the heat treatment.

The processes used to build multichip substrates are often repeated. Therefore it was interesting to study the effects of combinations of these surface treatments along with the process steps. During TEM investigation of different samples, voids were observed in all multiple-treatment samples in which RIE was one of the treatments, as evidenced in Fig. 6.

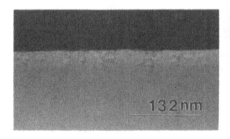 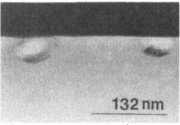

Fig. 6. a.) O_2/N_2 RIE + O Implant,   b.) O_2/N_2 RIE + Liftoff + O Implant

X-ray analysis during TEM examination revealed that these regions contained greater amounts of silicon than that present in the bulk of the BCB. It is possible that these artifacts are the result of the preferential removal of silicon/silicon dioxide agglomerates during the XTEM preparation. This phenomenon warrants further study.

Table III gives the results for some of these samples which received multiple treatments followed by the evaporation of Cr/Cu/Cr. The values shown are comparable to those obtained in a recent study utilizing sputter deposited materials.[9] The values obtained are considerably higher than those achieved with single treatments. Another interesting difference between samples receiving multiple versus single treatments was that samples which received multiple treatments were generally not adversely affected by subsequent heat treatment. The sample which received O_2/N_2 implant + heat treatment failed at the substrate BCB interface. The values for the O_2/N_2 + liftoff treated samples are listed under the 40KeV implant column for convenience, only. Those samples did not receive any implant treatment.

(ksl)	OXYGEN 40KeV		OXYGEN 70KeV	
		HEAT		HEAT
O$_2$/N$_2$ + IMPLANT	5.85	>6.81	5.59	8.67
O$_2$/N$_2$ + LIFTOFF	6.97	7.42	—	—
IMPLANT + LIFTOFF	8.34	—	7.73	2.83
O$_2$/N$_2$ + IMPLANT + LIFTOFF	9.82	—	6.92	8.82

Table III.
Pull Test Adhesion Results
for Multiple-Treatment
Samples

SUMMARY

Adhesion is improved through surface treatments. The use of O$_2$/N$_2$ plasma treatments improve adhesion for both Cr and Cu. However, the Cu agglomerates in the O$_2$/N$_2$ RIE treated BCB, which can be a reliability concern. The implant treated samples exhibit improved adhesion prior to heat treatment in the Cr/BCB structures and some improvement subsequent to heat treatment in the Cu/BCB structures. The effects of combining some of the surface treatments were studied to assess their applicability for processing Cu/BCB thin film high density interconnections.

ACKNOWLEDGEMENTS

The authors wish to acknowledge the contributions of S. Hofmeister for the Auger profiling, M. Ray and M. Denker for the SIMS verification and S. Hsia for the XTEM sample preparation.

REFERENCES

[1] Egitto, F. D., Emmi, F., Horwath, R.S., JVST B 3 (3) May/Jun (1985).

[2] Paik, K. W., Saia, R.J., Chera, J.J., MRS Conf. Nov. 1990.

[3] Baglin, J.E.E., Nuc. Instr. Methods Phys. Res. B39 (1989).

[4] Williams, J.S., Rep. Prog. Phys. 49 (1986).

[5] Adema, G.M., Turlik, I., Hwang, L.T., Rinne, G.A., Berry, M.J., IEEE Trans. (CHMT) 13,4,(1990).

[6] LeGoues, F.K., Silverman, B.D., Ho, P.S., JVST A 6 (4) Jul/Aug 1988.

[7] Jones, SK, Chapman, R.C., Dishon G., Pavelchek, E.K., Proc. Eighth Int'l Conf. on Photopolymers, Oct. 1988 (SPE).

[8] Flutie, R.E., Materials Research Society, vol. 62, 1986.

[9] Fong, S.O., Keister, F.Z., and Peters, J. W., 22nd. Int'l SAMPE Conference, Nov. 1990.

A STUDY OF THE INTERFACE BETWEEN POLYIMIDE AND PLATED COPPER

YOSHIYUKI MORIHIRO, KURUMI MIYAKE, MITSUMASA MORI, HIROSHI KUROKAWA, MASANOBU KOHARA, AND MASAHIRO NUNOSHITA
Materials & Electronic devices Laboratory, Mitsubishi Electric Corporation, 1-1, Tsukaguchi-Honmachi 8-chome, Amagasaki Hyogo, 661 Japan

ABSTRACT

A polyimide (PI) surface is analyzed by XPS at each electroless-copper-plating step. Zn ions are adsorbed on the PI surface, and subsequently replaced with Pd ions. Cu is deposited on the PI surface catalyzed by these Pd ions, As the surface concentration of Pd ions increases, the adhesive strength of the plated Cu to the PI surface increases.

In the case of PI fabrication onto a plated Cu, no significant PI degradation is observed.

INTRODUCTION

A Copper-polyimide system has many attractive features for electronic device applications. Applications of this system especially for high-speed multilayer circuit boards on which LSI's are mounted have been widely reported, because of the low dielectric constant of polyimide (PI) and the low resistance of Cu[1]. It is very important to understand the interfacial chemistry and adhesion properties between Cu and PI when considering the reliability of electronic devices. Poor adhesion and PI degradation at the Cu-PI interface have been reported as a result of interfacial reaction between Cu and PI[2,3]. Inserting a barrier metal such as Cr or Ti between Cu and PI has been one of the popular ways to suppress these problems[1].

On the other hand, we have developed a new hybrid IC system which was named MCPH (Mitsubishi Copper Polyimide Hybrids) as one of the applications of the Cu-PI system, and have confirmed its high reliability[4]. The cross-sectional structure of MCPH is shown in Figure 1. MCPH can be easily fabricated into a multilayer circuit board by utilizing a copper-plating method then a photolithographic method of photosensitive PI, alternatively. Although no barrier metal is inserted between Cu and PI, we have confirmed high adhesive strength at the Cu-PI interface in MCPH, because of our advanced copper-plating technology. In spite of these unique properties, there has been few works on interfacial studies between plated Cu and PI. In this study, a PI surface is analyzed at each electroless-copper-plating process, and the mechanism of high adhesive strength of the plated Cu to the PI surface is investigated.

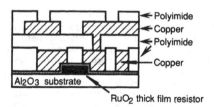

Fig. 1. Cross-sectional structure of MCPH

Mat. Res. Soc. Symp. Proc. Vol. 227. ©1991 Materials Research Society

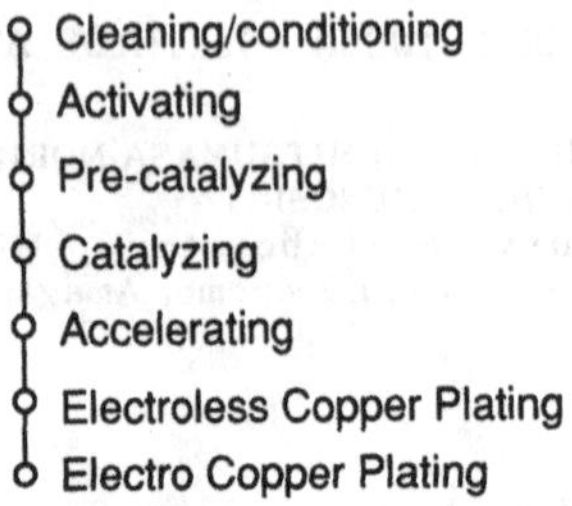

○ Cleaning/conditioning
○ Activating
○ Pre-catalyzing
○ Catalyzing
○ Accelerating
○ Electroless Copper Plating
○ Electro Copper Plating

Fig. 2. Copper plating process

In the case of PI fabrication onto a plated Cu, the degree of PI degradation is evaluated.

EXPERIMENTAL

The PI used in this study was a polyamic acid (PAA) of the pyromellitic dianhydride-oxydianiline (PMDA-ODA) type, which has photosensitive groups at the end of the carbonyl groups of PMDA. For the copper-on-polyimide (Cu/PI) study, PAA was spin-coated onto an alumina ceramic substrate (96% Al_2O_3), and cured at 350°C in a nitrogen atmosphere (O_2 content less than 5ppm) for imidization. The PI thickness was 7μm after curing. The Cu was fabricated to a 5μm thickness on the PI surface by a plating process shown in Figure 2. The changes of the PI surface structure after the activating step and the catalyzing step were characterized by X-ray photoelectron spectroscopy (XPS). In the following electro-plating step, the Cu patterns were fabricated on the PI surface in the shape of circles with 2mm diameters. After being annealed in the same condition as the PI curing, the Cu patterns were pulled to evaluate the adhesive strength of Cu to the PI surface. For the polyimide-on-copper (PI/Cu) study, electro-plated-copper was fabricated on a Cu substrate of 5μm thickness. Then PAA was spin-coated onto the Cu and cured in the same manner as the case in the Cu/PI study. The PI was lifted off the Cu by Cu dissolution in a nitric acid, and the PI surface at the Cu side was analyzed by XPS.

RESULTS AND DISCUSSION

(A) Cu/PI Interface

Change of PI Structure in the Activating Step

In the activating step, PI is dipped in a specially developed alkaline solution containing Zn ions before going into a conventional catalyzing step. The C1s spectra of the PI surface before and after the activating step is shown in Figure 3. Each peak can be associated with a carbon of the PI structure, as shown in Figure 4. The increase in the C-O peak and the decrease in the C=O peak after the activating step suggest a transition of the C=O bond to the C-O bond, because it has been reported that the C=O group is the most active portion in the PI structure and can easily form an interaction with other species[5]. Moreover, Zn ions are confirmed to be adsorbed on the PI surface in the activating step, as shown in Figure 5. These results suggest that an O-Zn bond is formed as a result of the covalent interaction between the activated C=O group and a Zn ion during the activating step.

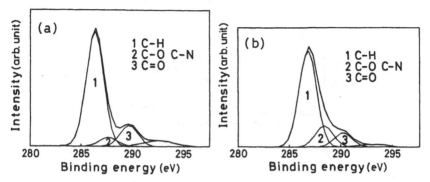

Fig.3 Carbon 1s spectra of polyimide surface (a)before and (b)after the activating step.

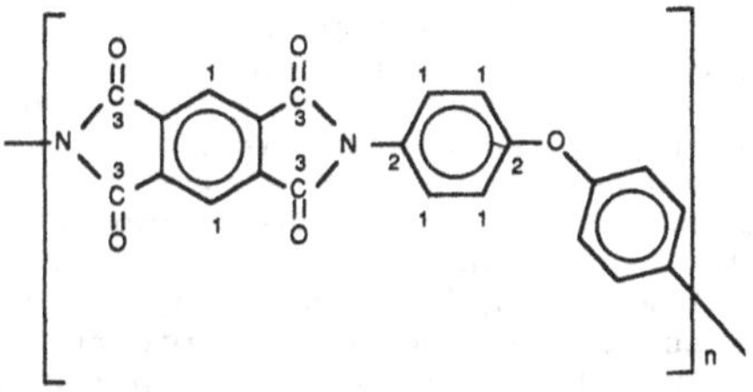

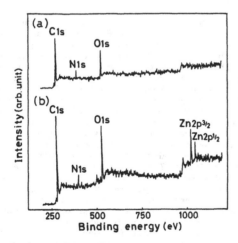

Fig.4 Structure of PMDA-ODA polyimide (one repeat unit). Numbers represent carbon positions referred to in Carbon 1s spectra.

Fig.5 XPS spectra in wide range (a)before and (b)after the catalyzing step.

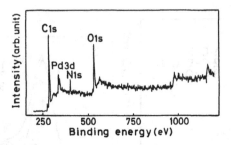

Fig.6 XPS spectrum of wide range after the catalyzing step.

Change of PI Structure in the Catalyzing Step

In the catalyzing step, Pd ions are adsorbed on the PI surface, and they act as catalysts for Cu precipitation in the subsequent electroless-plating step. Although Zn ions appear on the PI surface after the activating step, they disappear after the catalyzing step, as shown in Figure 6. In an attempt to clarify this phenomenon, the relative intensities of the Zn2p peak, after the activating step, and the Pd3d peak, after the catalyzing step in XPS spectra, are plotted as a function of the dipping time in the activator, as shown in Figure 7. Regarding the increase in the peak intensity in XPS spectra as the amount of adsorbed ions, it is realized that the increase in the amount of Pd ions shows the same tendency as that of Zn ions, i.e., the adsorption of Zn ions seems to contribute to the increase in the amount of adsorbed Pd ions. Considering the disappearance of Zn ions, it can be presumed that an adsorbed Zn ion is replaced with a Pd ion in the catalyzing step, resulting in the formation of an O-Pd bond at the C=O group of the PI structure.

Adhesion at the Cu/PI Interface

When Cu is deposited onto a PI surface by using the sputtering or evaporating method, the adhesion at the Cu/PI interface is very poor because of a lack of chemical interaction between Cu and PI[2]. On the other hand, when Cu is fabricated onto the PI surface by the electroless-plating process, higher adhesive strength can be obtained because the PI is activated chemically in an alkaline solution such as a catalyst. However, in the conventional pretreatment without the activating step, the adhesion at Cu/PI interface is not sufficient for the reliability standards of electronic devices (less than 5MPa). In this paragraph, the effect of our pretreatment on the adhesion at the Cu/PI interface is discussed. As shown in Figure 8, adhesive strength at the Cu/PI interface improves with the increase in the dipping time in activator, up to 1 minute. As this result shows almost the same change as the peak intensity of Pd ions shown in Figure 7, it is suggested that the amount of Pd ions adsorbed on the PI surface dominates the adhesion at the Cu/PI interface. Therefore, the effect of pretreatment on high adhesive strength is to create an environment where large amounts of Pd ions are adsorbed on the PI surface, making use of a large adsorption tendency of Zn ions on the PI surface and the subsequent replacement of Zn ions by Pd ions. When PI is dipped in the activator for more than 1 minute, the adhesive strength of more than 20MPa is achieved, and this is sufficient strength for the reliability standards of electronic devices.

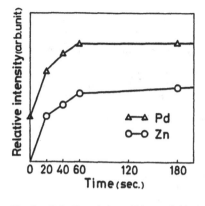

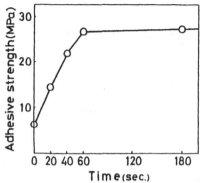

Fig.7 Relative intensities of the zinc 2p and palladium 3d peaks as a function of the dipping time in the activator.

Fig.8 Adhesive strength of copper to the polyimide surface as a function of the dipping time in the activator.

(B) PI/Cu INTERFACE

PI Structure on Electro-Plated-Copper

When PAA is coated onto Cu, it has been reported that Cu dissolves into PAA, and the dissolved Cu causes PI degradation during and after the PAA curing[3]. Reduction of C=O groups in the PI structure is one of the main features of the PI degradation at the PI/Cu interface. It is interesting, that in most of the cases in these reports, PI has been cured on Cu deposited by sputtering or evaporating. In this study, the PI structure cured on electro-plated-copper is analyzed to confirm the degree of PI degradation in our PI/Cu interface. As shown in Figure 9, no apparent reduction of the C=O peak appears in the C1s spectrum, compared to that in the PI which is cured without a reaction with metals (shown in Figure 3-a). This result suggests that no significant PI degradation occurs at the PI/plated-Cu interface. It has been reported that the

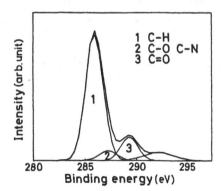

Fig.9 Carbon 1s spectrum of polyimide cured on electro-plated-copper. The spectrum is taken from the copper side.

interaction between PAA and Cu, such as the formation of a PAA-Cu complex, results in the high adhesive strength at the PI/Cu interface[6]. And it has also been pointed out that the degree of PI degradation at this interface is sensitive to the oxidation state of the Cu surface during PAA curing, which should be affected by the oxygen content in a curing atmosphere and the oxidation tendency of Cu[3]. As high adhesive strength of more than 50MPa has been confirmed at the PI/plated-Cu interface, the same kind of PI-Cu interaction should also be occurring at this interface. More detailed investigations are needed to clarify the mechanism of PI-Cu interaction and PI degradation.

CONCLUSION

In Cu fabrication onto the PI surface, we have developed a new copper-plating process, resulting in a high adhesive strength of more than 20MPa. We analyze the changes in the PI surface structure by XPS, at each pretreatment step. In the pretreatment, Zn ions are adsorbed on the PI surface, and these adsorbed Zn ions are replaced with Pd ions. A large amount of adsorbed Pd ions guarantees high adhesion at the Cu/PI interface. Therefore, the effect of pretreatment is to create an environment where large amounts of Pd ions are adsorbed on the PI surface, making use of a large adsorption tendency of Zn ions on the PI surface and the subsequent replacement of Zn ions by Pd ions. On the other hand, we have confirmed that there is no significant PI degradation at the PI/plated-Cu interface.

ACKNOWLEDGMENT

The authors would like to thank Mr. Yasuo Hashimoto and Dr. Tetsuo Ogama for their helpful discussions.

REFERENCES

[1] C. C. Chao, K. D. Scholz, J. Leibovitz, M. Cobarruviaz, and C. C. Chang, IEEE Trans. Comp., Hybrids, Manuf. Technol., vol. 12, no. 2, 180-184 (1989).

[2] J. Kim, S. P. Kowalczyk, Y. M. Kim, N. J. Chou, and T. S. Oh, Mat. Res. Soc. Symp. Proc., vol. 167, 137-145 (1990).

[3] D-Y. Shih, J. Paraszczak, N. Klymko, R. Flitsch, S. Nunes, J. Lewis, C. Yang, J. Cataldo, R. McGouey, W. Graham, R. Serino, and E. Galligan, J. Vac. Sci. Technol., A7(3), 1402-1412 (1989).

[4] M. Takada, T. Makita, E. Gofuku, K. Miyake, A. Endo, K. Adachi, Y. Morihiro, and H. Takasago, ISHM Proc., 540-544 (1988).

[5] S. C. Freilich and F. S. Ohuchi, POLYMER, vol. 28, 1908-1914 (1987).

[6] Y. H. Kim, G. F. Walker, J. Kim, and J. Park, J. Adhesion Sci. Tech., vol. 1, no. 4, 331-339 (1987).

ADHESION OF COPPER TO POLYTETRAFLUOROETHYLENE POLYMER FILM WITH ADHESIVE LAYERS

Chin-Jong Chan, Chin-An Chang and Curtis E. Farrel
IBM Research Division, T.J. Watson Research Center, Yorktown Heights, NY 10598

ABSTRACT

Enhanced adhesion of Copper to spin-coated polytetrafluoroethylene (PTFE) film is achieved by employing an intermediate adhesive layer consisting of fluorinated ethylenepropylene (FEP) and metallic titanium. The peel strength of Cu on PTFE is low (1-2 g/mm) without the adhesive layer. After incorporating the Ti/FEP adhesive layer, a peel strength as high as 80 g/mm can be achieved. The overall peel strength is found to increase with the adhesive layer thickness. Thermal annealing at high temperatures strongly affects its overall peel strength. Previous XPS study indicated the formation of TiC, at the Ti/FEP and Ti/PTFE interfaces. Rutherford backscattering spectrometry and cross-sectional transmission electron microscopy studies revealed that significant amount of Cu diffused into the FEP layer through the Ti layer when the specimens were heat treated at 375°C for various amount of time. Cu diffusion strongly affects the interfacial properties and hence results in the variation of overall peel strength. The results indicate that the thickness of Ti barrier as well as the heat treatment temperature are important for retaining the high peel strength.

INTRODUCTION

Thin film polymeric materials has been widely used in applications such as thin film multichip packaging and device fabrications as interdielectrics. Most attention has been focussing on polyimides [1,2]. However, polyimide has certain disadvantages such as high water absorption and relatively higher dielectric constant ($k' = 3.2$) than other polymers. For improved reliability and faster signal propagation, polymers with lower dielectric constant and lower water absorption will be required. Fluorocarbon polymers, commonly known as Teflon, are attractive in high performance electronic packaging [3] for their low dielectric constant ($k' = 2.1$), low dissipation factor ($\tan \delta = 0.0002$), low water absorption, high thermal stability, and excellent chemical resistance.

Adhesion between metal and polymer has been one of the key issues in electronic packaging applications. Adhesion of metals to fluorocarbon polymers has been studied to certain extent [4,5,6]. Research efforts have been primarily focussed on polytetrafluoroethylene (PTFE) and copper. PTFE has a relatively simple chemical structure and better high temperature properties among different fluorocarbon polymers. However, adhesion of Cu to PTFE is low without any surface modifications. The adhesion of Cu to bulk PTFE has been shown to increase dramatically when the PTFE surface is Argon ion presputtered prior to Cu deposition [4,5]. Other surface modification processes, such as ultraviolat irradiation of the polymer surface and the thermal treatment of the metal polymer interface, have also shown some increase in the Cu-PTFE adhesion [6]. More recent invesitigations have also shown that the adhesion was strongly dependent on the metal reactivity as well as the fluorocarbon polymer's chemical structure [7]. Among the combinations, Ti on fluorinated ethylenepropylene (FEP) was found to have the highest peel strength. This high peel strength is attributed to the strong titanium-carbon interaction from the largest electronegativity difference between the metal and carbon of FEP.

In this study, we have examined the possibility of using a thin Ti/FEP adhesive layer to improve the adhesion of Cu to PTFE. This is an alternative approach in improving the ad-

Mat. Res. Soc. Symp. Proc. Vol. 227. ©1991 Materials Research Society

hesive strength of Cu to PTFE without changing its interfacial morphorlogy. Effect of adhesive layer thickness, FEP and/or Ti, on the overall adhesion has been studied. The effect of thermal annealing on the adhesion strength was also investigated and presented in the following sections.

EXPERIMENTAL PROCEDURES

All polymer resins were obtained from E.I. du Pont De Nemours & Co. in the form of negatively charged hydrophobic colloids containing resin particles of 0.05 μm in diameter suspended in water. The fluorocarbon resins were spin-coated on the surfaces of SiO_2/Si wafers with 2000Å Cr layer. The film thickness under different spinning conditions was calibrated with respect to different spinnig rate and spinning time for both PTFE and FEP dispersions. The spin-coated resins were heated in an oven to 100°C for 10 min to remove water, and to 260°C to burn off the wetting agent. Multiple coatings were applied to obtain the desired polymer film thickness.

The adhesion strength between polymer-polymer (FEP and PTFE thin films) and metal-polymer were both studied. For polymer-polymer adhesion, approximately 10 μm or 20 μm of PTFE was first coated on the $Cr/SiO_2/Si$ wafers. A thin layer (500 Å) of Al_2O_3 was deposited onto one end of the substrate and followed by 500 Å Cu to initiate the peeling. The peel strip release area is about 10 mm in width. A thick layer of FEP was then spin-coated on top of the PTFE film and subsequently cured at 370°C in nitrogen atmosphere. Peel strip was then defined with a width of 1.6 mm.

The Cu/adhesive/PTFE samples were prepared according to the following procedures. An approximately 10 μm of PTFE film was spin-coated on $Cr/SiO_2/Si$ substrates. A thickness of 1000Å, 3000Å, 6000Å or 2 μm of FEP resin was then spin-coated on top of the 10 μm PTFE film by controlling the suspension's concentration. The final structures containing 10 μm of PTFE and a thin FEP overcoat were then cured at 370°C in a nitrogen atmosphere. Two reference samples were constructed with 10 μm FEP films on Cr coated SiO_2/Si wafers and cured under similar conditions described above.

Metal films were deposited on the cured polymers using electron-beam evaporation. Titanium films of 100 Å or 200 Å were deposited on the polymer, followed by 8 μm of Cu strips. For reference samples, 2000Å of Ti or Cu was deposited on the bulk FEP films, followed by 8μm of Cu peel strip. The geometry of Cu strips were defined using a metal mask with 1.6 mm strip width. A small amount of aquadag was applied to the edge of the polymer layer prior to the metal depostion to allow for subsequent peeling test of the Cu strips. The deposition rate for the metals was typically 25 Å/s. The base pressure of the deposition chamber was maintained at 2×10^{-6} Torr during the deposition process.

Adhesion was measured in peel strength using a 90 ° peel tester. A test specimen was loaded on the stage with an end of the metal strip attached to a sensitive load cell. During the peel test, the peeling angle was maintained at 90° to the substrate. The peel rate was kept constant at 7.5×10^{-2} mm/s. The peel strength was recorded as a quantitative measure of the force needed to peel off the metal strips in grams/mm. Each data point reported here represents an average value from 3 to 6 metal strips from two different samples.

Metallized samples were also annealed at 375°C in a nitrogen atmosphere to study the effect of thermal cycling on adhesion strength. Cross-sectional TEM specimens were prepared and selected area of the specimens were examined using energy dispersive spectroscopy (EDS) and selected area electron diffraction (SAD) techniques. Studies of the interdiffusion between polymers and metals were also carried out using Rutherford Backscattering Spectrometry (RBS) and X-ray Photoelectron Spectrometry (XPS).

RESULTS AND DISCUSSION

The adhesion strength between PTFE and FEP was found to be high (65 - 70 g/mm). During the peeling test, a FEP peel strip thickness of about 80 μm was necessary in order to minimize the work expenditure from the viscoelastic effect of the polymeric film during peeling. When the underlying PTFE thickness increased, the measured peel strength was found to be slightly higher than the thin one. The viscoelastic effect may result in lower observed peel strengths due to its effect on the peel rate. The true peel strength of the FEP-PTFE couple is believed to be slightly higher than the currently reported value. Since the adhesion between FEP and PTFE is good, a PTFE bulk film with thin FEP overcoat is hence a logical choice for a combination of good thermal properties, good dielectric properties and high adhesion strength.

As shown in Figure 1, the peel strengths of the reference samples, 2000Å titanium film on 10 μm spin-coated FEP and 2000Å Cu film on 10μm spin-coated FEP, were 80 g/mm and 9 g/mm respectively. The peel strength of Cu to PTFE was weak (1-2 g/mm). When a Ti/FEP adhesive layer was incorporated in the structure, the peel strengths were found to increase as the thickness of FEP overcoat increased (Figure 1). A similar trend was observed for both 100Å and 200Å of titanium in the adhesive layer. The overall peel strength approached the value of the reference Ti/FEP samples when the FEP overcoat thickness was in the μm range. The peel strength increase can be attributed to the strong interaction between Ti and carbon of FEP for their highest electronegativity difference, as discussed in the previous experiments [5]. SEM analyses on the peeled surfaces of both the substrate and Cu peel strip showed that fracture occured in the FEP layer. This might explain why the peel strength changed with the FEP thickness.

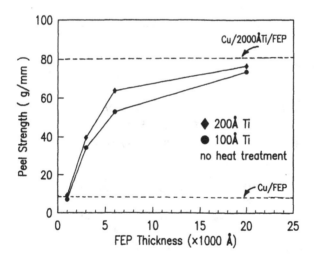

Figure 1. Peel strength of Cu to PTFE with intermediate Ti/FEP adhesive layer as a function of FEP overcoat thickness. Dashed lines are peel strength value of 2000Å of Cu or Ti on bulk FEP films.

When the samples were thermally cycled to 375°C, the peel strength of the reference Ti/FEP specimens did not change with the thermal treatments. The Cu/FEP reference samples, on the other hand, showed an increase in peel strength from 9 g/mm to 14 g/mm after one or more thermal cycles, as shown in Figure 2. The Cu/PTFE samples with adhesive layer, however, showed a decrease in peel strength from their previous level to about 15 g/mm after only one thermal cycle and remained constant throughout other cycles. This value corresponds to the peel strength of Cu/FEP reference samples after thermal cycling. RBS analyses on Ti/FEP samples showed no detectable interdiffusion between the two when the samples were heat treated to 375°C for 1 hour or longer as shown in Figure 3(a). On the contrary, a significant amount of Cu was detected diffusing into the FEP layer when the Cu/FEP samples were heat treated under a similar condition (Figure 3(b)). It is obvious that heat treatment of Cu/FEP interface promoted chemical reactions and consequently increase the peel strength. For Ti/FEP specimens, the initial metal-polymer chemical interaction was very strong as demonstrated by higher Ti-carbidelike and carbonlike structures in the previous XPS C 1s spectra [6]. Thermal treatment did not significantly increase the chemical reactions at the Ti/FEP interface. After heat treatment, little interdiffusion was detected and, indeed, no noticeable changes in peel strength were observed.

For a Cu/Ti/FEP trilayer structure significant interdiffusion between Cu and Ti were observed when the trilayer samples were heat treated under similar conditions (Figure 3(c)). A significant amount of Cu atoms can diffuse into the FEP layer when the Ti layer is thin. Cross-sectional TEM did reveal the occurence of Cu diffusion into the FEP layer through Ti layer (Figures 4(a), 4(b) and 4(c)). These results were further confirmed by EDS analyses. Cu was found as spherical or slightly faceted precipitates in FEP layer. The precipitates have a size ranging from 5 nm to 20 nm. Analyses of the SAD patterns (inserts of Figures 4(b)) indicate the presence of metallic Cu only. The diffusion of Cu reached about μm range. Figure 4(c) is the typical microstructure of area far away (greater than 1 μm.) from the Cu/Ti/FEP interface. Cu diffusion across the Ti layer is believed to be responsible for the decrease of the overall peel strength.

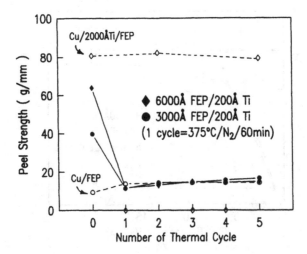

Figure 2. Peel strength of Cu to PTFE with intermediate Ti/FEP adhesive layer as a function of number of thermal cycling at 375°C in a nitrogen atmosphere. (Each thermal cycle = 375°C for 60 minutes)

Thermal treatment of the metal-polymer is generally believed to increase the metal-polymer chemical interactions and change the properties of the polymer to certain degree. Our results indicate that interdiffusion between metal and polymer is important in determining the peel strength of a metallized thin film structure. The thickness of the Ti barrier determines the ultimate temperature a Cu/PTFE structure with Ti/FEP adhesive layer can be exposed to. A finite thickness of Ti barrier layer is necessary for the bonding of the Cu to PTFE to remain at an acceptable level. At high temperatures, the diffusion of Cu through the barrier layer into the polymer cause variations in adhesion strength of the thin film structure. Understanding of the interdiffusion behavior is necessary in controlling the interfacial bonding strength.

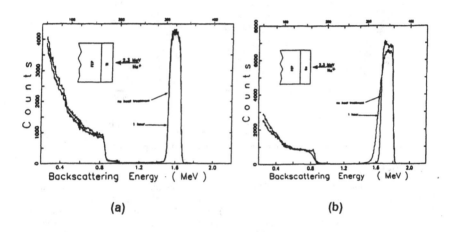

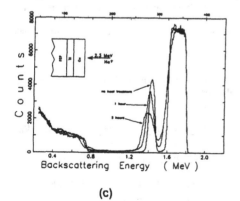

Figure 3. RBS spectra from (a) Ti deposited on FEP bulk film before and after heat treament at 375°C, (b) Cu deposited on FEP film before and after heat treatment, and (c) as-deposited Cu/Ti/FEP trilayer structure and samples being heat treated at 375°C for 1 or 5 hours.

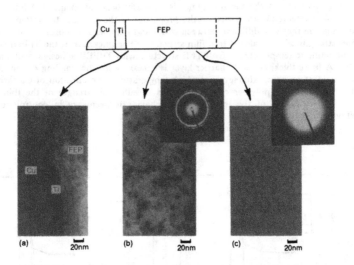

Figure 4. Cross-sectional TEM bright-field micrographs of Cu/adhesive/PTFE specimen after 5 thermal cycles at 370°C (a) Cu/Ti/FEP interface, (b) 3000Å from interface in FEP, (insert is the corresponding SAD pattern), (c) far from the interface showing no Cu precipitates.

CONCLUSIONS

Enhanced adhesion of copper to PTFE is achieved by employing a Ti/FEP intermediate layer. The overall peel strength was found to increase as the FEP thickness increased. Thermal annealing to high temperature (375°C) caused the peel strength to decrease to a constant value. This peel strength value corresponds to the value of Cu/FEP after thermal treatment. RBS and cross-sectional TEM analyses on the samples before and after thermal annealing revealed that significant diffusion of Cu into FEP has occured. Spherical or slightly faceted Cu precipitates with a size of 5 to 20 nm were observed in the FEP layer. The diffusion of Cu across the thin Ti layer into the FEP layer was identified to be responsible for the observed peel strength degradation during thermal cycling. Our results indicate that an optimum Ti layer thickness exists for retaining the observed high peel strength upon high temperature thermal treatments.

ACKNOWLEDGEMENTS

The authors would like to thank Grant Coleman for help in RBS analyses, Alex Schrott for XPS analyses, and Susan Lee for general assistances.

REFERENCES

1. R.R. Tummala and E.J. Rymaszewski, Microelectronic Packaging Handbook, Van Nostrand Reinhold, New York, (1989).

2. R.J. Jensen and J.H. Lai, in Polymers for Electronic Application, edited by J.H. Lai, (1989, CRC press), P. 33.

3. R.T. Traskos, W.D. Smith and S.C. Lockard, Proc. NEPCON East '90, 823 (1990).

4. C.-A. Chang, J.E.E. Baglin, A.G. Schrott, and K.C. Lin, Appl. Phys. Lett. 51, 103 (1987).

5. C.-A. Chang, Appl. Phys. Lett. 51, 1236 (1987).

6. Y.-K. Kim, C.-A. Chang, and A.G. Schrott, J. Appl. Phys., 67(1), 251 (1990).

7. C.-A. Chang, Y.-K. Kim and A.G. Schrott, J. Vac. Sci. Technol., A8, 3304 (1990).

PART VI

Late Papers Accepted

STRUCTURE AND PROPERTIES OF BINARY RODLIKE/FLEXIBLE POLYIMIDE MIXTURES.

S. ROJSTACZER, M. REE, D. Y. YOON,† AND W. VOLKSEN
IBM Almaden Research Center, 650 Harry Road, San Jose, California 95120

ABBSTRACT: Binary mixtures of a rodlike poly(p-phenylene pyromellitimide) (PMDA-PDA) and a flexible 6F-BDAF polyimide synthesized from hexafluoroisopropylidene diphthalic anhydride and 2,2-bis(4-aminophenoxy-p-phenylene) hexafluoropropane were prepared by solution-blending of the meta-PMDA-PDA poly(amic ethyl ester) and 6F-BDAF poly(amic acid) precursors, followed by solvent evaporation and thermal imidization. The size scale of the phase separation, as measured by light scattering, is ca. 1 μm or smaller in most cases. Dynamical mechanical thermal analysis measurements indicate that the glass transition temperature of 6F-BDAF is unaffected in all of the mixtures studied, indicating complete demixing of rodlike and flexible polyimides in agreement with theory. X-ray photoelectron spectroscopy results show a strong surface segregation of 6F-BDAF in mixtures containing as low as 10% by weight of the 6F-BDAF component in the bulk. The mixtures with PMDA-PDA as the major matrix component therefore maintain excellent bulk properties of rodlike polymers, i.e., high modulus to 500°C, and low coefficients of thermal expansion (< ca. 10 ppm/°C). On the other hand, the surface properties of the mixtures are dominated by the flexible 6F-BDAF, resulting in excellent polymer/polymer self-adhesion (lamination) properties between fully imidized films.

INTRODUCTION

Aromatic polyimides are increasingly used in microelectronic devices owing to their good dielectric properties, thermal stability, and easy processability [1,2]. The requirements of polymer properties for more advanced microelectronic applications are quite severe, however, such that they cannot be usually met by a single component polymer. For example, polyimides with rigid rodlike conformations provide the required thermo-mechanical properties, but these materials have inherently poor adhesion properties [3].

One approach to obtain a combination of desired properties is blending a rodlike polymer system with a flexible polymer which could provide enhanced adhesion and increased toughness. However, the intrinsic tendency towards phase separation between rodlike and flexible polymers [4] invariably leads to phase-separated morphologies which tend to be quite large in size. In the case of polyimides, the components can be mixed in the more flexible precursor forms, and in this way, the large scale phase separation can be potentially prevented. In this regard, Yokota et al. [5] investigated mixtures of rigid poly(p-phenylene biphenyltetracarboximide) (BPDA-PDA) with flexible poly(4,4′-diphenylether biphenyltetracarboximide) (BPDA-ODA), obtained by solution blending of the respective poly(amic acid) precursors. Based on the observation of a single T_g which was composition dependent and the transparency of the films, the authors concluded that the components are dispersed at a molecular level. However, studies by Ree et al. [6] and by Feger [7] showed that mixtures of poly(amic acid)s can lead to the

† To whom correspondence should be addressed.

formation of segmented blocky copolymers via transamidation exchange reactions rather than molecularly mixed blends of two homopolymers.

In our approach, polyimide/polyimide mixtures are obtained by using one component in its poly(amic ethyl ester) precursor form as a way to prevent such an exchange reaction. The system studied consists of a rodlike poly(p-phenylene pyromellitimide) (PMDA-PDA), and a flexible 6F-BDAF polyimide synthesized from hexafluoroisopropylidene diphthalic anhydride and 2,2-bis(4-aminophenoxy-p-phenylene) hexafluoropropane (see Fig. 1). The mixtures were obtained by blending the meta-PMDA-PDA poly(amic ethyl ester) (PMDA-PDA PEE) and the 6F-BDAF poly(amic acid) (6F-BDAF PA) precursors in N-methylpyrrolidone (NMP). The meta-isomer of the PMDA-PDA precursor was chosen specifically for increased flexibility and better solubility. Preliminary reports on this system [8,9] and another similar system of PMDA-PDA/ODPA-ODA [10] were presented previously.

SAMPLES

Fig. 1 shows the chemical structures of PMDA-PDA and 6F-BDAF precursors synthesized for the present study and the respective polyimide forms. The details of polymer sysnthesis are presented elsewhere [11].

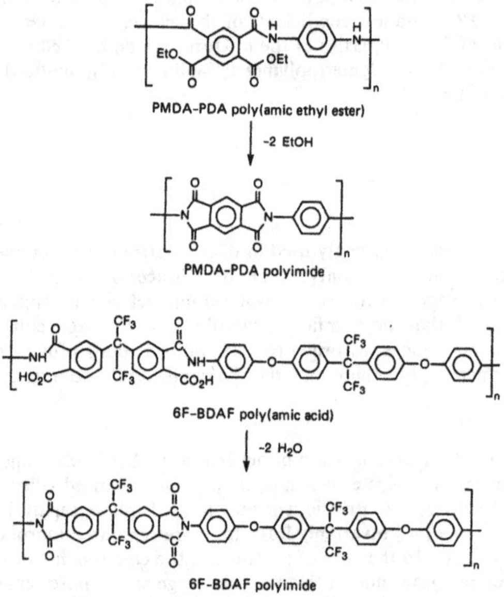

Fig. 1. Chemical structures of rodlike PMDA-PDA and flexible 6F-BDAF polyimides and their precursors.

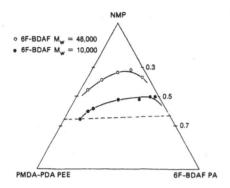

Fig. 2. Cloud-point curves of PMDA-PDA PEE/6F-BDAF PA/NMP mixtures measured at 75°C.

Miscible solutions of various PMDA-PDA PEE/6F-BDAF PA weight ratios were prepared by mixing different amounts of the homopolymers in NMP solutions. (All the polymer compositions in this paper refer to the weight ratios of the precursor polymers.) Films of PMDA-PDA/6F-BDAF precursors were prepared either by doctorblading or by spin-casting, followed by drying at 75°C for 90 min, all under ambient conditions. Thermal imidization was performed at 400°C for one hour in nitrogen atmosphere. The thicknesses of the precursor films for the light scattering measurements were in the range of 10-13 μm. After imidization, the thickness is reduced by 40-45%. The thickness of the polyimide films used for the rest of the measurements was in the range of 20 to 40 μm.

RESULTS AND DISCUSSION

Miscibility of Precursor Polymers

The miscibility of the precursor polymers was characterized by the biphasic curve which separates the single phase from the two phase regions of the ternary phase diagram of the system comprising the solvent NMP, PMDA-PDA PEE and 6F-BDAF PA. The compositions corresponding to the biphasic curve were measured by means of a cloud point technique. Fig. 2 shows two cloud-point curves corresponding to mixtures containing different molecular weights of 6F-BDAF PA. The compositions are in units of weight fraction, and the numbers on the right side of the ternary diagram represent the solid contents of the compositions along the respective horizontal lines. Fig. 2 indicates a significant effect of the 6F-BDAF PA molecular weight on the area of the single phase region.

Blend Morphology

The domain sizes measured by light scattering for the various mixtures are summarized in Table 1. For films containing the higher molecular weight 6F-BDAF polymer, the domain size shows only a slight composition dependence, with smaller domain sizes measured for mixtures with increasing PMDA-PDA content. More significant is the effect of molecular weight, as evidenced by comparing the mixtures of equal composition.

TABLE 1

Domain Sizes Measured by Light Scattering

M_w	PMDA-PDA/6F-BDAF		
6F-BDAF PA	25/75	50/50	75/25
48,000 (48K)	1.1 μm	1.0 μm	0.8 μm
10,000 (10K)	0.8 μm	0.6 μm	<0.5 μm

While the 75/25 (10K) mixtures became turbid during drying in the cloud-point measurements, films of up to ca. 15 μm in thickness prepared by spin casting and their imidized films were transparent, with a very low scattering which was angularly independent. This indicates that in these films, if any phase separation occurs, it is confined to a size scale smaller than ca. 0.5 μm.

One important characteristic of semirigid and rodlike polyimides is the fact that when prepared on substrates, the films exhibit a significant extent of molecular orientation parallel to the plane of the film [12,13]. The molecular orientation can be examined by comparing the wide-angle X-ray diffraction (WAXD) measured with the diffraction vector oriented along the film plane (transmission runs) and the film thickness (reflection runs), respectively. Figs. 3a and 3b present the WAXD curves measured using 1.53 Å radiation in transmission and in reflection runs, respectively, from a PMDA-PDA polyimide film prepared from PMDA-PDA PEE. The large difference in these patterns demonstrate a very strong in-plane molecular orientation: the sharp peaks at 7.2°, 14.4°, 21.6°, 29.0°, and 36.5° 2θ in the transmission pattern, are indexed as the 001 to 005 peaks of the 12.3Å monomer repeat distance. Figs. 3c and 3d show the diffraction patterns obtained from a 75/25 (48K) mixture measured in transmission and reflection runs, respectively. Comparison between Figs. 3a-b and 3c-d indicates that the molecular orientation of PMDA-PDA polyimide along the film plane is nearly retained even after the addition of

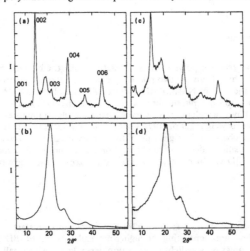

Fig. 3. X-ray diffraction patterns of PMDA-PDA polyimide and PMDA-PDA/6F-BDAF (75/25) mixtures: (a) transmission pattern of PMDA-PDA; (b) reflection pattern of PMDA-PDA; (c) transmission pattern of 75/25 mixture; (d) reflection pattern of 75/25 mixture.

25% 6F-BDAF. The WAXD results for a 50/50 mixture were very similar to those of 75/25 mixture except the change in peak intensities, but a 25/75 mixture exhibited no sharp peaks in either transmission or reflection patterns.

Bulk Properties

The high temperature dynamic mechanical properties and the glass transition (softening) temperatures of the homopolymers and their mixtures were studied by dynamical mechanical tensile analyzer (DMTA) measurements. Fig. 4 shows the dynamic storage and loss tensile moduli at 10 Hz for some representative samples as a function of temperature. PMDA-PDA shows only a small gradual decrease in modulus up to 500°C, without any indication of a transition occurring in the temperature range studied. On the other hand, 6F-BDAF (48K) shows a distinct glass transition at ca. 260°C. All of the mixtures studied showed a relaxation at a nearly constant temperature around the glass transition determined for 6F-BDAF (48K) homopolymer. This constant glass transition temperature of the 6F-BDAF component in the mixtures indicates that no significant amount of PMDA-PDA is included in the 6F-BDAF rich phase. The drastic difference in the dynamic modulus between the 70/30 and 30/70 samples above ca. 280° reflects the fact that in the former sample the matrix is the rigid PMDA-PDA, whereas in the latter 6F-BDAF forms the matrix. The relaxation corresponding to the glass transition of 6F-BDAF was also observed in the optically transparent mixtures containing the flexible component in its low molecular weight form, indicating the presence of submicron 6F-BDAF domains dispersed in the PMDA-PDA matrix.

The tensile properties of the polyimide mixtures were studied as a function of composition and molecular weight. The mechanical properties of the homopolymers and their mixtures are summarized in Table 2. The extent of reduction in the values of the modulus observed with increasing 6F-BDAF content is independent of its molecular weight. The toughening effect of 6F-BDAF, however, is only observed for the mixtures containing the 48K molecular weight polymer. Addition of 6F-BDAF of $M_w = 10K$ does not lead to any significant increase in the values of the elongation at break, probably due to the fact that this homopolymer forms very brittle films which could not be handled in the Instron tests without breaking them. The brittleness of this material suggests that its molecular weight is below the entanglement molecular weight.

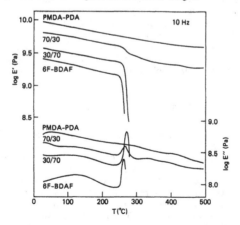

Fig. 4. Temperature dependence of the dynamic storage (E') and loss (E") modulus at 10 Hz as a function of composition.

TABLE 2

Tensile Properties and Coefficients of Thermal Expansion(CTE) of
PMDA-PDA/6F-BDAF Polyimide Mixtures

Composition PMDA-PDA/ 6F-BDAF $(M_w)^a$	Modulus (GPa)	Elongation (%)	CTE $(10^{-6}/°C)$
100/0	8.0	3.0	5
90/10 (48K)	6.4	4.5	-
90/10 (10K)	6.5	3.5	-
75/25 (48K)	4.4	5.0	12
75/25 (10K)	4.1	3.6	-
50/50 (48K)	4.0	6.5	-
25/75 (48K)	2.3	9.4	50
0/100 (48K)	2.3	25.0	55

a The number within the parenthesis denotes the weight-average molecular weight of the 6F-BDAF PA precursor.

Table 2 includes the coefficients of thermal expansion (CTE) measured for the homopolymers and some representative mixtures. The thermal expansivity of polymeric materials is directly related to the chain conformation/stiffness as well as the extent of molecular orientation. The low coefficient measured for PMDA-PDA (5 ppm/°C) should be considered in terms of its rodlike character in conjunction with the strong in-plane chain orientation discussed in the previous section. The CTE's exhibit a drastic increase to ca. 50 ppm/°C as the flexible 6F-BDAF polyimide becomes the matrix, consistent with the DMTA results.

Surface Composition and Adhesion Properties

The surface composition of the polyimide mixtures was characterized by X-ray photoelectron spectroscopy (XPS) in order to determine the extent of any surface segregation of one of the components. The results for various mixtures are presented in Table 3 in terms of the 6F-BDAF content at the surface region. The results include measurements performed at incidence angles of 10° and 60°, corresponding to detection depths of ca. 15 and 75Å, respectively. The results indicate a very strong surface segregation of 6F-BDAF. The top 15Å layers of mixtures containing as low as 10% 6F-BDAF in the bulk are composed of pure 6F-BDAF. Similar results were obtained for the 75Å layer, except that the 90/10 (10K) mixtures shows a 50% 6F-BDAF content.

TABLE 3

Surface Composition of 6F-BDAF in PMDA-PDA/6F-BDAF Polyimide Mixtures and Self-Adhesion (Lamination) Strength

6F-BDAF Content (%)			Peel Strength (g/mm)	
Bulk (%)	top 15Å	top 75Å	cast	laminated
0	-	-	≃1	0
10 (10K)	100	≃50	3.8	>3
10 (48K)	100	100	6.0	>5
25*	100	100	failed	>10
50*	100	100	failed	-
100	-	-	failed	30

* for both 10K and 48K molecular weight.

Also included in Table 3 are the measured values of adhesion strength. For specimens prepared by casting on the fully imidized bottom layers, the mixtures containing 10% and more 6F-BDAF exhibit a very significant enhancement of the adhesion as compared to that for the PMDA-PDA homopolymer. The 75/25 mixtures exhibited excellent adhesion which could not be evaluated quantitatively, as the top strip failed on tension during peeling. The efficiency in improving the adhesion properties is greater for the higher molecular weight 6F-BDAF. In view of the XPS results, this observation may be interpreted mainly in terms of the effect of molecular weight on the thickness of the surface segregated layer. The higher adhesion strength of 6F-BDAF over the values of PMDA-PDA is due to its flexible conformation that allows for interdiffusion to occur at the polymer-polymer interface at elevated temperatures.

While the peel test of polyimide bilayer samples prepared by casting is commonly used for characterizing the polyimide/polyimide self-adhesion properties, a more valid test of polyimide self-adhesion should be obtained by welding or laminating previously imidized films at temperatures above the softening transition. Such process, however, is not feasible for rigid polyimides such as PMDA-ODA or PMDA-PDA due to the absence of any softening temperature below the decomposition temperature. As expected, in an experiment in which two imidized films of PMDA-PDA were pressed at 320°C under a nominal stress of ca. 1000 psi for 30 min, the films could not be consolidated into one specimen. This was not the case with 6F-BDAF films, which could be laminated at 320°C, resulting in a bilayer specimens with peel strength of ca. 30 gr/mm. All of the mixtures containing 10% or more 6F-BDAF could be laminated.

The surface segregation of 6F-BDAF is most likely due to the lower surface energy associated with the fluorinated segments. In this regard, it is to be noted that for another rodlike/flexible polyimide mixture system comprising PMDA-PDA and flexible ODPA-ODA polyimides no enhancement of self-adhesion was observed when the flexible ODPA-ODA remained as the minor component [10]. Hence, the chain flexibility does not seem to be an important factor in determining the surface composition.

CONCLUSION

In the mixtures of a rodlike PMDA-PDA polyimide and a flexible 6F-BDAF polyimide, the size scale of the phase-separated structures is ca. 1 μm or smaller in most cases. Mixtures of PMDA-PDA containing up to 25% low molecular weight 6F-BDAF (10K) were obtained as transparent films. However, the glass transition (softening) temperature of the pure 6F-BDAF polyimide is found in all the mixtures that contain 6F-BDAF component, regardless of composition and domain size, indicating complete demixing of rodlike and flexible polyimides as predicted by theory [4]. The mixtures with PMDA-PDA as the major component exhibited excellent dimensional stability up to 500°C and low coefficients of thermal expansion, comparable to PMDA-PDA. On the other hand, the surface properties and adhesion strength of the mixtures were dominated by the flexible 6F-BDAF, due to the segregation of this component to the surface region, resulting in excellent self-adhesion (lamination) properties.

ACKNOWLEDGMENT

The authors would like to thank P. Iannelli and W. Parrish for carrying out the X-ray measurements, D. Miller for the XPS data, and R. Siemens for the CTE results, and P. Cotts for the molecular weight data. We also thank B. Smith for assistance with the light scattering measurements and B. Fuller for the assistance with DMTA, dielectric, and Instron experiments. It is also a pleasure to acknowledge the support and encouragement of IBM-STD, Endicott and IBM-GTD, Fishkill.

REFERENCES

1. K.L. Mittal, Ed., *Polyimides* (Plenum Press, New York, 1984).
2. J.H. Lupinski and R.S. Moore, Eds., *Polymeric Materials for Electronic Packaging and Interconnection*, ACS Symposium Series 407, 1989.
3. H.R. Brown, A.C. M. Yang, T.P. Russell, W. Volksen and E.J. Kramer, *Polymer* 29, 1807 (1988).
4. P.J. Flory, *Macromolecules* 11, 1138 (1978).
5. R. Yokota, R. Horiuchi, M. Kochi, H. Soma and I. Mita, *J. Polym. Sci.: Part C* 26, 215 (1988).
6. M. Ree, D.Y. Yoon and W. Volksen, *ACS Polym. Mat. Sci. Eng.* 60, 179 (1989).
7. C. Feger, see page 114 in ref. 2.
8. M. Ree, S.A. Swanson, W. Volksen, and D.Y. Yoon, U.S. Patent 4,954,578, Sept. 4, 1990.
9. S. Rojstaczer, D.Y. Yoon, W. Volksen and B.A. Smith, *Mater. Res. Soc. Proc.* 171, 171 (1990).
10. M. Ree, D.Y. Yoon and W. Volksen, *ACS Polym. Prep.* 31(1), 613 (1990).
11. S. Rojstaczer, M, Ree, D.Y. Yoon, and W. Volksen, *J. Polym. Sci., Polym. Phys. Ed.*, in press.
12. N. Takahashi, D.Y. Yoon and W. Parrish, *Macromolecules* 17, 2583 (1984).
13. T.P. Russell, H. Gugger and J.D. Swalen, *J. Polym. Sci. Polym. Phys. Ed.* 21, 1745 (1983).

STIFF POLYIMIDES: CHAIN ORIENTATION AND ANISOTROPY OF THE OPTICAL AND DIELECTRIC PROPERTIES OF THIN FILMS

D. BOESE, S. HERMINGHAUS, D. Y. YOON,† J. D. SWALEN, AND J. F. RABOLT

IBM Almaden Research Center, 650 Harry Road, San Jose, California 95120

ABSTRACT: Thin films of poly(p-phenylene biphenyltetracarboximide), prepared by thermal imidization of the precursor poly(amic acid) on substrates, have been investigated by optical waveguide, UV-visible, infrared (IR), and dielectric spectroscopies. The polyimide films exhibit an extraordinarily large anisotropy in the refractive indices with the in-plane index $n_\parallel$ = 1.852 and the out-of-plane index $n_\perp$ = 1.612 at 632.8 nm wavelength, indicating a strong preference of polymer chains to orient along the film plane. No discernible effect of the film thickness on this optical anisotropy is found in the range of ca. 0.4 μm to 7.8 μm in thickness. The frequency dispersion of the in-plane refractive index to 1.06 μm wavelength is consistent with the results calculated by the Lorentz-Lorenz equation from the UV-visible spectrum. The contribution from the entire IR range from 7000 to 200 cm,$^{-1}$ computed by the Spitzer-Kleinmann dispersion relations from the measured spectra, adds ca. 0.07 to the in-plane refractive index $n_\parallel$. Approximately the same increase is assumed for the out-of-plane index $n_\perp$, based on the tilt-angle dependent IR results. Application of the Maxwell relation leads to the out-of-plane dielectric constant $\varepsilon_\perp \simeq 2.8$ at ca. 10^{13} Hz, as compared with the measured value of ca. 3.0 at 10^6 Hz. Assuming this small difference to remain the same for the in-plane dielectric constants $\varepsilon_\parallel$, we obtain a a very large anisotropy in the dielectric properties of these polyimide films with the estimated in-plane dielectric constant $\varepsilon_\parallel \simeq 3.5$ at ca. 10^{13} Hz, and $\varepsilon_\parallel \simeq 3.7$ at 10^6 Hz.

INTRODUCTION

The unique features of polyimides - thermal and chemical stability, mechanical and electrical properties, and easy processability, for example - have made possible their applications in many high performance devices from aerospace to microelectronics [1]. For the latter they have become increasingly important as interlevel dielectrics for multilevel interconnections in microchips and electronic packaging structures [2 – 4]. In these applications, the dielectric material should have as low a dielectric constant, ε, as possible [4], since the interconnections are designed primarily to minimize signal propagation delay, resistive losses, and noise due to reflection and crosstalk. In this regard, the lower dielectric constants of polyimides present a definite advantage, as compared with those of inorganic materials such as alumina ($\varepsilon \approx 9.5$) and sputtered quartz ($\varepsilon \approx 4.2$) [3]. In addition, a low coefficient of thermal expansion (CTE) is needed in order to minimize the stress resulting from the mismatch in thermal expansion between the polymer and the other materials such as oxides, metals, ceramics, etc.

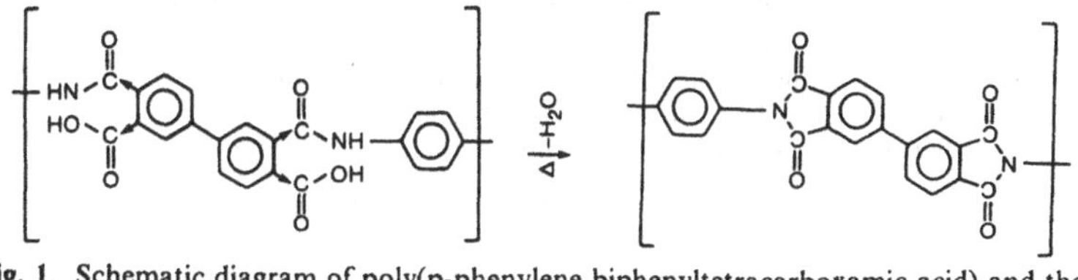

Fig. 1. Schematic diagram of poly(p-phenylene biphenyltetracarboxamic acid) and the imidized BPDA-PDA polyimide.

† To whom correspondence should be addressed.

Mat. Res. Soc. Symp. Proc. Vol. 227. ©1991 Materials Research Society

Poly(*p*-phenylene biphenyltetracarboximide) usually denoted as BPDA-PDA (see Fig. 1) after its precursor monomers, biphenylene tetracarboxylicdianhydride (BPDA) and *p*-phenylene diamine (PDA), is considered most promising due to recent reports of low CTE and stress, excellent thermal and mechanical properties [5,6]. Moreover, its dielectric constant measured along the film thickness direction is ca. 3.0 (see below), lower than the typical value of ca. 3.5 for most polyimides [3]. However, these measurements were carried out with the electric field normal to the film plane. Since polyimide chains tend to align along the surface of coated substrates during the imidization process [6 – 8], the dielectric properties are expected to be anisotropic, with the in-plane dielectric constants, $\varepsilon_\parallel$, generally larger than the normally measured out-of-plane values, $\varepsilon_\perp$, along the film thickness direction. Such in-plane dielectric properties have not been studied in any detail, despite the fact that the in-plane dielectric constants are more relevant to some important device design considerations such as the cross-talk noise between signal lines in the same wiring plane.

The degree of chain orientation and the resultant anisotropy in dielectric properties in thin polyimide films can be estimated readily by optical measurements. It is the purpose of this study to investigate the anisotropy of BPDA-PDA films over a wide range of film thickness and frequency and then to estimate the refractive indices at the lowest possible optical frequncies (ca. 10^{13} Hz) from the combination of UV-visible and infrared (IR) absorptions by appropriate dispersion relations. The estimated out-of-plane refractive index is then compared with the experimental values of out-of-plane dielectric constants, $\varepsilon_\perp$, at radio frequencies to deduce the additional contributions between the two frequencies. Adding this contribution to the in-plane refractive index then leads to a reasonable estimate of the desired in-plane dielectric constants, $\varepsilon_\parallel$, at radio frequencies.

EXPERIMENTAL

Polyimide films were prepared by dissolving the precursor poly(amic acid) (PYRALIN PI-5810D from DuPont Co.) in N-methylpyrrolidone (NMP). The solution was filtered through a 1 μm Teflon filter and spin-coated onto suitable substrates. The coated films were then dried at 125°C for 30 min in air and thermal imidization was performed at 400°C for one hour under nitrogen.

Samples for the attenuated total reflectance waveguides (ATR) were spun onto gold-coated glass slides. The gold layer was necessary for the optical experiments [9,10]. The resulting polyimide film thicknesses ranged from 404 nm to 3950 nm by varying the polymer concentrations and the spinning speeds of coating solutions. Samples for Metricon waveguides, UV-visible (UV-VIS), and IR measurements were prepared by spin-coating the polymer solution onto quartz wafers or sodium chloride substrates, respectively. The free-standing film for the tilt-angle dependent IR measurement was obtained by releasing the cured polymer film from a glass substrate in a water bath. The released sample of 1.55 μm thickness was then dried at 150°C for at least 2 hours.

For the dielectric measurements the solution was spun onto microscope slides with a two layer base electrode (bottom layer of 500Å thick Cr and top layer of 1500Å thick Au), which had been deposited in advance. After imidization, top electrodes of gold of 1500Å thickness were evaporated.

The optical characterization of the polymer films as a function of film thickness was done by attenuated total reflectance (ATR) spectroscopy of the waveguide modes, using the incident light of 632.8 nm wavelength. We were able to determine two refractive index values, namely, the ordinary ($n_\parallel$) and extraordinary ($n_\perp$) refractive index, to an accuracy of ±0.001, while the thickness could be determined to ±1nm [10].

Frequency dependence of the refractive indices were measured at four different wavelengths (543.5, 594.1, 632.8, and 1064.2 nm) at room temperature with a Metricon PC 2000 prism coupler [9]. Two seperate samples with thickness of 2.16 μm and 7.8 μm were tested.

UV-visible spectra were recorded with a Perkin-Elmer UV-VIS-NIR spectrophotometer (Perkin-Elmer Lambda 9). The infrared measurements were made in an evacuated IBM IR 98 Michelson interferometer on films supported on alkali halide substrates or free standing films.

Permittivity measurements were carried out with two Hewlett-Packard Multifrequency LCR meters (models 4274A & 4275A). Before starting the measurements the sample was kept dry at 130°C for 1 hour, and the sample cell was flushed with dry nitrogen during the measurements. The sample thickness was ca. 2.1 μm.

RESULTS AND DISCUSSION

Sixteen samples of various thicknesses were prepared for the ATR measurements by changing the polymer concentrations and the spinning speeds of coating solutions. On each sample, ATR spectra were measured at different spots and analyzed by fitting the measured data to the theoretical curves. Although the film thickness varied substantially (more than 30%) on some of the samples, the refractive indices were found to be constant within the experimental accuracy across each sample. All refractive index data collected are presented in Fig. 2; the experimental uncertainty is much smaller than the symbol size. Aside from slight variations for very thin films, the refractive indices are fairly independent of the film thickness over the whole range covered.

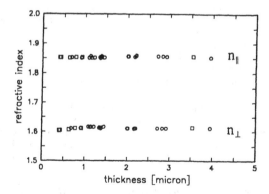

Fig. 2. The refractive indices $n_\parallel$ and $n_\perp$ vs. the thickness of BPDA-PDA polyimide films at 632.8 nm wavelength; circles denote samples parepared on glass substrate, and squares are for the samples prepared on rutile substrate.

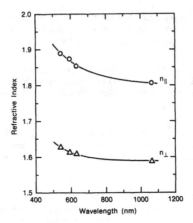

Fig. 3. Dispersion of refractive indices $n_\parallel$ and $n_\perp$ for BPDA-PDA.

The refractive index dispersion of a 2.16 μm thick BPDA-PDA film, measured by Metricon prism coupler, is shown in Fig. 3. The results at 632.8 nm are in excellent agreement with the results in Fig. 2 by attenuated total reflectance (ATR) spectroscopy. The results show a very large optical anisotropy: the difference Δn of the (in-plane) ordinary ($n_\parallel$) and (out-of-plane) extraordinary ($n_\perp$) refractive index is 0.262 at 543.5 nm and decreases to 0.217 at 1064.2 nm. A rather thick film of ca. 7.8 μm thickness was also investigated at 632.8 nm, and the results obtained were the same as those of the 2.16 μm thick sample. Hence, this optical anisotropy, indicating a strong tendency of these molecules to align along the film surface, is nearly independent of the film thickness up to at least ca. 8 μm.

The optical anisotropy of the BPDA-PDA films is remarkable: the difference of the ordinary and the extraordinary refractive index is ca. 0.240 at 632.8 nm. This is to be compared with the maximum values 0.007 and 0.04, reported for spin-coated polystyrene and poly(vinyl carbazole) films [11,12], respectively, and 0.078 found for the conventional polyimide films of poly(N,N'-phenoxyphenyl pyromellitimide) (PMDA-ODA) [7]. The observed anisotropy of 0.24 is even larger than that of most small liquid crystalline materials, for which Δn falls in the range 0.1-0.3 [13].

The frequency dispersion of the refractive index plotted in Fig. 3 is due to absorption in the UV-VIS region. The inset of Fig. 4 shows the UV-VIS absorption spectrum for a very thin film (0.057 μm) of BPDA-PDA. The index of refraction can be calculated from the measured absorbance by two common methods: the Kramers-Kronig dispersion relations, plus some sum rules, or by a Lorentz-Lorenz fit to each individual absorption line and summing each contribution. While the infrared portion was analyzed by a variation of the latter technique fitting all of the many absorption bands, the ultraviolet, visible and near infrared were analyzed by both methods. We compared the results of both methods and finally favored the Lorentz-Lorenz method because it gave a longer wavelength tail more consistent with the experimental values.

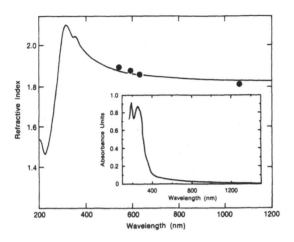

Fig. 4. The in-plane refractive index $n_\parallel$ of BPDA-PDA vs. wavelength, calculated by the Lorentz-Lorenz method [9] from the UV-VIS absorption spectrum shown in the inset. The circles denote the experimental results of $n_\parallel$ in Fig. 3.

For the Lorentz-Lorenz calculations, the experimental absorption curve was fitted by using six lines: one at 196 nm, three around 260 nm, with a small shoulder at 350 nm, and one transition arbitrarily placed at 100 nm to give the background level. The details of the adopted Lorentz-Lorenz method are described in reference 9. The calculated dispersion curve plotted in Fig. 4 reproduces the results at the measured frequencies (circles) reasonably well, and thus allows one to deduce values at longer wavelengths. From this plot it is obvious that there is little significant change in the refractive index values at wavelengths greater than 1064.2 nm, attributable to the the UV-VIS absorptions. Hence, we assume a constant value at long wavelengths.

Next, the contributions from the IR spectral region to the real part of the refractive index was considered by applying the Spitzer-Kleinmann dispersion analysis. In this method which is essentially the same as that applied to the UV-VIS absorption, the sample is thought to consist of a collection of Lorentzian oscillators whose band centers, widths and strengths are obtained from an infrared absorption spectrum. The oscillator parameters are used to compute the complex dielectric function and subsequently, equated to the complex refractive index [14]. Cast in this format the calculation was performed by using the REFMAT Fortran program developed by Pacansky et al. [14]. The calculated in-plane refractive index included all the contributions from the near infrared to the far infrared between ca. 7000 and 200 cm.$^{-1}$ The highest contributing wavenumber is at ca. 3000 cm,$^{-1}$ and the lowest one occurs at ca. 400 cm.$^{-1}$ The in-plane refractive index at ca. 300 cm $^{-1}$ is thus estimated to be 1.876, which is increased by ca. 0.07 over the entire IR range.

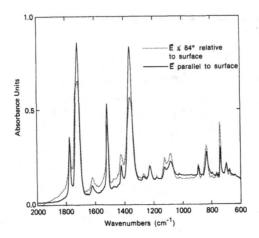

Fig. 5. Polarized IR spectra of an unsupported BPDA-PDA film of 1.55 μm thickness. The solid line spectrum was obtained for the normal incidence and the dotted line spectrum resulted when the film plane was tilted to 84° from the normal incidence.

In order to estimate the out-of-plane refractive indices one needs to have the corresponding absorption spectra with the the electric field of the incident light along the film thickness direction. Direct experiments of this kind are not practical, and hence we carried out polarized IR measurements as a function of tilt-angle of the film plane from the normal incidence. The results, obtained from a free-standing film of 1.55 μm thickness, with the electric field vector of the incident beam parallel to the surface and by tilting the sample almost perpendicular (84°) to the surface are shown as solid and dotted lines, respectively, in Fig. 5. From these results one can conclude that the overall contribution from the IR spectral range does not exhibit a large difference between the in-plane and out-of-plane directions. That is, the major absorbing units (imide planes, for example) are not aligned perfectly along the film plane.

This is most clearly seen in the 736 cm⁻¹ band arising from the out-of-plane bending of the C = O in the imide moieties [15]. Since the transition dipole of this band is along the normal to the imide plane, this absorption would have disappeared if the imide plane had been aligned perfectly along the film plane. The presence of this band, therefore, indicates that the imide planes are instead tilted. However, the increase in this absorption at 84° tilt-angle indicates the imide planes are preferentially oriented along the film plane. This preferential orientation of the imde planes is also corroborated by the strongly decreased absorptions for the tilted sample at 1360 cm⁻¹ (C-N stretching in the imides) and at 1720 cm⁻¹ (out-of-phase stretching of C = O), both with their transition dipoles along the imide planes [15].

Assuming that the oscillator strength for the out-of-plane refractive index $n_\perp$, contributed from the entire IR region, is nearly equal to that (ca. 0.07) of the in-plane refractive index $n_\parallel$, this yields the value of $n_\perp \simeq 1.66$ at 300 cm⁻¹ Application of the Maxwell relation, $\varepsilon = n^2$, then leads to an estimated dielectric constant of $\varepsilon_\perp \simeq 2.76$ at ca. 10^{13} Hz.

TABLE 1

Out-of-Plane Dielectric Constant $\varepsilon_\perp$ of BPDA-PDA Polyimide Films at 25°C at Various Frequencies

Frequency (Hz)	Dielectric Constant
100	3.051
1k	3.037
10k	3.021
100k	3.007
1M	2.980

The experimental results of $\varepsilon_\perp$ at 100 to 10^6 Hz are listed in Table 1. There is a slight dispersion in this frequency range, indicating some contribution from the segmental dipoles. The small difference (0.22) between the experimental value of 2.98 (± 0.05) at 10^6 Hz and the estimated value of 2.76 at 10^{13} Hz is likely to arise mainly from the remaining dispersion of the molecular dipoles, since the presence of water, a normal source of contribution in this range, has been eliminated by drying the sample at 130°C before starting the measurements.

By applying the Maxwell relation to the in-plane refractive index of $n_\parallel = 1.876$ at 300 cm^{-1} and assuming that the addtional contribution from the molecular dipoles at the 10^6 Hz is the same as that for the out-of-plane value, we estimate that $\varepsilon_\parallel \simeq 3.5$ at 10^{13} Hz and $\varepsilon_\parallel \simeq 3.7$ at 10^6 Hz.

CONCLUSION

The ordinary and extraordinary refractive indices of BPDA-PDA polyimide films, prepared by thermal imidization on substrates, were measured revealing a large optical anisotropy with the in-plane refractive index $n_\parallel = 1.852$ and the out-of-plane value $n_\perp = 1.612$ at 632.8 nm. This remarkable optical anisotropy is found to be practically the same in the film thickness range of ca. 0.4 μm to 8 μm. The frequency dispersion of the refractive index follows closely the calculated curve by the Lorentz-Lorenz method from the measured UV-VIS spectra. The dispersion relations from the combination of UV-VIS and the entire IR range is estimated to add ca. 0.07 to the measured value at 1064.2 nm for the in-plane refractive index $n_\parallel$. From the tilt-angle dependent polarized IR spectra, a simliar contribution is assumed for the out-of-plane refractive index $n_\perp$. Applying the Maxwell relation, the estimated dielectric constant along the film thickness direction $\varepsilon_\perp \simeq 2.76$ at 10^{13} Hz compares favorably with the measured value of 2.98 at 10^6 Hz. Assuming the small difference at these two frequencies, most likely due to molecular dipolar contributions, to remain the same for the in-plane dielectric constants, we estimate $\varepsilon_\parallel \simeq 3.5$ at 10^{13} Hz and $\varepsilon_\parallel \simeq 3.7$ at 10^6 Hz.

The estimated dielectric anisotropy, ca. 25%, should be a major factor to consider for device designs and optimizations. One should, therefore, take caution in describing the dielectric properties of stiff polymers in general without considering the effect of molecular orientation. Moreover, one needs to devise the experimental methods to measure directly the in-plane dielectric properties of thin films of stiff polymers.

ACKNOWLEDGMENT

We acknowledge the support of IBM GTD-Fishkill, and thank M. Jurich for assistance with the waveguide measurements, R. Waltman for help with the REFMAT program, and B. Fuller for assistance with the dielectric set-up.

REFERENCES

1. C.E. Sroog, *J. Polym. Sci.: Macromol. Rev.* 11, 161 (1976).
2. K.L. Mittal, *Polyimides: Synthesis, Characterization and Applications*, Vol. I & II, (Plenum Press, New York, 1984).
3. R.J. Jensen and J.H. Lai, in *Polymers for Electronic Applications*, J.H. Lai, Ed. (CRC Press, Boca Raton, 1989).
4. R.J. Jensen, in *Polymers for High Technologies, Electronics and Photonics*, M.J. Bowden and S.R. Turner, Eds. (ACS, Washington, D.C., 1987).
5. S. Numata and N. Kinjo, *Polym. Eng. Sci.* 28, 906 (1988).
6. M. Ree, D.Y. Yoon, L.E. Depero and W. Parrish, submitted to: *J. Polym. Sci., Polym. Phys. Ed.*
7. T.P. Russell, H. Gugger, J.D. Swalen, *J. Polym. Sci., Polym. Phys. Ed.* 21, 1745 (1983).
8. N. Takahashi, D.Y. Yoon and W. Parrish, *Macromolecules* 17, 2583 (1984).
9. J.D. Swalen, *J. Molec. Electron.* 2, 155 (1986), and references therein
10. S. Herminghaus, D. Boese, D.Y. Yoon and B.A. Smith, *J. App. Ph. ... .t.*, in press.
11. J.D. Swalen, R. Santo, M. Take, and J. Fischer, *IBM J. Res. Dev.* 21, 168 (1977).
12. W.M. Prest, Jr. and D.J. Luca, *J. Appl. Phys.*, 51, 5170 (1980).
13. J.A. Castellano and K.J. Harrison, in *The Physics and Chemistry of Liquid Crystal Devices*, G.J. Sprokel, Ed. (Plenum Press, New York, 1980).
14. J. Pacansky, C. England and R.J. Waltman, *J. Poly. Sci., Polym. Phys. Ed.* 25, 901 (1987).
15. R.W. Snyder, C.W. Sheen and P.C. Painter, *Appl. Spectrosc.* 42, 503 (1988).

CHAIN CONFORMATIONS OF AROMATIC POLYIMIDES AND THEIR ORDERING IN THIN FILMS.

D. Y. YOON, W. PARRISH, L. E. DEPERO, and M. REE
IBM Almaden Research Center, 650 Harry Road, San Jose, California 95120

ABSTRACT: We have investigated the conformation - order relationships of four aromatic polyimides prepared by thermal imidization of the precursor poly(amic acids) in thin films. They are: (i) PMDA-4,4'-ODA polyimide from pyromellitic dianhydride (PMDA) and 4,4'-oxydianiline (ODA) monomers; (ii) PMDA-PDA polyimide from PMDA and p-phenylene diamine (PDA); (iii) PMDA-Benzidine polyimide from PMDA and benzidine and (iv) BPDA-PDA polyimide from biphenylene tetracarboxylicdianhydride (BPDA) and PDA. X-ray diffraction results and their analyses by molecular modeling show that all the four polyimides exhibit extended chain conformations in various smectic-type ordered structures that form monomer repeat layers but differ in the details of interchain packing. Furthermore, the polyimide chains are highly aligned along the film plane.

INTRODUCTION

Aromatic polyimides have become increasingly important in a number of aerospace and microelectronic applications, owing to their superior high temperature properties that can be attributed to the thermal and oxidative stability of aromatic heterocyclic structures.[1] In addtion to the thermal stabiltiy, aromatic polyimides often exhibit a combination of exceptional physical properties such as mechanical, thermo-mechanical, electrical, and solvent-resistance properties. For example, poly(4,4'-oxydiphenylene pyromellitimide) (PMDA-4,4'-ODA), which is well known as Kapton films, exhibits an elongation at break of more than 50 % and a softening/flow tempeature of above 400 °C. However, the origin of such physical properties are not understood yet.

One prominet aspect common to most aromatic polyimides is that they do not have a sufficient chain flexibility, as can be expressed by the equivalent Kuhn length. For example, the Kuhn length of the PMDA-4,4'-ODA polyimide is ca. 62 Å, and it increases to ca. 110 Å for the BPDA-PDA polyimide. Therefore, their Kuhn axial ratios defined as the Kuhn length divided by the average cross-sectional diameter (ca. 4.5 Å) normally exceed the critical value of 6.7, that marks the critical flexibility beyond which a disordered random amorphous state is impossible owing to the steric packing problems, according to the prediction of Flory's lattice theory [2]. Furthermore, the highly aromatic characteristics of the chain segments add the orientation-dependent thermotropic interactions that strongly favor an ordered state with the adjacent aromatic units aligned parallel to each other [3,4]. Hence, aromatic polyimide chains in soild films are expected to assume a highly ordered state.

Takahashi, Yoon and Parrish [5] carefully studied the structure of PMDA-4,4'-ODA poly(amic acid) precursor in a condensed state and its thermally cured polyimide in thin films. They found that the poly(amic acid) precursor in the condensed state prior to imidization exhibits extended-chain confomations and smectic-like lateral packing. This molecular and packing order is slightly improved by thermal imidization and subsequent high temperature annealing. However, the coherence length for the lateral molecular packing normal to the chain axis is quite short, and consequently appreciable long range crystalline order does not exist in thermally cured PMDA-4,4'-ODA polyimide films. Hence, the presence of such a mesomorphic order is likely to be a key factor responsible for the unusual physical properties of thin films of this polyimide.

In the present study, we extended the investigations of the ordering in thin films of aromatic polyimides to three other polymers: PMDA-PDA polyimide prepared from pyromellitic dianhydride (PMDA) and p-phenylene diamine (PDA); PMDA-Benzidine polyimide from PMDA and benzidine; and BPDA-PDA polyimide from biphenylene tetracarboxylicdianhydride (BPDA) and PDA. The ordering in these polyimides are then compared with that of the PMDA-4,4'-ODA polyimide.

EXPERIMENTAL

Samples

The precursor poly(amic acid) solutions were prepared in NMP solution. The polyimide films were prepared from the poly(amic acid) solution on glass substrates. In this method the precursor solution was cast on microscope slides using a doctor blader, dried at 80 °C for about 1 hr, subsequently imidized by raising the temperature to 400 °C in about 30-40 min, and held at this curing temperature for 0.5 to 1 hr. Both drying and imidization were performed under moderate nitrogen gas flow. The imidized films were subsequently immersed in distilled water for 30 min and then could be easily removed from the substrate. The films were subsequently dried in a vacuum oven at about 50°C and kept in the ambient conditions; the final thickness was about 25 μm.

Measurements

Wide angle X-ray diffraction (WAXD) patterns were obtained using two vertical-scanning powder diffractometers with θ-2θ scanning [6]. One diffractometer was set up for measurements in the transmission mode (i.e., reflections from lattice planes normal to the film surface) with a thin asymmetric cut (10.1) quartz plate monochromator bent to a section of a logarithmic spiral and located in the diffracted beam. The other was used for measurements in the reflection mode (reflections from lattice planes parallel to the film surface) with a curved graphite monochromator in the diffracted beam. A comparison of the relative intensities of the transmission and the reflection patterns gives a good measure of the degree of preferred orientation with respect to the film surface. The source was a standard sealed long fine focus X-ray tube operated at 50 kV, 25 mA. The transmission diffractometer was used with 2° entrance slit and anti-scatter slits. The reflection diffractometer was operated with 1° entrance slit, 0.11° receiving slit and anti-scatter slits. Vertical divergence was limited by 4° Soller slits in the incident beam of both diffractometers. The patterns were recorded using step-scanning with 0.02° steps and 2 sec count time per step.

RESULTS AND DISCUSSION

PMDA-4,4'-ODA Polyimide

Fig. 1 shows the WAXD patterns of PMDA-4,4'-ODA films obtained with the transmission (dashed trace line) and the reflection (normal signal line) geometry, respectively. As discussed in detail previously [5], the reflection pattern comprises mainly featureless amorphous halos. Only the transmission pattern exhibits a single sharp peak at $5.7°(2\theta)$, representing the monomer repeat distance of ca. 15.5 Å along the extended chain axis. the relative sharpness and the symmetric shape of this peak is indicative of smectic-like layer of monomer repeat units, as drawn schematically in Fig.2. Within the monomer repeat layer the registration of neighboring segments in the lateral directions is relatively poor, analogous to the situation in smectic A liquid crystals.

The appearance of this smectic-like layered order in PMDA-4,4'-ODA films indicates that the polymer chains assume locally extended conformations. Also, the fact that this monomer layer peak is nearly absent in the reflection pattern shows a very strong tendency of polymer chains to align along the film surface. Hence, one should recognize that the polyimide chains adopt both the conformational order to form an extended local conforamtion and the orientational order to align along the film surface in considering the properties of thin polyimide films.

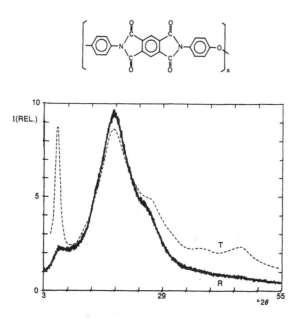

Fig. 1. WAXD patterns of PMDA-4,4'-ODA polyimide films in transmission (dashed trace line) and in reflection (normal signal line) runs.

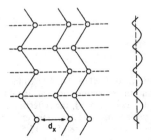

Fig. 2. Schematic of the smectic-like order of PMDA-4,4′-ODA polyimide films. The oxygen kink in the ODA unit is denoted by the open circle, and the electron density projection along the chain axis is drawn in the right side of the figure.

PMDA-PDA Polyimide

Fig. 3 shows the transmission pattern (solid line) and the reflection pattern of PMDA-PDA polyimide films. The reflection pattern has one broad amorphous halo, indicating no sign of any order. In contrast, the transmission pattern exhibits five sharp peaks at 7.2°, 14.4°, 21.6°, 29.0°, and 36.5° 2θ, that can be indexed as the 001 to 005 peaks of the 12.3 Å monomer repeat distance. The numbers in the parentheses in Fig. 3 denote the calculated powder pattern intensities for pefectly layered **crystalline** order of monomer repeats along the chain axis. Hence, the weak intensities of the odd numbered 001, 003, and 005 peaks are consistent with the calculated structure factors.

The PMDA-PDA polyimide films therefore can be cosidered to have a highly ordered smectic Λ crystal order comprised of monomer repeat layers, with the polymer chains aligned parallel to the film surface.

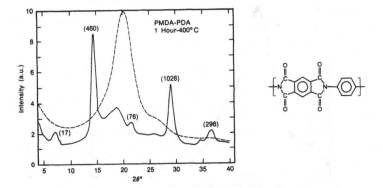

Fig. 3. WAXD patterns of PMDA-PDA polyimide films in transmission (solid line) and in reflection (dashed) runs. The numbers within the parentheses denote the calculated powder diffraction intensities for a perfect smectic crystal along the chain axis.

PMDA-Benzidine Polyimide

The WAXD patterns of PMDA-Benzidine films, plotted in Fig.4, show a more dramatic smectic crystal order. The reflection pattern (dashed line) is featureless with one amorphous halo. On the other hand, the transmission pattern exhibits seven very sharp symmetric peaks at 5.34°, 10.66°, 16.0°, 21.4°, 26.8°, 32.3°, and 38.0° 2θ, that can be indexed as the 001 to 007 peaks of the 16.6 Å monomer repeat distance. Moreover, the sharpness and the symmetric shape of the peaks do not decrease appreciably in the high-order peaks: the full width at half height changes from 0.65° for the 002 peak, 0.7 for the 003, 0.75 for the 005, to 0.80 for the 006 peak. Taking into account the instrumental broadening of 0.15°, the coherence length is estimated to be ca. 159 Å for the 001 peak, and decreases only slightly to ca. 127 Å for the 006 peak, according to the Sherrer equation [7].

The PMDA-Benzidine polyimide films therefore exhibit a nearly perfect crystal order along the chain axis, with nearly perfect matching of monomer units along the chain axis. On the other hand, the lateral packing order within the monomer layer is rather poor. This may be described as a very highly ordered smectic A crystal in this regard. Again, the chain axis is aligned strongly along the film surface.

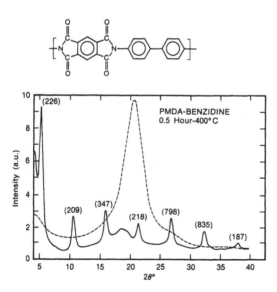

Fig. 4. WAXD patterns of PMDA-Benzidine polyimide films in transmission (solid line) and in reflection (dashed) runs. The numbers within the parentheses denote the calculated powder diffraction intensities for a perfect smectic crystal along the chain axis.

BPDA-PDA Polyimide

The experimental WAXD patterns are shown in Fig 5b (transmission) and 5c (reflection). In addition to the four 00*l* peaks in the transmission pattern, there are also other relatively sharp peaks which can be indexed as *hk0* peaks [8]. According to atomistic model calculations [8], the model which best fitted the experimental data (see Fig. 5a) could be interpreted as having an orthorhombic unit cell, a = 8.620Å, b = 6.270Å, and c = 31.986Å, with space group Pba2, that requires a twofold rotational disorder by 180° around the chain axis. It is therefore analogous to the the smectic-E order of rodlike molecules. The reflection pattern has only a very weak 004 reflection on the low 2θ side of the profile, and the broad *hk0* reflections 110 and 210 which also appear in the transmission pattern.

The virtual absence of *00l* peaks in the reflection pattern indicates a strongly preferred orientation of the molecular chains along the film plane. The coherence length, i.e., the average "crystallite" size along the chain axis, was determined to be ca. 155 Å from the Sherrer equation.

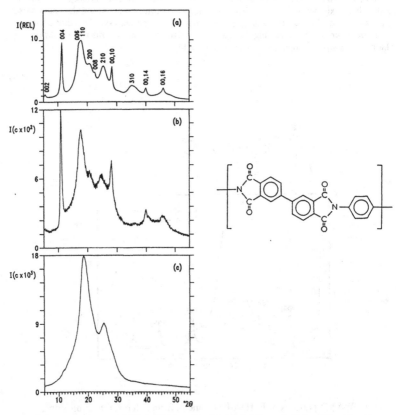

Fig. 5. WAXD patterns of BPDA-PDA polyimide films in transmission (b) and reflection (c) runs. The top section (a) is the fitted pattern according to the atomistic model calculations [8].

CONCLUSION

All the four aromatic polyimides investigated here exhibit a smectic-like order with sharp and symmetric 00ℓ peaks representing the monomeric repeat layer and a strong preference for the polyimide chains to orient along the film surface. The smectic layer consists of monomeric repeat units in a fully extended conformation. While the PMDA-4,4'-ODA exhibits a relatively poor smectic-like order with only one diffraction peak, both the rodlike PMDA-PDA and PMDA-Benzidine chains show a very highly ordered smectic crystal order, as indicated by multiple 00ℓ peaks that do not show any appreciable peak broadening at higher angles. Despite such a very high order along the monomer repeat axis, the packing order lateral to the chain axis is rather poor, analogous to the smectic A liquid crystals. The BPDA-PDA polyimide is unique in this regard, showing a significant lateral interchain (smectic E type) packing order in addition to the excellent monomer repeat packing order along the extended chain axis.

ACKNOWLEDGMENT

We thank D. Hofer and S. Swanson for providing the PMDA-PDA and PMDA-Benzidine poly(amic acid) precursor samples.

REFERENCES

1. K.L. Mittal, *Polyimides: Synthesis, Characterization, and Applications,*. Vol. I & II, Plenum, New York, 1984.
2. P.J. Flory, *Advances in Polymer Science* **59**, 1 (1984).
3. W. Maier and A. Saupe, *Z. Naturforschg.* **14a**, 882 (1959); *Ibid.* **15a**, 287 (1960).
4. D.Y. Yoon and P.J. Flory, *Mat. Res. Soc. Symp. Proc.* **134**, 11 (1989).
5. N. Takahashi, D.Y. Yoon, and W. Parrish, *Macromolecules* **17**, 2584 (1984).
6. W. Parrish, *X-Ray Analysis Papers*, Centrex Publ. Co., Eindhaven, 1965.
7. P. Sherrer, *Nachr. Göttinger Gesell.*, 98 (1918).
8. M. Ree, D.Y. Yoon, L.E. Depero, and W. Parrish, *J. Polym. Sci., Polym. Phys. Ed.*, submitted.

Author Index

Subject Index

Printed in the United States
By Bookmasters